SCIENCE & CULTURE In the Western Tradition

Sources and Interpretations

SCIENCE & CULTURE In the Western Tradition

Sources and Interpretations

Senior Editor
JOHN G. BURKE

Contributing Editors
Mark B. Adams, Judith V. Grabiner, Frederick Gregory,
Russell C. Maulitz, Richard Olson, Robert S. Westman

Gorsuch Scarisbrick, Publishers
Scottsdale, Arizona

Science and Culture in the Western Tradition was developed by Coast Community College District and International Video Connection, Inc., with partial funding by BBC Enterprises, London Writers, and RKO Pictures.

Coast Community College District

David A. Brownell, Chancellor
Coast Community College District

William M. Vega, President
Coastline Community College

Leslie N. Purdy, Director
Alternative Learning Systems

David P. Stone, Assistant Dean
Alternative Learning Systems

Edward Hirsch, Instructional Designer

Michael S. Werthman, Publications Editor

International Video Connection, Inc.

George A. Colburn, Director of Operations
Jane L. Scheiber, Director of Education Programs
Gail Tirana, Research Associate/Permissions Editor

Gorsuch Scarisbrick, Publishers

Gay L. Orr, Production Manager
Carlisle Graphics, Typesetting
Bill Nebel, Graphic Arts Services, Layout
BookCrafters, Printing and Binding

Gorsuch Scarisbrick, Publishers
8233 Via Paseo del Norte, Suite E-400
Scottsdale, Arizona 85258

10 9 8 7 6 5 4

ISBN 0-89787-811-6

Printed in the United States of America.

Preface

This Reader provides an introduction to the history of the Western world, from ancient Greece to the present, in terms of the interplay between scientific discoveries and cultural developments. Going beyond traditional histories of science and traditional histories of Western civilization, this volume explores scientific breakthroughs and technological developments in terms of their cultural contexts and their impact on how Western societies have viewed themselves and their universe.

The readings include both primary sources and secondary selections. Also included are original essays by the editors that provide historical, cultural, and scientific contexts for the topics discussed in the readings.

The book, originally designed for a telecourse produced by Coast Community College District and International Video Connection, Inc., in association with South Carolina Educational Television, was developed over a period of more than two years. Its basic structure evolved from conferences in June 1985 and June 1986, which brought together a group of content consultants under the leadership of Robert S. Westman, Professor of History, and John G. Burke, Professor Emeritus of History, both at the University of California, Los Angeles. These and the other scholars who subsequently served as contributing editors of the Reader, conferred on the selections for each chapter. Individual editors edited the readings, wrote original introductory essays and headnotes, and compiled annotated bibliographies for further reading. Burke served as senior editor, shaping, coordinating, and further refining the book. Jane Scheiber of IVC and Michael S. Werthman of Coastline Community College served as supervisory and manuscript editors, respectively.

Appreciation is expressed to Gail Tirana for her work in obtaining permissions for the readings; to Lois Havens for her skill in preparing various drafts of the manuscript and completing permissions acquisitions; to Dorothy McCollom and Marjorie Kustra for extensive clerical work on portions of the Study Guide; to Florence Randolph and Sandra Leslie for their careful proofreading; and to Gay Orr, Production Manager at Gorsuch Scarisbrick, for her expertise in seeing the book through all stages of production.

Contents

Introduction

Science may be defined narrowly as the systematic knowledge of the physical world which has been amassed by the activity of scientists. In this endeavor scientists conduct investigations of natural phenomena of all kinds and seek to derive theories or general laws of nature from the logical assessment of the observations, experiments, or measurements which their research may require. In this view, what counts is what scientists do, what methods they employ, and what results they achieve.

A much broader definition of science includes, in addition to the above, many other aspects. Among them are the educational background of scientists, the specialized training that each one must experience to enter a particular scientific discipline, the professional societies with which each is affiliated, and the scientific journals with which each must be familiar. It also incorporates the uses to which their investigations lead—for example, the technology developed as a consequence of new knowledge or the application of a scientific theory in a completely different area, such as the employment of a biological theory to support a social or political point of view. It is important to note that this definition of science also takes into account the cultural context in which scientists are immersed; that is, the beliefs—political, social, economic, religious, or philosophical—which scientists share with their nonscientific contemporaries.

Historians of science, until early in the twentieth century, generally adopted the narrow view of science. They could, thereby, explain scientific progress as the result of more refined observations, more accurate instrumentation, more ingenious experimentation, more precise measurements, or more cogent analysis. Isotopes of the chemical elements were discovered, for example, when chemists were able to make much more accurate determinations of atomic weights. This finding, in turn, influenced theories of the physical constitution of the atom. The history of science, then, consisted of reporting when, by whom, and by what means important discoveries were made, and how this knowledge led to contemporary science. Historians dutifully recorded the sources of previous errors. They labeled such aberrations as astrology and alchemy the pseudosciences, which were the products of fertile imaginations and hardly worthy of mention. Some authors roundly condemned organized religions as enemies of science, and many scoffed at philosophical convictions or religious beliefs that had actually or apparently hindered progress toward scientific truth.

During the course of the twentieth century, historians became increasingly uncomfortable with this view of the history of science. They came to the realization that more satisfactory and more accurate explanations of the rise and development of the various scientific disciplines could be achieved only if one adopted the broader definition of science. Factors external to the scientific activity and scientific thought of a scientist had to be taken into account. Such relevant factual information as education, associations, friendships, and the beliefs and speculations of a scientist with respect to philosophy, the arts, religion, or politics had to be considered.

To write a faithful narrative of the scientific activity of any period such as the late middle ages or the Enlightenment, then, historians must be cognizant not only of the contemporary scientific theories and practices, but of the cultural milieu in which scientists worked. Only then will the historian comprehend why particular natural phenomena attracted the attention of scientists, why certain problems were chosen for investigation rather than others, and why preconceptions colored theories or errors impaired them. This approach to the history of science also sheds light on the occurrence of so-called scientific revolutions, that is, the replacement or transformation of a prevailing or dominant scientific theory by a new one.

This is not to say that historians must agree in their accounts of how and why science developed at a particular time as it did, or why the efforts of one individual should be considered as more important than those of another. Disagreements occur usually because one historian bases an explanation on a piece or body of evidence that another historian may consider trivial or inconsequential. In some instances more exhaustive research has resolved the conflict; in others the student must decide which argument is more persuasive or convincing.

This book of readings exemplifies the broader perspective of the history of science. The contributing editors, in their introductory essays to the chapters and headnotes for the excerpted reading selections, have attempted to place the most significant scientific activity of a historical period into a cultural context and to describe at the same time how scientists approached the problems posed by the phenomena of nature and what those scientists accomplished.

No single book, of course, can encompass adequately the developments through centuries in all of the scientific disciplines. Most of the chapters, therefore, concentrate on a particularly significant area of science in which a marked or radical change in scientific thought or practice occurred—for example, the introduction and reception of Copernican astronomy in the sixteenth and early seventeenth centuries, or the genesis and impact of Darwinian evolutionary theory in the nineteenth and early twentieth centuries.

Attempts to understand nature by rational means and to bring the forces of nature under control for human benefit have not been limited to Western civilization. The Chinese, as Joseph Needham and his associates have demonstrated in their definitive multivolume work *Science and Civilization in China,* have pursued science for well over 2,000 years. In fact, many of the technological by-products of Chinese science, among them gunpowder and the magnetic compass, were introduced into Western Europe centuries after they appeared in China. Similarly, scholarly articles devoted to the history of science in Islamic countries and in India centuries ago are now beginning to be published in increasing numbers. In Chapter 2, a reading selection describes briefly the contributions that Islamic science made to medieval European scholarship. However, science as it has developed in Western civilization and the scientifically based technology that commenced in the West have become dominant throughout the world: hence this concentration on the Western tradition.

The results of science, during the 2,500 years that it has been a feature of Western culture, have changed our views of our planet, of the cosmos, of the intimate constitution of matter, of the functioning of the human body, and of the development of life. The technological products of science, moreover, have in the past two centuries transformed our material culture. There is no reason to believe that this process will not continue, such that two and a half millennia in the future the world inhabited by our descendants will be a much different place. They will undoubtedly have a much greater store of knowledge, and their perspectives of nature will unquestionably differ substantially from ours. The final chapter of this book is intended to emphasize the fact that science has a future as well as a history, and also that humans will undoubtedly employ, as in the past, scientific technology for purposes that, in their judgment, will ensure the prolonged existence of the human species.

John G. Burke

CHAPTER 1

Greek Science and Society

Introduction by Richard Olson

The ancient cultures of Egypt and Mesopotamia produced highly sophisticated and culturally important mathematical, astronomical, biological, and medical knowledge long before Greek-speaking people migrated into the Mediterranean basin. Moreover, the content of Egyptian and Babylonian knowledge was frequently appropriated by the earliest Greek "scientists" as a foundation for their own work; and Greek historians, such as Herodotus, were generally well aware of the debt owed to these earlier cultures. Yet the Greek speculators about nature felt that their efforts signaled a radical break with earlier traditions, and many of those who study the history of science in the late twentieth century are still inclined to place the origins of Western science in Ionian Greece between 600 and 500 B.C., rather than in the Nile or Tigris and Euphrates river basins a millennium or two earlier.

If the mathematics and astronomy of ancient Babylon and the medical knowledge of ancient Egypt were often technically more sophisticated and more useful for the practices of accountants, engineers, calendar makers, and physicians than the comparable "sciences" of the classical Greeks—as they almost certainly were—why do we persist in attributing the beginnings of science to the Greeks? Is it because we learned about Greek science first and, like all humans, we are trying to hang on to our traditional beliefs in the face of contrary arguments and evidence? Or was there *really* something new, different, and important about the ways in which the Greeks approached questions about the natural world which made them "scientific" whereas the approaches of earlier peoples had not been?

How we answer these questions will depend heavily on what point of view or perspective we bring to a key question: What should we count as scientific, or how do we define "science"? Moreover, the perspective we bring to this question in the twentieth century is likely to be influenced by political considerations. For most Marxist scholars, including Benjamin Farrington, author of one of the selections in this chapter, one of the central defining characteristics of science lies in its practical aims. Science *is* for Farrington "the system of behavior by which man acquired mastery of his environment." It follows from the very nature of science that "it has its origin in techniques . . . in various activities by which man keeps body and soul together. Its source is experience, its aims practical, its *only* test, that it works.'' Farrington and the Marxists are thus inclined to see the astronomy, mathematics and medicine of pre-Greek Mediterranean cultures as scientific and to say that we overidealize the Greeks when we claim that they invented science.

At least during the first two-thirds of the twentieth century, most non-Marxist historians of science, concerned much more with issues of intellectual freedom than with considerations of social relevance, viewed science in a very different way. With a few notable exceptions they saw science as a disinterested search for knowledge of nature. For these historians, the ties of science to productive activity are thus fundamentally incidental to its prime goal, which is to satisfy a basic human psychological need to discover or create pattern and order amid the apparent chaos of experience. From this perspective, knowing takes precedence over doing, and the Greeks must be counted as the first scientists because they offered the first self-conscious commitment to this position.

Regardless of what stance historians take regarding the relationship between knowing and doing and whether or not science ''begins'' with the Greeks, they generally agree that around 600 B.C. the Greek approach to natural topics began to differ from those of earlier cultures—including that of earlier Greeks—in a series of ways that have given a particular shape to the subsequent Western scientific tradition. First, the Ionian Greeks were among the earliest to offer general accounts or explanations of natural objects and events, accounts which are naturalistic rather than theological—that is, which seek to understand whole classes of events by appealing to causes which lie within the realm of nature rather than outside of it in a supernatural realm.

For example, Homer, sometime around the eighth century B.C., explained the lightning that destroyed Odysseus' ship by fashioning a story in which the sun god, Helios (Hyperion), enraged by the unauthorized slaughter of his sacred cows by Odysseus' men, demanded that Zeus punish them. Homer clearly showed that it was their particular act that provoked divine retribution in the form of thunder and lightning aimed specifically at them. As Odysseus recalled what happened:

> 'Lampetia of the light robes [daughter of the sun god and keeper of his flocks]
> ran swift with the message
> to Hyperion the Sun God, that we had killed his cattle,
> and angered at the heart he spoke forth among the immortals:
> ''Father Zeus, and you other everlasting and blessed
> gods, punish the companions of Odysseus, son of Laertes;
> for they outrageously killed my cattle, in whom I always
> delighted, on my way up into the starry heaven,
> or when I turned back again from heaven toward earth. Unless
> these are made to give me just recompense for my cattle,
> I will go down to Hades' and give my light to the dead men.''
> 'Then in turn Zeus who gathers the clouds answered him:
> ''Helios, shine on as you do, among the immortals
> and mortal men, all over the grain-giving earth. For my part
> I will strike these men's fast ship midway on the open
> wine-blue sea with a shining bolt and dash it to pieces.'' . . .
> 'Six days thereafter my own eager companions feasted
> on the cattle of Helios the Sun God, cutting the best ones
> out; but when Zeus the son of Kronos established the seventh
> day, then at last the wind ceased from its stormy blowing,
> and presently we went aboard and put forth on the wide sea,
> and set the mast upright and hoisted the white sails on it.
> 'But after we had left the island and there was no more
> land in sight, but only the sky and the sea, then Kronian
> Zeus drew on a blue-black cloud, and settled it over

> the hollow ship, and the open sea was darkened beneath it;
> and she ran on, but not for a very long time, as suddenly
> a screaming West Wind came upon us, stormily blowing,
> and the blast of the stormwind snapped both the forestays that were
> holding
> the mast, and the mast went over backwards, and all the running gear
> collapsed in the wash; and at the stern of the ship the mast pole
> crashed down on the steersman's head and pounded to pieces
> all the bones of his head, so that he like a diver
> dropped from the high deck, and the proud life left his bones there.
> Zeus with thunder and lightning together crashed on our vessel,
> and, struck by the thunderbolt of Zeus, she spun in a circle,
> and all was full of brimstone. My men were thrown in the water,
> and bobbing like sea crows they were washed away on the running
> waves all around the black ship, and the god took away their
> homecoming.'

By contrast, two centuries later the Milesian nature philosopher, Anaximander, gave an account of thunder and lightning that typifies the newly emerging demand for naturalism and also for generality:

> These things occur as a result of wind: for whenever it is shut up in a thick cloud and then bursts out forcibly, through its fineness and lightness, then the twisting makes the noise, while the rift against the blackness of the cloud makes the flash.

Homer's so-called mythopoeic (myth-creating), poetic, or theogonic (concerned with the gods and their origins) approach focuses on a particular event, the single bolt of lightning that destroyed Odysseus' ship, and that event is explained in terms of a human emotional reaction supposed to exist in a superhuman, supernatural, divine being. These two characteristics are typical of prescientific accounts of natural phenomena. Anaximander, on the other hand, seeks to account for all cases of lightning with one explanation; that is, he assumes that there is what we call a *universal* cause for lightning and he seeks that cause in terms of the regular behavior of objects of our ordinary sensory experience. Both of these characteristics are what we call "scientific."

In addition, the classical Greeks were the first to evidence a characteristic that some have called "organized skepticism" and others a "critical intelligence" in connection with accounts of natural phenomena: They began to demand that before any proposed account of natural phenomena be accepted it should be justified or defended by appeal to arguments drawn from sensory experience and/or from reason. They developed formal rules of logical inference and required that mathematical propositions be proved, rather than simply accepted because of their usefulness. Indeed, many Greek philosophers became so impressed with mathematical reasoning that they sought to cast accounts of all natural phenomena in the form of a demonstrative science—initiating a pervasive feature of much subsequent Western science. Some Greek thinkers even began to suggest that all legitimate human reasoning must partake of the character of mathematics—that is, must deal with objects which, like those of mathematics, are unchanging and eternal. In Plato's words, "mathematics is the science of the intelligible." In this view, what we can really know for certain, in contrast to what we can only have uncertain opinions about, must be treatable mathematically.

If the long-standing demands for logical consistency and reasoned argument which characterize Western science emerged largely in connection with Greek mathematical developments, the parallel demands for conformity to sensory experience or for empirical testing emerged largely, though not exclusively, in connection with Greek medical and biological writings. Explaining how his own subject is superior to others, for example, the author of *On Ancient Medicine,* writing sometime around 440 B.C., bragged that medicine had no need of empty hypotheses like those needed by astronomers, because the physician could appeal directly to his experiences, whereas, "with regard to things above and below the earth; if anyone should treat of these and undertake to declare how they are constituted, the reader and hearer could not find out whether what is delivered is true or false; for there is nothing which can be referred to [to] discover the truth." This passage demonstrates two interesting characteristics of Greek "scientific" writing. First, it is aggressively and typically critical in seeking to raise the status of one kind of knowledge by attacking the purported grounds for belief in another. Second, it emphasizes the claim that scientific knowledge should be subject to some kind of direct empirical testing.

All modern scientists would agree that scientific theories must be both logically coherent and open to empirical testing in some way. Each of these demands first emerged self-consciously in the Greek world, though only a few Greek natural philosophers incorporated both of them. Moreover, in general, Greek attempts to develop techniques for establishing logical coherence were more successful than their attempts to establish canons for assessing empirical evidence. For this reason early Greek science seems highly speculative by our twentieth-century standards.

So far we have claimed that the following features associated with the Western scientific tradition emerged in connection with Greek attempts between c. 600 and c. 350 B.C. to account for natural phenomena:

1. the search for universal explanations,
2. the search for naturalistic explanations,
3. the insistence upon logical coherence,
4. the demand for "proofs" in mathematics,
5. the general presumption that nature can be described using mathematics, and
6. the insistence that sensory evidence (and/or empirical testing) may or must be used to support the truth claims of a credible theory.

Although there was no one Greek intellectual prior to 300 B.C. whose thought incorporated every one of the above features, by the time that Greek science reached its apex at the Alexandrian museum during the first three centuries of the Christian era, all six were universally accepted. The astronomical works of Claudius Ptolemy, the writings of Archimedes on mechanics and hydrostatics, and the medical writings of Galen illustrate the dramatic achievement those six features made possible.

One of the fundamental aims of this chapter is to provide an understanding of what Greek science was like. The selections from Aristotle's *Physics* and *Metaphysics* include not only his own explanation of the four kinds of causes which should be sought in accounting for natural events, but also one of the first historical accounts of early Greek science. Although he does have his own axes to grind, Aristotle provides an introduction to his major scientific predecessors and their doctrines. From him we learn of the Milesians, such as Thales and Anaximenes, who first attempted to explain the world in terms of some single basic substance of which it is composed; of the Eleatics, who initiated discussions of logical inference; of the Pythagoreans, who tried to understand everything in terms of numbers; and the Atomists, who were the first to explain events in terms of indivisible and indestructible particles moving in a void. The brief passage from *On The Sacred Disease* illustrates the rejection of divine in favor of naturalistic causes; the selection about Thales illustrates the character and importance of mathematics in Greek science; and Adam Smith's essay identifies an ordered society that provides the leisure to engage in scientific inquiry as a prerequisite for the development of science in a civilized culture such as that of ancient Greece.

Once we know what Greek science was like we can turn to a set of questions that are specifically related to the themes of this book. Why did a science with those characteristics we have identified emerge in ancient Greece? Were there special characteristics of Greek life that stimulated these developments, or was it a kind of miracle or accident? For anyone with a scientific cast of mind it is unthinkable that a phenomenon as important as the scientific tradition should be inexplicable; and though there is still no universal agreement about the social and cultural factors that shaped Greek science, several of the following selections address this issue. Both Aristotle and Adam Smith emphasize the notion that the emergence of a leisured class of intellectuals in Greece was a central factor in the emergence of science. Although Farrington admits the importance of Greek class structure in giving Greek science its special speculative cast, he is inclined to view that characteristic as not only inessential to the emergence of science, but as positively harmful in fundamental ways. G. E. R. Lloyd brings the perspective of recent cultural anthropology to bear on this question and argues that Western scientific demands for logical coherence and the appeal to sensory evidence were both stimulated by the competition and argumentation that stemmed from the special nature of Greek political life.

There is another question even more critical to us than that of how Greek culture shaped science, the question of how Greek science in turn helped to shape the culture in which it emerged—that is, of how "the universe changed" as a consequence of Greek scientific developments. The short selection from Plato's *Republic* illustrates one dramatic long-term cultural consequence of Greek scientific thought—the emergence of the hypothetico-deductive structure of thought associated with mathematics as one of the dominant models for *all* thought. And the selection from Olson's *Science Deified and Science Defied* illuminates the impact which scientific thought had on Greek religious beliefs and points out that when our universe changes we often feel threatened and act to protect our traditional view of the world.

As you read the selections of Chapter 1, try to keep one final issue in mind. We have already pointed out that what we "see" or count as science depends on the cultural perspective we bring to our investigation. And the following selections illustrate that historical perspectives on the nature and origins of science have changed over time. Most of the selections at least implicitly accept the claim that what counts as an acceptable scientific explanation in any particular culture will be influenced by factors peculiar to that culture. If this is true, then even our own modern science might be presumed to reflect the characteristics of our own culture, and science cannot ever be expected to produce an objectively "correct" understanding of some presumed autonomous "reality." If what people see or count as an adequate scientific theory is clearly culture dependent, it may also be the case that even what we accept as the "facts" to be explained, the natural objects and events themselves, are as sensitive to the cultural context as is what we count as science itself.

It has been suggested that every scientific theory, regardless of how successful and sophisticated it may be, must be counted as merely provisional and subject to improvement. While this view of the nature of science does not demand the so-called "cultural relativist" perspective taken in this book, it does at least suggest how it can be made consistent with the widespread modern belief that successive scientific theories are truly progressive rather than simply different. That is, we may well be willing to trade in one set of scientific beliefs, commitments, or theories for another only when the latter encompasses a greater range of phenomena, or when it accounts for a given set of phenomena in a simpler and more satisfying way. In this sense scientific theories may be said to be progressive. But such "progress" does not guarantee that successive theories are gradually approaching some "truth" about an independent and "real" natural world.

Henri Frankfort

Explanations of Natural Phenomena in the Ancient Near East

One of the pervasive themes of this book is that scientific developments have frequently wrought important changes in the intellectual "universes" shared by the members of the cultures in which they occurred. In order to recognize the basic changes which grew out of the first Greek developments in science, we need to have some feeling for what the prescientific universe of early Mediterranean cultures was like. The following passage from Henri Frankfort et al., Before Philosophy, *finds the essence of prescientific cultures in the personal and particular character of human interactions with the natural world.*

The ancients, like . . . modern savages, saw man always as part of society, and society as imbedded in nature and dependent upon cosmic forces. For them nature and man did not stand in opposition and did not, therefore, have to be apprehended by different modes of cognition. . . . Natural phenomena were regularly conceived in terms of human experience and . . . human experience was conceived in terms of cosmic events. We touch here upon a distinction between the ancients and us which is of the utmost significance. . . .

The world appears to primitive man neither inanimate nor empty but redundant with life; and life has individuality, in man and beast and plant, and in every phenomenon which confronts man—the thunderclap, the sudden shadow, the eerie and unknown clearing in the wood, the stone which suddenly hurts him when he stumbles while on a hunting trip. It is experienced as life confronting life, involving every faculty of man in a reciprocal relationship. Thoughts, no less than acts and feelings, are subordinated to this experience. . . .

We are here concerned particularly with thought. It is likely that the ancients recognized certain intellectual problems and asked for the 'why' and 'how', the 'where from' and 'where to.' Even so, we cannot expect in the ancient Near Eastern documents to find speculation in the predominantly intellectual form with which we are familiar and which presupposes strictly logical procedure even while attempting to transcend it. We have seen that in the ancient Near East, as in present-day primitive society, thought does not operate autonomously. . . .

. . . Early man does, in fact, view happenings as individual events. An account of such events and also their explanation can be conceived only as action and necessarily take the form of a story. In other words, the ancients told myths instead of presenting an analysis or conclusions. We would explain, for instance, that certain atmospheric changes broke a drought and brought about rain. The Babylonians observed the same facts but experienced them as the intervention of the gigantic bird Imdugud which came to their rescue. It covered the sky with the black storm clouds of its wings and devoured the Bull of Heaven, whose hot breath had scorched the crops.

In telling such a myth, the ancients did not intend to provide entertainment. Neither did they seek, in a detached way and without ulterior motives, for intelligible explanations of the natural phenomena. They were recounting events in which they were involved to the extent of their very existence. They experienced, directly, a conflict of powers, one hostile to the harvest upon which they depended, the other frightening but beneficial: the thunderstorm reprieved them in the nick of time by defeating and utterly destroying the drought.

Editor's note: *By comparison, the Greeks, who lived in a region that did not often experience the extreme violence of searing heat and flash floods which characterized Mesopotamia, viewed rain in terms that were not less personal, but which were certainly more closely tied to gentler emotional states. Aeschylus gives us a typical Greek account of the drought-breaking storms in Greece: "Holy sky passionately longs to penetrate the earth, and desire takes hold of the earth to achieve this union. Rain, from her bedfellow, sky, falls and impregnates the earth, and she brings forth for mortals pasturage for flocks and Demeter's livelihood."[RO]*

Plato

The Allegory of the Cave

Plato (c. 427–c. 348 B.C.) stands as the only serious rival to his student, Aristotle, for the title of the Greek philosopher-scientist who had the greatest long-term impact. Although he was interested primarily in political and moral issues, two circumstances make him virtually impossible to ignore in any attempt to understand the place of science in Western culture. Certain features of Plato's single writing on natural science, the Timaeus, *made that document particularly appealing to Christian scholars; so the* Timaeus *became* the *chief embodiment of Greek science in Europe for over a thousand years. More important, Plato's philosophy is characterized by such a deep-seated distrust of sensory experience that he insists that true knowledge can be had only of some transcendent eternal reality approachable by the intellect alone. Mathematics stands for Plato both as an illustration of the kind of knowledge which should be sought and as that discipline which draws our attention away from the imperfections of the sensory world to the perfection of the eternal intelligible reality. The following passages from Plato's* Republic *present the famous "cave analogy" through which Plato seeks to explain his general attitude toward the relation between the sensory and real worlds, as well as the beginning of his discussion of how the study of mathematics—arithmetic, plane geometry, solid geometry, and astronomy—prepares the soul to apprehend "The Good."*

[*Plato, in the character of Socrates, speaks to his pupil Glaucon.*]

Here is a parable to illustrate the degrees in which our nature may be enlightened or unenlightened. Imagine the condition of men living in a sort of cavernous chamber underground, with an entrance open to the light and a long passage all down the cave. Here they have been from childhood, chained by the leg and also by the neck, so that they cannot move and can see only what is in front of them, because the chains will not let them turn their heads. At some distance higher up is the light of a fire burning behind them; and between the prisoners and the fire is a track with a parapet built along it, like the screen at a puppet-show, which hides the performers while they show their puppets over the top.

I see, said he.

Now behind this parapet imagine persons carrying along various artificial objects, including figures of men and animals in wood or stone or other materials, which project above the parapet. Naturally, some of these persons will be talking, others silent.

It is a strange picture, he said, and a strange sort of prisoners.

Like ourselves, I replied; for in the first place prisoners so confined would have seen nothing of themselves or of one another, except the shadows thrown by the fire-light on the wall of the Cave facing them, would they?

Not if all their lives they had been prevented from moving their heads.

And they would have seen as little of the objects carried past.

Of course.

Now, if they could talk to one another, would they not suppose that their words referred only to those passing shadows which they saw?

Necessarily.

And suppose their prison had an echo from the wall facing them? When one of the people crossing behind them spoke, they could only suppose that the sound came from the shadow passing before their eyes.

No doubt.

In every way, then, such prisoners would recognize as reality nothing but the shadows of those artificial objects.

Inevitably.

Now consider what would happen if their release from the chains and the healing of their unwisdom should come about in this way. Suppose one of them set free and forced suddenly to stand up, turn his head, and walk with eyes lifted to the light; all these movements would be painful, and he would be too dazzled to make out the objects whose shadows he had been used to see. What do you think he would say, if someone told him that what he had formerly seen was meaningless illusion, but now, being somewhat nearer to reality and turned towards more real objects, he was getting a truer view? Suppose further that he were shown the various objects being carried by and were made to say, in reply to questions, what each of them was. Would he not be perplexed and believe the objects now shown him to be not so real as what he formerly saw?

Yes, not nearly so real.

And if he were forced to look at the fire-light itself, would not his eyes ache, so that he would try to escape and turn back to the things which he could see distinctly, convinced that they really were clearer than these other objects now being shown to him?

Yes.

And suppose someone were to drag him away forcibly up the steep and rugged ascent and not let him go until he had hauled him out into the sunlight, would he not suffer pain and vexation at such treatment, and, when he had come out into the light, find his eyes so full of its radiance that he could not see a single one of the things that he was now told were real?

Certainly he would not see them all at once.

He would need, then, to grow accustomed before he could see things in that upper world. At first it would be easiest to make out shadows, and then the images of men and things reflected in water, and later on the things themselves. After that, it would be easier to watch the heavenly bodies and the sky itself by night, looking at the light of the moon and stars rather than the Sun and the Sun's light in the daytime.

Yes, surely.

Last of all, he would be able to look at the Sun and contemplate its nature, not as it appears when reflected in water or any alien medium, but as it is in itself in its own domain.

No doubt. . . .

Every feature in this parable, my dear Glaucon, is meant to fit our earlier analysis. The prison dwelling corresponds to the region revealed to us through the sense of sight, and the fire-light within it to the power of the Sun. The ascent to see the things in the upper world you may take as standing for the upward journey of the soul into the region of the intelligible; then you will be in possession of what I surmise, since that is what you wish to be told. Heaven knows whether it is true; but this, at any rate, is how it appears to me. In the world of knowledge, the last thing to be perceived and only with great difficulty is the essential Form of Goodness. Once it is perceived, the conclusion must follow that, for all things, this is the cause of whatever is right and good; in the visible world it gives birth to light and to the lord of light, while it is itself sovereign in the intelligible world and the parent of intelligence and truth. Without having had a vision of this Form no one can act with wisdom, either in his own life or in matters of state. . . .

If this is true, then, we must conclude that education is not what it is said to be by some, who profess to put knowledge into a soul which does not possess it, as if they could put sight into blind eyes. On the contrary, our own account signifies that the soul of every man does possess the power of learning the truth and the organ to see it with; and that, just as one might have to turn the whole body round in order that the eye should see light instead of darkness, so the entire soul must be turned away from this changing world, until its eye can bear to contemplate reality and that supreme splendour which we have called the Good. Hence there may well be an art whose aim would be to effect this very thing, the conversion of the soul, in the readiest way; not to put the power of sight into the soul's eye, which already has it, but to ensure that, instead of looking in the wrong direction, it is turned the way it ought to be. . . .

Well now, number is the subject of the whole art of calculation and of the science of number; and since the properties of number appear to have the power of leading us towards reality, these must be among the studies we are in search of. The soldier must learn then in order to marshal his troops; the philosopher, because he must rise above the world of change and grasp true being, or he will never become proficient in the calculations of reason. Our Guardian is both soldier and philosopher; so this will be a suitable study for our law to prescribe. Those who are to take part in the highest functions of state must be induced to approach it, not in an amateur spirit, but perseveringly, until, by the aid of pure thought, they come to see the real nature of number. They are to practise calculation, not like merchants or shopkeepers for purposes of buying and selling, but with a view to war and to help in the conversion of the soul itself from the world of becoming to truth and reality.

Excellent.

Moreover, talking of this study, it occurs to me now what a fine thing it is and in how many ways it will further our intentions, if it is pursued for the sake of knowledge and not for commercial ends. As we were saying, it has a great power of leading the mind upwards and forcing it to reason about pure numbers, refusing to discuss collections of material things which can be seen and touched. Good mathematicians, as of course you know, scornfully reject any attempt to cut up the unit itself into parts: if you try to break it up small, they will multiply it up again, taking good care that the unit shall never lose its oneness and appear as a multitude of parts.

Quite true.

And if they are asked what are these numbers they are talking about, in which every unit, as they claim, is exactly equal to every other and contains no parts, what would be their answer?

This, I should say: that the numbers they mean can only be conceived by thought: there is no other way of dealing with them.

You see, then, that this study is really indispensable for our purpose, since it forces the mind to arrive at pure truth by the exercise of pure thought.

Aristotle

Inquiries into the Causes of Changes in Nature

Aristotle (384–322 B.C.) is among the greatest scientific thinkers of all time. His writings provided the foundations of several modern sciences: our physics, astronomy, biology, and psychology. In the following selections from his Physics *and* Metaphysics, *Aristotle first lists what he thinks are the four causes of phenomenal changes in the world—the material, formal, efficient, and final causes. He then states that philosophy had its start when his predecessors began to wonder about the causes of events in nature, and he presents a historical overview of the scientific opinions of earlier Greek philosophers—from Thales to Plato—to demonstrate that they discerned, but only "vaguely," the causes he enumerated.*

W. D. Ross, in his commentary Aristotle *[5th ed., p. 75], points out:*

> It will be noted that of Aristotle's four causes only two, the efficient and the final, answer to the natural meaning of "cause" in English. We think of matter and form not as relative to an event which they cause but as static elements which analysis discovers in a complex thing. This is because we think of cause as that which is both necessary and sufficient to produce a certain effect. But for Aristotle none of the four causes is sufficient to produce an event; and speaking generally we may say that in his view all four are necessary for the production of any effect. We have, then, to think of his "causes" as conditions necessary but not separately sufficient to account for the existence of a thing; and if we look at them in this way we shall cease to be surprised that matter and form are called causes.

We must proceed to consider causes, their character and number. Knowledge is the object of our inquiry, and men do not think they know a thing till they have grasped the 'why' of it (which is to grasp its primary cause). So clearly we too must do this as regards both coming to be and passing away and every kind of physical change, in order that, knowing their principles, we may try to refer to these principles each of our problems.

In one sense, then, (1) that out of which a thing comes to be and which persists, is called 'cause', e.g. the bronze of the statue, the silver of the bowl, and the genera of which the bronze and the silver are species.

In another sense (2) the form or the archetype, i.e. the statement of the essence, and its genera, are called 'causes' (e.g. of the octave the relation of 2 : 1, and generally number), and the parts in the definition.

Again (3) the primary source of the change or coming to rest; e.g. the man who gave advice is a cause, the father is cause of the child, and generally what makes of what is made and what causes change of what is changed.

Again (4) in the sense of end or 'that for the sake of which' a thing is done, e.g. health is the cause of walking about. ('Why is he walking about?' we say. 'To be healthy', and, having said that, we think we have assigned the cause.) The same is true also of all the intermediate steps which are brought about through the action of something else as means towards the end, e.g. reduction of flesh, purging, drugs, or surgical instruments are means towards health. All these things are 'for the sake of' the end, though they differ from one another in that some are activities, others instruments.

This then perhaps exhausts the number of ways in which the term 'cause' is used. . . .

Now, the causes being four, it is the business of the physicist to know about them all, and if he refers his problems back to all of them, he will assign the 'why' in the way proper to his science—the matter, the form, the mover, 'that for the sake of which'. The last three often coincide; for the 'what' and 'that for the sake of which' are one, while the primary source of motion is the same in species as these (for man generates man), and so too, in general, are all things which cause movement by being themselves moved. . . .

It is owing to their wonder that men both now begin and at first began to philosophize; they wondered originally at the obvious difficulties, then advanced little by little and stated difficulties about the greater matters, e.g. about the phenomena of the moon and those of the sun and of the stars, and about the genesis of the universe. And a man who is puzzled and wonders thinks himself ignorant (whence even the lover of myth is in a sense a lover of Wisdom, for the myth is composed of wonders); therefore since they philosophized in order to escape from ignorance, evidently they were pursuing science in order to know, and not for any utilitarian end. And this is confirmed by the facts; for it was when almost all the necessities of life and the things that make for comfort and recreation had been secured, that such knowledge began to be sought. Evidently then we do not seek it for the sake of any other advantage; but as the man is free, we say, who exists for his own sake and not for another's, so we pursue this as the only free science, for it alone exists for its own sake. . . .

Of the first philosophers, then, most thought the principles which were of the nature of matter were the only principles of all things. . . .

Yet they do not all agree as to the number and the nature of these principles. Thales, the founder of this type of philosophy, says the principle is water (for which reason he declared that the earth rests on water), getting the notion perhaps from seeing that the nutriment of all things is moist, and that heat itself is generated from the moist and kept alive by it (and that from which they come to be is a principle of all things). He got his notion from this fact, and from the fact that the seeds of all things have a moist nature, and that water is the origin of the nature of moist things. . . .

Anaximenes and Diogenes make air prior to water, and the most primary of the simple bodies, while Hippasus of Metapontium and Heraclitus of Ephesus say this of fire, and Empedocles says it of the four elements (adding a fourth—earth—to those which have been named). . . .

From these facts one might think that the only cause is the so-called material cause; but as men thus advanced, the very facts opened the way for them and joined in forcing them to investigate the subject. However true it may be that all generation and destruction proceed from some one or (for that matter) from more elements, why does this happen and what is the cause? For at least the substratum itself does not make itself change; e.g. neither the wood nor the bronze causes the change of either of them, nor does the wood manufacture a bed and the bronze a statue, but something else is the cause of the change. And to seek this is to seek the second cause, as *we* should say—that from which comes the beginning of the movement. . . .

These thinkers, as we say, evidently grasped, and to this extent, two of the causes which we distinguished . . .—the matter and the source of the movement—vaguely, however, and with no clearness. . . .

Leucippus and his associate Democritus say that the full and the empty are the elements, calling the one being and the other non-being—the full and solid being being, the empty non-being (whence they say being no more is than non-being, because the solid no more is than the empty); and they make these the material causes of things. And as those who make the underlying substance one generate all other things by its modifications, supposing the rare and the dense to be the sources of the modifications, in the same way these philosophers say the differences in the elements are the causes of all other qualities. These differences, they say, are three—shape and order and position. For they say the real is differentiated only by 'rhythm' and 'inter-contact' and 'turning'; and of these rhythm is shape, inter-contact is order, and turning is position; for A differs from N in shape, AN from NA in order, ⌶ from H in position. The question of movement—whence or how it is to belong to things—these thinkers, like the others, lazily neglected.

Contemporaneously with these philosophers and before them, the so-called Pythagoreans, who were the first to take up mathematics, not only advanced this study, but also having been brought up in it they thought its principles were the principles of all things. Since of these principles numbers are by nature the first, and in numbers they seemed to see many resemblances to the things that exist and come into being—more than in fire and earth and water (such and such a modification of numbers being justice, another being soul and reason, another being opportunity—and similarly almost all other things being numerically expressible); since, again, they saw that the modifications and the ratios of the musical scales were expressible in numbers;—since, then, all other things seemed in their whole nature to be modelled on numbers, and numbers seemed to be the first things in the whole of nature, they supposed the elements of numbers to be the elements of all things, and the whole heaven to be a musical scale and a number. And all the properties of numbers and scales which they could show to agree with the attributes and parts and the whole arrangement of the heavens, they collected and fitted into their scheme; and if there was a gap anywhere, they readily made additions so as to make their whole theory coherent. E.g. as the number 10 is thought to be perfect and to comprise the whole nature of numbers, they say that the bodies which move through the heavens are ten, but as the visible bodies are only nine, to meet this they invent a tenth—the 'counter-earth'. . . .

After the systems we have named came the philosophy of Plato, which in most respects followed these thinkers, but had peculiarities that distinguished it from the philosophy of the Italians. For, having in his youth first become familiar with Cratylus and with the Heraclitean doctrines (that all sensible things are ever in a state of flux and there is no knowledge about them), these views he held even in later years. . . .

Plato . . . held that the problem applied not to sensible things but to entities of another kind—for this reason, that the common definition could not be a definition of any sensible thing, as they were always changing. Things of this other sort, then, he called Ideas, and sensible things, he said, were all named after these, and in virtue of a relation to these; for the many existed by participation in the Ideas that have the same name as they. Only the name 'participation' was new; for the Pythagoreans say that things exist by 'imitation' of numbers, and Plato says they exist by participation, changing the name. But what the participation or the imitation of the Forms could be they left an open question.

Further, besides sensible things and Forms he says there are the objects of mathematics, which occupy an intermediate position, differing from sensible things in being eternal and unchangeable, from Forms in that there are many alike, while the Form itself is in each case unique.

Hippocratic Collection

On the Sacred Disease

Most of what we know about early Greek medicine comes from a collection of about seventy treatises written by a number of different authors. In antiquity these works were attributed to Hippocrates, the most famous physician in the ancient world, who seems to have practiced and taught during the late fifth century B.C. *on the island of Cos, then a major center of medical study. The treatise,* On the Sacred Disease *does not deny that the gods may influence human life, but it does insist that no disease is more divine than any other. Most important, it demands that diseases be understood and treated in naturalistic and rationalistic ways rather than in theological terms and through religious rituals.*

But this disease [epilepsy] is in my opinion no more divine than any other; it has the same nature as other diseases, and the cause that gives rise to individual diseases. It is also curable, no less than other illnesses, unless by long lapse of time it be so ingrained as to be more powerful than the remedies that are applied. Its origin, like that of other diseases, lies in heredity. For if a phlegmatic parent has a phlegmatic child, a bilious parent a bilious child, a consumptive parent a consumptive child, and a splenetic parent a splenetic child, there is nothing to prevent some of the children suffering from this disease when one or the other parent suffered from it; for the seed comes from every part of the body, healthy seed from the healthy parts, diseased seed from the diseased parts. Another strong proof that this disease is no more divine than any other is that it affects the naturally phlegmatic, but does not attack the bilious. Yet, if it were more divine than others, this disease ought to have attacked all equally, without making any difference between bilious and phlegmatic.

This disease styled sacred comes from the same causes as others, from the things that come to and go from the body, from cold, sun, and from the changing restlessness of winds. These things are divine. So that there is no need to put the disease in a special class and to consider it more divine than the others; they are all divine and all human. Each has a nature and power of its own; none is hopeless or incapable of treatment. Most are cured by the same things as caused them. One thing is food for one thing, and another for another, though occasionally it does it harm. So the physician must know how, by distinguishing the seasons for individual things, he may assign to one thing nutriment and growth, and to another diminution and harm. For in this disease as in all others it is necessary, not to increase the illness, but to wear it down by applying to each what is most hostile to it, not that to which it is accustomed. For what is customary gives vigour and increase; what is hostile causes weakness and decay. Whoever knows how to cause in men by regimen moist or dry, hot or cold, he can cure this disease also, if he distinguish the seasons for useful treatment, without having recourse to purifications and magic.

B. L. Van Der Waerden

The Logical Structure of Greek Geometry

About 300 B.C. Euclid, working at Alexandria, produced his famous Elements of Geometry. *Starting from a set of twenty-three definitions, five postulates, and five "common notions," he was able to demonstrate almost every mathematical proposition known to his contemporaries and to create a textbook which represented the pinnacle of logical elegance and mathematical learning well into the nineteenth century.*

It is possible, however, that the logical structure of Greek geometry displayed in Euclid's work was the result of a long process of development. B. L. Van Der Waerden, in the following selection, takes the position that Thales, who lived almost three centuries before Euclid, may have logically constructed the so-called "congruence theorem," which was used to determine the distance of a ship at sea.

The following information concerning Thales is given by Proclus [c. A.D. 410–485], the commentator of the first book of Euclid's Elements, who obtained it from the History of Mathematics of Eudemus [fourth century B.C.], itself unfortunately lost:

1. He was the first to prove that a circle is divided into two equal parts by its diameter (Proclus, p. 275).

2. Besides several other theorems, he had obtained the equality of the base angles in an isosceles triangle; in ancient fashion, he called these angles not equal, but similar (Proclus, p. 341).

3. According to Eudemus, he discovered that when two straight lines intersect, angles are equal (Proclus, p. 374).

4. The congruence proposition concerning two triangles, in which a side and two angles are equal, was ascribed by Eudemus to Thales, with the remark that, in order to demonstrate the validity of his method for determining the distance between two ships at sea, Thales had to make use of this congruence theorem (Proclus, p. 409).

How might Thales have determined the distance between ships at sea? According to Tannery, the most ancient method that has been brought down to us is that of the Roman surveyor Marcus Junius Nipsius, who gives the following, indeed very primitive, rule:

In order to find the distance from *A* to the inaccessible point *B*, one erects in the plane a perpendicular *AC* to *AB*, of arbitrary length and determines its midpoint *D*. In *C* one constructs a line *CE* perpendicular to *CA*, in a direction opposite to that of *AB*, and one extends it to a point *E*, collinear with *D* and *B*. Then *CE* has the same length as *AB*.

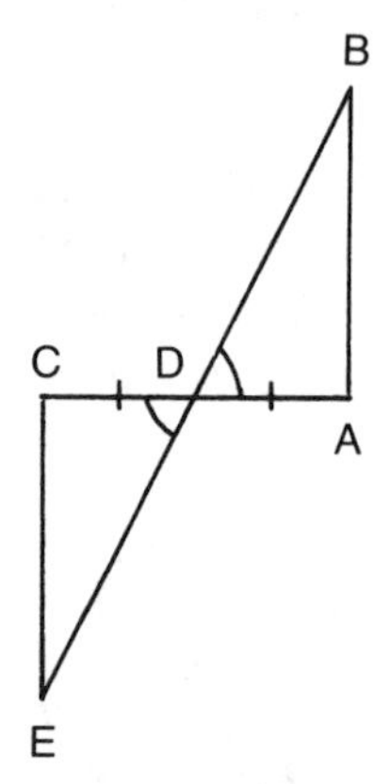

The congruence theorem (4), referred to by Eudemus, is indeed used in the proof of this rule, as well as proposition (3), concerning the equality of vertical angles, also known to Thales. It is therefore possible that this was Thales' method. . . .

The absolute accuracy of statements (1) and (4) has been drawn into question, even though they come from the best source. It is thought that the structure of old Thales' mathematics can not have been so strictly logical, that he would undertake to *prove* such obvious things as the equality of the parts into which a diameter divides a circle. Heath, the eminent English historian of Greek mathematics, observes in this connection, that even in Euclid this proposition is not proved. It has been held that the statements of Eudemus should be accounted for in this way, that he incorrectly assumed for the mathematics of Thales the same external structure as had been given to the mathematics of his own time (about 330–300), in which every proposition was derived by strictly logical steps from previous propositions, definitions and axioms. Thus Eudemus might, for instance, have reasoned as follows: Purely logically, the measurement of the distance between ships at sea, depends upon the congruence theorem mentioned in (4); therefore Thales must have known this theorem and have formulated it explicitly. It is thought that, in reality, Thales did perhaps apply this congruence proposition without being aware of it. Some people even believe that Thales did in no way prove his discoveries, but that he established them empirically, the very opposite of what Eudemus explicitly says in (1)!

The . . . objection to this view is that Eudemus not only knew the results of the mathematics of Thales, but also their external form, at least to a certain extent; he even knew the terminology which Thales used for the equality of angles. We are hardly justified therefore in simply ignoring his judgment that the geometry of Thales was constructed logically in the same way as that of the later mathematicians—and that evidently is his judgment because otherwise he would not have drawn the conclusion (4)—and certainly not his explicit statement that Thales had proved proposition (1). Only on the basis of better knowledge can one fairly correct an antique historian.

Adam Smith

The Leisure to Study and Understand Nature

Adam Smith (1723–1790) is best known for his The Wealth of Nations, *one of the founding texts of classical economics. Few people know that Smith's approach to economic topics owed a great deal to his fascination with the history of science, which he explored as a classics student in his youth. The following passages from "The principles which lead and direct philosophical inquiries: illustrated by the History of Astronomy" illustrates a widely held eighteenth-century version of the "leisure class" interpretation of the origins of Greek science.*

Nature, after the largest experience that common observation can acquire, seems to abound with events which appear solitary and incoherent with all that go before them, which therefore disturb the easy movement of the imagination; which make its ideas succeed each other, if one may say so, by irregular starts and sallies. . . . Philosophy, by representing the invisible chains which bind together all these disjointed objects, endeavours to introduce order into this chaos of jarring and discordant appearances, to allay this tumult of the imagination, and to restore it, when it surveys the great revolutions of the universe, to that tone of tranquility and composure, which is both most agreeable in itself, and most suitable to its nature. Philosophy, therefore, may be regarded as one of those arts which address themselves to the imagination. . . . Let us endeavour to trace it, from its first origin, up to that summit of perfection to which it is at present supposed to have arrived, and to which, indeed, it has equally been supposed to have arrived in almost all former times. It is the most sublime of all the agreeable arts, and its revolutions have been the greatest, the most frequent, and the most distinguished of all those that have happened in the literary world. Its history, therefore, must, upon all accounts, be the most entertaining and the most instructive. Let us examine, therefore, all the different systems of nature, which, in these western parts of the world, the only parts of whose history we know any thing, have successively been adopted by the learned and ingenious; and, without regarding their absurdity or probability, their agreement or inconsistency with truth and reality, let us consider them only in that particular point of view which belongs to our subject; and content ourselves with inquiring how far each of them was fitted to sooth the imagination, and to render the theatre of nature a more coherent, and therefore a more magnificent spectacle. . . .

When law has established order and security, and subsistence ceases to be precarious, the curiosity of mankind is increased, and their fears are diminished. The leisure which they then enjoy renders them more attentive to the appearances of nature, more observant of her smallest irregularities, and more desirous to know what is the chain which links them all together. . . .

Greece, and the Greek colonies in Sicily, Italy, and the Lesser Asia, were the first countries which, in these western parts of the world, arrived at a state of civilized society. It was in them, therefore, that the first philosophers, of whose doctrine we have any distinct account, appeared.

Benjamin Farrington

Greek Philosophy and Technology

Benjamin Farrington was one of a group of brilliant British Marxist thinkers who became deeply concerned during the 1930s by what they perceived to be the perversion of science in capitalist society. In a series of works published during the next two decades, Farrington located one of the major sources of this perversion in the anti-practical orientation of Platonic and Aristotelian thought and claimed that that orientation grew out of the fact that Greek society was built on slave labor. Though subsequent Marxists have produced more refined and sophisticated arguments, Farrington clearly expresses their basic perspective.

In the following passage Farrington focuses his critical attention on Plato's writings. (Note, where Farrington used the term technique modern American usage would substitute the term technology.)

When we come to . . . the connexion between philosophy and the techniques, which had proved so fruitful in an earlier period, we find that Plato has nothing to contribute. Preoccupied with theological, metaphysical, or political problems, and disbelieving in the possibility of a science of nature, Plato has little appreciation of the connexions between Greek thought and Greek practice which were clear to an earlier age. These connexions are many. Astronomy was, of course, not studied out of mere curiosity. It was studied in order to solve those very problems concern with which Plato deprecates—the exact relations of the lengths of day and night, of both to the month, and of the month to the year. On the solution of these problems depended the improvement of the calendar. On the improvement of the calendar depended improvements in agriculture, navigation, and the general conduct of public affairs. Neither was geometry studied, outside the Academy, purely for the good of the soul. It was studied in connexion with land-surveying, navigation, architecture, and engineering. Mechanical science was applied in the theatre, the field of battle, the docks and dockyards, the quarries, and wherever building was afoot. Medicine was a conspicuous example of applied science. It was a scientific study of man in his environment with a view to promoting his well-being. But the political programme put forward by Plato in the *Republic* and the *Laws* is all but barren of understanding of the rôle of applied science in the improvement of the lot of humanity. In his *Republic* and *Laws* Plato is wholly occupied with the problem of managing men, not at all with the problem of the control of the material environment. Accordingly the works, if full of political ingenuity, are devoid of natural science.

Plato carries this hostility, or indifference, to the science implicit in the techniques to very great lengths. A characteristic of the Ionian scientists had been the honour paid to great inventors, such as Anacharsis, who invented the bellows and made an improvement in the design of the anchor, or Glaucus of Chios, who invented the soldering-iron. These were examples of human ingenuity to an older age. Plato, however, did not think a human craftsman could originate anything; he had to wait for God to invent the Idea or Form of it. A carpenter, says Plato, could only make a bed by fixing his mind's eye on the Idea of the bed made by God. Theodorus of Samos, who invented the level, the lathe, the set-square, and the key, was thus shorn of his originality and of his title to honour; and Zopyrus, who invented the *gastrophetes,* or cross-bow held against the belly, had stolen the patent from God. The propounders of the modern theory of evolution found themselves embarrassed by the teaching of the Old Testament, that the various species of plants and animals, as they now exist, had been created by God. The technicians of the ancient world must have found it still more embarrassing to be told to wait upon the divine initiative before originating, or even improving, any technical device, since the present stage of technical development represented the divine plan.

But Plato went further than this in depressing the intellectual status of the technician. Not only is the technician robbed of the credit of inventiveness, he is also denied the possession of any true science in the art of manufacture. By an ingenious piece of sophistry Plato proves, in the same passage of the *Republic,* that it is not the man who *makes* a thing, but the man who *uses* it, who has true scientific knowledge about it. The user, who alone has true science, must impart his science to the maker, who then has 'correct opinion.' This doctrine effectually exalts the position of the consumer in society and reduces the status of the producer. Its political importance, in a slave-owning society, is obvious. A slave who made things could not be allowed to be the possessor of a science superior to that of the master who used them. But it constitutes an effective bar to technical advance or a true history of science. Plato has here, in fact, prepared the way for the grotesquely unhistorical opinion later current in antiquity, that it was philosophers who invented the techniques and handed them over to slaves.

Why did Plato think in this way? Plato had one of the best brains of which human history holds record. Why do his arguments lead sometimes to such wrong-headed conclusions? The answer is not difficult to give. . . . Suffice it here to suggest that Plato's thought was corrupted by his approval of the slave society in which he lived. Plato and Aristotle regretted the fact that any free labour still survived. In his *Politics* Aristotle remarks: 'The slave and his master have a common existence; whereas the artisan stands to his master in a relation far less close and participates in virtue only in so far as he participates in slavery.' In his *Laws* Plato organizes society on the basis of slavery, and, having done so, puts a momentous question: 'We have now made arrangements to secure ourselves a modest provision of the necessities of life; the business of the arts and crafts has been passed on to others; agriculture has been handed over to slaves on condition of their granting us a sufficient return to live in a fit and seemly fashion; how now shall we organize our lives?' A still more pertinent question would have been: 'How will our new way of life reorganize our thoughts?' For the new way of life did bring a new way of thinking, and one that proved inimical to science. It was henceforth difficult to hold to the view that true knowledge could be arrived at by interrogating nature, for all the implements and processes by which nature is made to obey man's will had become, if not in fact yet in the political philosophy of Plato and Aristotle, the province of the slave.

G. E. R. Lloyd

Greek Science and Society

G. E. R. Lloyd, the most outstanding contemporary interpreter of Greek science, addresses the general question of why self-conscious demands for logical coherence and empirical evidence to support claims about nature first emerged in classical Greece. In the following selection he argues that the critical or skeptical dimension of Greek science was closely linked with a focus on rhetoric that emerged as a consequence of the broadening participation of Greek citizens in political life.

In the admittedly speculative business of attempting to elucidate why it was that certain kinds of intellectual inquiry came to be initiated in ancient Greece, we must first take stock of certain of the economic, technological and other factors . . . affecting not only Greece itself, but also one or more of her ancient Near Eastern neighbours, notably (1) the existence of an economic surplus and of money as a medium of exchange, (2) access to, and curiosity about, other societies, and (3) changes in the technical means of communication and the beginnings of literacy. Without the first of these, the development of the institutions of the city-state—so expensive in time and manpower—is inconceivable. The second had its positive contribution to make to the widening of mental, as well as geographical, horizons, while without the third it is hardly an exaggeration to say that the new knowledge (which must in any case have been expressed quite differently) would have been stillborn.

Nevertheless the distinctive additional factors that must also be taken into account are, broadly speaking, political. Ancient Greece is marked not just by exceptional intellectual developments, but also by what is in certain respects an exceptional political situation: and the two appear to be connected. In four fundamental ways aspects of Greek political experience may be thought either to have directly influenced, or to be closely mirrored in, key features of the intellectual developments we are concerned with. First there is the possibility of radical innovation, second the openness of access to the forum of debate, third the habit of scrutiny, and fourth the expectation of justification—of giving an account—and the premium set on rational methods of doing so. . . .

While philosophy and science did not involve a different mentality or a new logic, they may be represented as originating from the exceptional exposure, criticism and rejection of deep-seated beliefs. Whereas limited scepticism about traditional schemata can be paralleled in other societies readily enough, the generalised scepticism about the validity of magical procedures we find in ancient Greek authors was unprecedented. If the concepts of 'nature' and of 'causation' develop from certain implicit assumptions, those ideas had, again, to be made explicit and generalised. These conceptual moves sound simple: but they could not be made without allowing fundamental aspects of traditional beliefs to come under threat. Philosophy and science can only begin when a set of questions is substituted for a set of vaguely assumed certainties. It is true that, the questions once posed, the answers given were sometimes not just schematic, but contained . . . elements of pure bluff. Yet while the Greeks' confidence in the rightness of their methods often outran their actual scientific performance—particularly in the matter of the collection of empirical data—those methodological ideals not only permitted, but positively promoted the further growth of the inquiry. The investigability of nature was explicitly recognised, even while the epistemological debate covered a wide spectrum of opinions on the character, aims and limits of that investigation.

The society in which these inquiries were first pursued was far from a primitive one. The level of technology and that of economic development were far in advance of those of many modern non-industrialised societies: above all literacy presents a difference not just of degree but of kind. Yet a comparison with Greece's ancient Near Eastern neighbours suggests that none of these three factors individually, nor all of them collectively, can be used to account fully for the developments we are interested in. So far as an additional distinctively Greek factor is concerned, our most promising clue (to put it no more strongly) lies in the development of a particular social and political situation in ancient Greece, especially the experience of radical political debate and confrontation in small-scale, face-to-face societies. The institutions of the city-state called for new qualities of leadership, put a premium on skill in speaking and produced a public who appreciated the exercise of that skill. Claims to particular wisdom and knowledge in other fields besides the political were similarly liable to scrutiny, and in the competition between the many and varied new claimants to such knowledge those who deployed evidence and argument were at an advantage compared with those who did not, at least—to repeat our proviso once again—so far as some audiences and contexts were concerned.

Moreover if this hypothesis helps to account for the strengths of Greek science, it also throws some light on some of its weaknesses. Although eventually Greek scientists produced lasting (if often elementary) results in areas of astronomy, mathematical geography, statics and hydrostatics, anatomy and even physiology, Greek science down to Aristotle is more notable for its achievements in second-order inquiries, in epistemology, logic, methodology and philosophy of science, in, for instance, the development of the concepts of an axiomatic system and of an exact science, and in that of the notion of empirical research. Whatever their limitations in the implementation of their ideas, the Greeks provided science with its essential framework, asserting the possibility of the inquiry and initiating the debate that continues today on its aims and methods. Yet several of the shortcomings of Greek science correspond closely to its strengths. The quest for certainty in an axiomatic system—itself in part a reaction against what was represented as the seductiveness of merely plausible arguments—was sometimes bought at the cost of a lack of empirical content. More generally, the way in which evidence and 'experiment' were often used to support, rather than to test, theories, a certain over-confidence and dogmatism, above all a certain failure in self-criticism, may all be thought to reflect the predominant tendency to view scientific debate as a contest like a political or a legal *agon*. Aristotle noted in the *De Caelo* . . . that 'we are all in the habit of relating an inquiry not to the subject-matter, but to our opponent in argument.' This remains true, no doubt, today, but the observation appears especially relevant to early Greek science. The sterility of much ancient scientific work is, we said, often a result of the inquiry being conducted as a dispute with each contender single-mindedly advocating his own point of view. This is easy to say with hindsight: but an examination of the Greek evidence suggests that this very paradigm of the competitive debate may have provided the essential framework for the growth of natural science.

Richard Olson

Science and Religion in Pre-Socratic Greece

Richard Olson, in the following passage, emphasizes the new attitudes toward the nature of the divine, which emerged in classical Greece as a result of the scientific arguments of pre-Socratic natural philosophers, and suggests why those new attitudes met with hostile conservative reactions.

Most discussions and classifications of the pre-Socratic philosopher-scientists have followed lines established by Aristotle, who was interested in how they accounted for the apparent flux of experience in terms of some unchanging principles. According to his classification scheme, the early Milesian philosophers, Thales, Anaximander, and Anaximenes, are best understood as monists; for each posited a single material substrate underlying the sensed variety of the cosmos. For Thales it was water; for Anaximander it was an indefinite amorphous substance (the apeiron); and for Anaximenes it was air. Other pre-Socratics, like Anaxagoras,

Empedocles, and the early atomist Democritus, are best understood as pluralists because they argued that there were two or more primordial material substances, and that nature demanded not only a material substrate but a cause of motion as well. Pythagoras and his followers form a special class because they believed that "the elements of numbers are the elements of everything, and that the whole cosmos is harmony and number."

. . . I would like to focus on certain common characteristics of pre-Socratic thought. Many of the fragments attributed to or about most pre-Socratics are unrelated, or only marginally related, to grand cosmogonic [concerned with the origin of the universe] speculations. Most of them offer explanations of particular natural phenomena, and many of the phenomena discussed deal with heavenly bodies, the weather, mathematics, and the nature of man—though there is an important class of fragments dealing with human society in post-Milesian times.

For the purpose of later reference, here is a short and characteristic selection of fragments dealing with a variety of topics:

> [1] [on thunder, lightning, thunderbolts, whirlwinds and typhoons] Anaximander says that all these things occur as a result of wind: for whenever it is shut up in a thick cloud and then bursts out forcibly, through its fineness and lightness then the twisting makes the noise, while the rift against the blackness of the cloud makes the flash.
>
> [2] Anaximenes says that the earth, through being drenched and dried off, breaks asunder, and is shaken by the peaks that are thus broken off and fall in. Therefore, earthquakes happen in periods both of droughts and again of excessive rains: for in droughts, as has been said, it dries up and cracks, and being made overmoist by the waters it crumbles apart.
>
> [3] [According to Anaxagoras] . . . the sun, the moon, and all the stars are red hot stones which the rotation of the aether carries around with it. . . . The sun exceeds the Peloponnesus in size. The moon has not any light of its own but derives it from the sun. . . . Eclipses of the moon are due to its being screened by the earth . . . those of the sun to screening by the moon when it is new. . . . He held that the moon was made of earth, and had plains and ravines on it. . . .
>
> [4] [Democritus explained man's belief in divine power as follows] . . . the belief arose from alarming natural occurrences such as thunder, lightening, thunderbolts, and eclipses, which men in their terror imagined to be caused by the Gods.

The final passage of this group has a different character than the others, but it may help to explain why the others had an important bearing on religious questions. Traditionally, Zeus was god of the sky, responsible for such occurrences as thunder and lightning; Poseidon was the "earth-shaker" whose anger manifested itself in earthquakes and volcanic eruptions; and Helios, the sun, was the ubiquitous witness of all sacred vows. To account for such events as thunder, lightning, and earthquakes without calling upon the gods was, by itself, to challenge or erode a very substantial element of religious belief. But even more importantly, if one accepted the naturalistic explanations of such phenomena, one might easily be drawn into atheistic arguments like those of Democritus and lose faith in other aspects of divinity.

Such implications of the naturalists' ideas were recognized and deeply resented by substantial numbers of Athenians. Euripides pointed out that Anaxagoras had reduced "the all seeing Helios, who traversed the sky every day in his flashing chariot and was the awful witness of men's most sacred oaths, to the status of a lifeless lump of glowing stone," and this made such an impression that it was said to have caused an attempted expulsion of Anaxagoras from Athens. From the challenge to Anaxagoras until the trial of Socrates about fifty years later it was dangerous to be known in Athens as a man who speculated about the heavens; astronomical and meteorological thought were tied to impiety and atheism in the popular mind. Plutarch tells us that Athenians "did not tolerate the natural philosophers and chatters about things in the sky, as they called them, dissolving divinity into irrational causes, blind forces, and necessary properties. Protagoras was banished, Anaxagoras put under restraint and with difficulty saved by Pericles, and Socrates, though in fact he had no concern in such matters, lost his life through devotion to philosophy."

Athenian intolerance showed up, as Plutarch suggests, not only in literary complaints but in the form of direct legal action. In about 440 B.C. a law that would impeach or exile "those who denied the Gods or taught about celestial phenomena" was introduced by Diopeithes. And in 411 B.C. Protagoras, who had been charged by the comic writer Eupolis of being an "imposter about the phenomena of the heavens," was brought to trial for impiety. Though we have only Plutarch's word for the outcome of that trial we can be reasonably sure that his report was correct, for Protagoras died in a shipwreck on leaving Athens in about 410 B.C., and exile would have been the natural sentence had he been convicted. Moreover, we know that the Athenians ordered copies of his books to be collected and burned in the marketplace. Far and away the most infamous persecution of ostensible impiety associated with "meddling in the affairs of the heavens" occurred in 399 B.C. when Socrates was tried and condemned to death.

In each of these cases (i.e., those of Anaxagoras, Protagoras, and Socrates) there were no doubt political reasons for the prosecutions, which were only marginally related to the avowed charges of impiety. But in each case an Athenian jury, which was probably responding at least in part to the charges rather than to the prosecution's hidden motives, felt that conviction was warranted because of the impious dangers of scientific speculation about the causes of heavenly phenomena.

Most pre-Socratic nature philosophers were neither atheistic nor agnostic, but their religious views tended to be very unorthodox. The two direct attributions that we have from Anaximander's cosmological writings demonstrate this clearly.

> [Anaximander] says that it is neither water nor any other of the so-called elements, but some other *apeiron* nature from which come into being all the heavens and the worlds in them. And the source of coming-to-be for existing things is that into which destruction, too, happens "according to necessity: for they pay penalty and retribution to each other for their injustice according to the assessment of time," as he describes it in these rather poetical terms.

> This seems to be the beginning of the other things and to surround all things and steer all, as all those say who do not postulate other causes, such as mind or love, above and beyond the infinite. And this is the divine; for it is immortal and indestructible, as Anaximander says.

Anaximander's divine is quite unlike the traditional gods. It has no human form and cannot be appealed to for special favors. Its governance of the world is inexorable. In these ways Anaximander's divine apeiron is abstract and remote from human experience. On the other hand, it has an important compensating virtue when compared with the Homeric gods because it is, above all, absolutely just—neither capricious nor immoral—and therefore is well suited to be a guiding principle for the new polis society in which equity in law and in commercial transactions was a crucially important issue.

Though Anaximander's religious ideas were more profound, and ultimately more influential, those of Anaximenes were of more immediate significance. Anaximenes and his near follower, Diogenes of Apollonia, asserted that air was the primary constituent of the universe, justifying their attitude in part by reference to the old notion of the breath soul. "Just as our souls, which were made of air, hold us together, so does breath and air encompass the world." According to their cosmologies the various things that we sense are caused by the felting or compacting of air. But Diogenes carries this idea two steps further to argue that what we cannot sense (i.e., the soul) must be composed of the rarest and purest air, to which men give the name aether, and that this purest of all things is the divine.

The notions of Anaximenes and Diogenes were adopted widely in connection with the so-called Orphic religions. W. C. K. Guthrie suggests that

> the religious mind, supported by the edifice of theory, which the philosophers had erected on the foundations of ancestral belief, could, without difficulty, find comfort in some such train of thought as this. The air is of kindred substance to our souls, that is, it is alive. At its purest, it must be the highest form of our soul or life. Untrammeled by mortal bodies, it is clearly immortal, and the purity which it attains in the upper

> reaches . . . is such that it must be a very high and intelligent form of life indeed. What else can one call such life but God? The mystically minded could deduce from these same premises a satisfying promise of immortality for man. . . . It seems therefore that the air God of the Ionian Philosophers, with which we started, was not so far removed as one might suppose from the religious aspirations of men whose minds were very different.

Whether or not we accept Guthrie's reconstruction of the Orphic mentality, there is substantial evidence of a semipopular association of both the soul and Zeus with the air or aether in the late fifth century, and following the natural philosophers' speculations. A funeral inscription dedicated to the Athenian soldiers who died fighting at Potidaea in 432 B.C., for example, reads, "Aether received their souls, their bodies, Earth."

Moreover, Euripides associates Zeus with the aether in several passages. And Phelemon, writing in the early fourth century, finds the air-Zeus association a fit subject for parody: "I am he from whom none can hide in any act which he may do, or be about to do, or have done in the past, be he God or man. Air is my name, but one might also call me Zeus."

Pre-Socratic science did, then, seem to authorize some religious beliefs. But these beliefs tended to be radically different from those of the Homeric tradition—beliefs in abstract, impersonal gods, or in natural gods so stripped of their old mystery and personality that they were subject to ridicule on the stage. During a period of terrifying rapid change and instability in such things as political institutions, moral standards, and economic circumstances, any challenge to old beliefs was likely to appear dangerous to the conservative element in society. And any person whose religious conceptions were at variance with those of tradition was likely to be charged with atheism and impiety. Thus, even where natural philosophy had a substantial theological bias it was often seen as being antireligious, and to seem antireligious was to seem dangerous.

CHAPTER 2

Medieval Contexts of Natural Knowledge

Introduction by Richard Olson

Greek science reached levels of mathematical elegance and conceptual complexity at Alexandria between 300 B.C. and A.D. 150 which were not approached, much less exceeded, in Europe for nearly a millennium and a half. As we shall see, for a brief period between A.D. 900 and A.D. 1300, Islamic scholars were able to add to the Greek legacy—especially in medicine, mathematics, and those physical sciences, like optics and astronomy, which drew heavily on mathematical techniques. But generally speaking, there was a dramatic decline in the Mediterranean region in both qualitative and quantitative levels of scientific activity during the period spanning the rise and fall of the Roman Empire and the first half millennium of the Christianization of Europe.

To a large extent the decline of ancient science preceded, or at least developed independently of, the growth of Christianity. Roman intellectuals never demonstrated much interest in knowledge for its own sake, and even those like Vitruvius, whose *Ten Books on Architecture* (c. A.D. 50) provides one of the best Roman encyclopedic surveys of the mathematical and physical sciences, generally warned students against getting carried away into a concern with nonutilitarian aspects of their studies: He writes that though architects should be "instructed in geometry—have followed the philosophers with attention, have some knowledge of medicine, . . . and be acquainted with astronomy and the theory of the heavens, . . . it appears . . . that he has done enough and to spare who in each subject possesses a fairly good knowledge of those parts with their principles, which are indispensable for architecture, so that if he is required to pass judgement and to express approval in the case of those things or arts, he may not be found wanting."

Totally aside from the fact that a pervasive antitheoretic attitude among Roman intellectuals encouraged superficial knowledge of scientific subjects even at the height of Roman prestige and power, there is little doubt that the rapid deurbanization of the western empire which accompanied the fall of Rome effectively destroyed almost all secular education in the West between the fall of Rome and the Carolingian Renaissance of the ninth century. Against this general background of Roman anti-intellectualism and cultural decline during the early middle ages, early Christian attitudes toward natural science and the preservation of learning in Christian monasticism appear relatively enlightened. As David Lindberg's article demonstrates, Christian attitudes toward science were complex and they varied from place to place; but with a few notable exceptions, Christian scholars advocated the study of nature and natural science on the grounds that it could provide valuable assistance in both understanding Scripture and defending the faith against pagan attack. Plato's *Timaeus* was particularly well thought of because its emphases on a single creator God and on the design of every entity in the universe toward some "good" end seemed to offer a philosophical parallel to the Genesis account of creation.

Of course, even in the so-called Hexamera—or commentaries on the six days of creation—scientific knowledge was generally subordinated to Christian ends, just as in the architectural tradition flowing from Vitruvius, scientific knowledge was clearly subordinated to productive ends. But because Christian intellectuals did maintain some interest in natural science while they had little but hostility to most pagan literature, their educational schemes emphasized the logical, mathematical, and scientific sides of the seven liberal arts that had provided the major structure to late Roman education. Consequently, arithmetic, geometry, music, and astronomy as well as logic provided a key part of monastic learning, and scientific concerns were kept alive even though scientific research was not encouraged.

Without the presence of an autonomous scientific community to promote scientific knowledge, and in the absence of printing to stabilize knowledge and expand access to it, scientific learning in Western Europe gradually deteriorated. Errors in scientific drawings and texts were copied and multiplied from one generation of manuscripts to the next in a process that was reversed only during the twelfth and thirteenth centuries, when the strengthening European economy gave rise to a new institution of learning—the university—which in turn recreated a more-or-less autonomous scientific community.

The Eastern Mediterranean and North African regions never experienced the extreme destruction of urban intellectual life that characterized Europe during the early middle ages; and as Islam swept through this area during the seventh century it was rapidly transformed into what Brian Stock characterizes as an "intensively urbanized culture, largely dependent on paper, writing, and administration." Mohammed had demonstrated both a deep respect for learning and a sincere love of the natural world in his own writings, and as Islam matured and became urbanized, its emphasis on learning intensified.

It is true that in Islam as in Christendom, "worldly" knowledge was, in principle, subordinated to religious purposes. But as Stock makes clear in "The Growth of Islamic Science," Islamic support for learning in general led to a situation in which "science in one form or another was the part-time or full-time occupation of a large number of intellectuals." And this situation in turn produced a corpus of scientific works which both incorporated and advanced upon knowledge drawn from the Greeks, from the ancient Near Eastern tradition, and from the Indian tradition, which became accessible when Islam spread eastward. Thus, when Western contacts with Islam became increasingly frequent and intense—especially in Muslim Spain and Sicily during the eleventh and twelfth centuries—Europeans were exposed to a vital and exciting scientific tradition.

Initially, European interest in Islamic science was largely limited to the applied sciences. Works on mathematics and such mathematical sciences as astronomy were translated by Europeans visiting Islamic Spain because the Arabic number system made commercial arithmetic easier to do than it was with Roman numerals and because Islamic astronomy made calendar construction and navigation easier and more reliable. Thus, at the very beginning of the eleventh century a monk named Gerbert of Aurillac—who would eventually become Pope Sylvester II—brought back a number of translated mathematics and astronomy texts from the Spanish monastery of Santa Maria del Rippole, where he had studied as a novice, and began to teach from them at the cathedral school at Reims. Similarly, Europeans who came to recognize the advanced state of Islamic medicine in

Spain and Muslim-controlled Sicily began translating medical works for the education and use of physicians.

Three sets of circumstances, however, soon led to increased European interest in the "philosophical" or theoretical side of Islamic science, to an intense interest in recovering Greek—especially Aristotelian—scientific knowledge, and to the revitalization of science in general in Latin Christendom.

One of these sets of circumstances was connected with the nature of Islamic scientific ideas and institutions. Within Islam the distinction between theoretical and practical knowledge, which had been so important for Plato and Aristotle, was minimized. So Europeans, searching Islamic libraries for useful information, were inevitably brought face to face with theoretical materials. And they often found these fascinating. Moreover, their fascination was enforced and expanded by the arguments of such Islamic scholars as Avicenna (980–1037), author of the most advanced and popular medical work, *Canon of Medicine*. Avicenna was constantly referring to and praising the virtues of a Greek philosopher named Aristotle. His attitude toward Aristotle is beautifully portrayed by his follower Nazami Arudi:

> For four thousand years the wise men of antiquity travailed in spirit and melted their very souls in order to reduce the science of philosophy to some fixed order, yet could not effect this, until . . . that incomparable philosopher and most powerful thinker, Aristotle, weighed this coin in the balance of Logic, assayed it with the touchstone of definitions, measured it by the scale of analogy, so that all doubt and ambiguity departed from it, and it became established on a sure and critical basis. And during these fifteen centuries which have elapsed since his time, no philosopher hath won to the inmost essence of his doctrine, nor travelled the high road of his method, save that most excellent of moderns, the Philosopher of the East, the Proof of God unto his creature, Avicenna. He who finds fault with these two great men will have cut himself off from the company of the wise, placed himself in the category of madmen, and exhibited himself in the ranks of the feeble minded.[1]

In the face of such enthusiasm for Aristotle it is hardly surprising that we should find Avicenna's first translator, Gerard of Cremona, turning to translate Aristotle's *Meteorology, Physics, On the Heavens,* and *On Generation and Corruption* from Arabic into Latin, initiating a rash of translating activity which made virtually all of Aristotle's works available in Latin by the middle of the thirteenth century.

A second set of circumstances that encouraged European Christian interest in natural science in general and Islamic science in particular was the emergence of a concern for natural phenomena in their own right, rather than as mere symbols of Christian doctrines. As Lynn White discusses in his article on "Natural Science and Naturalistic Art in the Middle Ages," this changed attitude developed first among members of the Franciscan order and spread rapidly among Christian intellectuals and artisans during the twelfth and thirteenth centuries.

The third set of circumstances leading to the reemergence of science—long dominated by Islamic and Aristotelian elements—as the central concern of many European intellectuals is related to the growth of medieval universities. These institutions came into existence during the twelfth and thirteenth centuries as a rapidly expanding and urbanizing European economy and society demanded an increasing number of educated clerics, lawyers, physicians, and bureaucrats. But they soon became the unintended institutional setting for a renewal of scientific study throughout Europe. In his discussion of "The Emergence of the Professional University Teacher in the Medieval University," Joseph Ben-David explains how a new class of intellectuals, professionally committed to learning and teaching what was known about the natural world, emerged in the universities and began to challenge long-standing European assumptions about the subordination of scientific to theological knowledge.

Through their commitment to studies that had a strong Islamic and Aristotelian orientation rather than a Christian and Platonic one, and through their emphasis on nature for its own sake rather than as Christian symbol, many of the new arts masters were clearly attempting to change the traditional Western view of the universe. Their attempts quite naturally met strong resistance; and this resistance was intensified both by a doctrinal conflict between Islamic-Aristotelian doctrines and Christian beliefs and by a parallel institutional conflict between some of the new arts masters and the old faculties of theology for control of the university curriculum.

It is important to recognize that very few medieval intellectuals were either pure philosophers or pure theologians. Almost all of the greatest thirteenth-century thinkers taught in the arts faculty (where philosophy was the central subject) while they earned their degree in theology and then moved on to the theology faculty as positions opened up. This was the pattern followed by Robert Grosseteste at Oxford and by Thomas Aquinas at Paris, for example. Thus, most medieval scholars were deeply concerned with finding a way to reconcile new scientific developments with their theological commitments. In the process of accommodation between the emerging scientific tradition and traditional Christian theology *both* traditions were dramatically transformed, as A. C. Crombie explains in the following excerpt from *Medieval and Early Modern Science*. On the one hand, Christian theologians began to move away from their long-standing Platonic emphasis on God's omniscience and rationality and toward an emphasis on his unfathomable and unlimited will and power. On the other hand, Christian scientists gradually abandoned the Islamic interpretations of Aristotle, which viewed his doctrines as necessarily true, and focused increasingly on empirical studies of natural phenomena and on critical, though seldom hostile, analyses of Aristotelian interpretations of nature.

Although he was undoubtedly oversimplifying the situation, the medieval historian C. Warren Hollister put his finger on one of the important consequences of the thirteenth-century interactions between science and theology when he claimed that they opened up "two [new] paths into the future: pietism uninhibited by reason and science uninhibited by revelation."[2]

1. Seyyed Hossein Nasr, *Science and Civilization in Islam* (Cambridge, Mass.: Harvard University Press, 1968), 210–211.
2. C. Warren Hollister, *Medieval Europe,* 4th ed. (New York: John Wiley, 1978), 273.

David C. Lindberg

Science and the Early Church

From the mid-nineteenth to the mid-twentieth century almost all attempts to interpret the relationships between science and Christianity were shaped by a perspective which is accurately reflected in the very titles of the books in which it was first expressed. In 1874 John William Draper's History of the Conflict Between Religion and Science *appeared, to be followed in 1876 by Andrew Dickson White's "Warfare of Science," which was expanded over twenty years into his massive* A History of the Warfare of Science With Theology *(2 vols., 1896). Within a very few years these two works, which tended to see a continuous history of "enlightened" science battling with superstitious religion, were met with such Christian defenses as Gavin Carlyle's* The Battle of Unbelief *(1878) and Nevison Loraine's* The Battle of Belief *(1891). Regardless of which side historians chose to support, they continued until recently to view the interactions between science and Christianity as generally hostile. During the past twenty years, however, there has been an attempt to get beyond the warfare and battle metaphor and to approach the relations between science and Christianity with less bias and greater subtlety. David Lindberg's analysis of science and early Christianity is a good example of this new approach.*

Although science was neither an autonomous discipline nor a profession during the patristic period [c. 95–604, when Church fathers developed the concepts and doctrines to guide the Christian faithful], we can nonetheless investigate the relationship between Christianity and those aspects of the philosophical enterprise which were concerned with nature. Did science in this sense benefit or suffer from the appearance and triumph of Christianity? Did Christianity, with its otherworldliness and its emphasis on biblical authority, stifle interest in nature, as the old stereotype proclaims? Or was there a more ambiguous and subtle relationship?

We must begin our inquiry by briefly surveying the state of science in late antiquity. Was there, in fact, a decline of science for which Christianity might be held responsible? The answer is not simple. Surely there are instances of important scientific work in the early centuries of the Christian era. Ptolemy's work in astronomy and Galen's in medicine (both in the second century A.D.) and Diophantus's mathematical efforts (in the third century) are outstanding examples. And . . . John Philoponus presented a major and important reassessment of Aristotelian physics and cosmology as late as the sixth century. Nevertheless, it is agreed by most historians of ancient science that creative Greek science was on the wane, perhaps as early as 200 B.C., certainly by A.D. 200. Science had never been pursued by very many people; it now attracted even fewer. And its character shifted away from original thought toward commentary and abridgment. Creative natural science was particularly scarce in the Roman world, where scholarly interests leaned in the direction of ethics and metaphysics; such natural science as Rome possessed was largely confined to fragments preserved in handbooks and encyclopedias.

Can Christianity be held responsible in any way for this decline? Let us first consider Christian otherworldliness. In antiquity there was a broad spectrum of attitudes toward the material world. At one end of the spectrum was pagan cosmic religion, constructed from a mixture of Pythagorean, Platonic, Aristotelian, and Stoic doctrines. This cosmic religion saw the material cosmos, or at least its upper, heavenly part, as a perfect expression of divine creativity and providence, "the supreme manifestation of divinity," and indeed itself a divine being. Moreover, study and contemplation of the cosmos were judged the only ways to God; natural philosophy and theology had been merged. At the other end of the spectrum was the Gnostic attempt to equate the material world with evil. The cosmos was viewed by "pessimistic" Gnostics as a disastrous mistake, the scene of disorder and sin, the product of evil forces, the antithesis of the divine, and a prison from which the soul must escape in order to make its way to its true home in the spiritual realm. Finally, between these extremes there was Platonic philosophy (or, in the hands of certain Neoplatonists, Platonic religion), which distinguished clearly between the transcendent world of eternal forms and their imperfect replication in the material cosmos. Neoplatonists by no means considered the world to be evil; it was the product of divine intelligence and, as A. H. Armstrong puts it, the "best possible universe that could be produced under difficult circumstances." Contemplation of it was even held to play a positive, albeit small, role in leading the soul upward to the eternal forms. Nevertheless, Neoplatonism was fundamentally otherworldly; the material world, for all its beauty, remained the scene of imperfection and disorder; and it had to be escaped before humanity could achieve its highest good, the contemplation of eternal truths.

There was, of course, no unitary Christian view of the material world. But orthodox Christianity, as it developed, emphatically rejected the extremes; nature was neither to be worshiped nor to be repudiated. Christianity was deeply influenced by Neoplatonic philosophy, and most Christian thinkers adopted some form of the Neoplatonic attitude. Gregory of Nyssa (ca. 331–ca. 396) believed deeply in the unreality and deceitfulness of the material world and yet recognized that it could provide signs and symbols that would lead mankind upward to God. Augustine insisted that sin is situated not in the body but in the will. This was a point of extraordinary importance, because it helped to liberate western Christendom from the notion that the soul is contaminated by its contact with the body—and therefore that matter and flesh must be inherently evil. Nevertheless, in a well-known passage from the *Enchiridion,* Augustine expressed serious doubt about the value of natural science:

> When it is asked what we ought to believe in matters of religion, the answer is not to be sought in the exploration of the nature of things, after the manner of those whom the Greeks called "physicists." Nor should we be dismayed if Christians are ignorant about the properties and the number of the basic elements of nature, or about the motion, order, and deviations of the stars, the map of the heavens, the kinds and nature of animals, plants, stones, springs, rivers, and mountains; about the divisions of space and time, about the signs of impending storms, and the myriad other things which these "physicists" have come to understand, or think they have. . . . For the Christian, it is enough to believe that the cause of all created things, whether in heaven or on earth, whether visible or invisible, is nothing other than the goodness of the Creator, who is the one and the true God.

Yet, insofar as scientific knowledge is required, it must be taken from the pagan authors who possess it:

> Usually, even a non-Christian knows something about the earth, the heavens, and the other elements of this world, about the motion and orbit of the stars and even their size and relative positions, about the predictable eclipses of the sun and moon, the cycles of the years and the seasons, about the kinds of animals, shrubs, stones, and so forth, and this knowledge he holds as certain from reason and experience. Now it is a disgraceful and dangerous thing for an infidel to hear a Christian, presumably giving the meaning of Holy Scripture, talking nonsense on these topics; and we should take all means to prevent such an embarrassing situation, in which people show up vast ignorance in a Christian and laugh it to scorn.

A view broadly the same as Augustine's was presented by Pope Leo the Great (440–461). Leo argued that the material world is not to be denigrated:

> Man, awake, and recognize the dignity of your own nature. Remember that you were made in the image of God; and though it was spoilt in Adam, it has been remade again in Christ. Use these visible creatures as they ought to be used, as you use earth, sea, sky, air, springs and rivers; and praise and glorify the Creator for everything fair and wonderful in them.

But neither should the material creation be allowed to occupy the center of attention:

> Do not devote yourself to the light in which birds and snakes, beasts and cattle, flies and worms delight. Feel bodily light with your bodily senses and clasp with all the strength of your mind that true light which "lightens every man coming into this world." . . . For if we are the temple of God, and the Spirit of God dwells in us, what everyone of the faithful has in his own soul is more than what he admires in the sky. We are not, of course . . . telling you this to persuade you to despise the works of God, or to think that there is anything against your faith in the things which the good God has made good; but so that you may use every kind of creature, and all the furniture of this world, reasonably and temperately. . . . So, since we are born to things of this present life but reborn to those of the future life, let us not devote ourselves to temporal goods but be set on eternal ones. . . .

The material world is not to be loved but to be used; it is not an end in itself but a means to the contemplation of higher things.

What were the implications of this attitude for the scientific enterprise? If we employ as a standard of comparison some sort of ideal world, a scientist's paradise, in which social values and resources are all marshaled in support of scientific research, Christianity may be judged harshly: the church was certainly not calling for the establishment of scientific research institutions nor urging able young men to undertake scientific careers. Most of the pejorative pronouncements regarding the early church in relation to science seem to spring from the anachronistic application of precisely such a standard. But what we must realize is that the early church was expressing values obtained from the pagan environment. On a spectrum of *pagan* values, from cosmic religion to Gnostic repudiation of the cosmos, the church fathers chose a middle position. There can be no doubt that biblical teaching about the creation as God's handiwork was influential in determining where on the spectrum Christians would land, and therefore it is clear that their Christianity was highly relevant to the issue; but it must be recognized that the alternatives from which they chose were of pagan origin.

It seems unlikely, therefore, that the advent of Christianity did anything to diminish the support given to scientific activity or the number of people involved in it. The study of nature held a very precarious position in ancient societies; with the exception of medicine and a little astronomy, it served no practical function and was rarely seen as a socially useful activity. As a result it received little political patronage or social support but depended on independent means and individual initiative. With the declining economic and political fortunes of the Roman Empire in late antiquity, people of independent means decreased in number, and initiative was directed elsewhere. Moreover, changing educational and philosophical values were diverting attention from the world of nature. Inevitably the pursuit of science suffered. Christianity did little to alter this situation. If anything, it was a little less otherworldly than the major competing ideologies (Gnosticism, Neoplatonism, and the mystery religions) and offered slightly greater incentive for the study of nature. Christians regarded science as important only insofar as it served the faith; but at least it did, on occasion, serve the faith. . . .

. . . Thus the church fathers used Greek natural science, and in using it they transmitted it. We must count this transmission as one of the major Christian contributions to science. Until the twelfth century, when a wave of translation brought an abundance of new sources to the Latin-speaking West, patristic writings constituted a major repository of scientific learning.

What the church transmitted, it also altered—and had its own doctrines altered in return. Christian doctrine and Greek natural philosophy must be viewed not as independent, unchangeable bodies of thought, situated side by side in the patristic period, with an occasional exchange of fisticuffs, but as interacting and mutually transforming views of the world. Christianity transformed the philosophical tradition, first, by performing a selective function. Because the church fathers had a strong preference for Platonic philosophy, they helped to determine that Plato's view of the world would prevail for a thousand years, until direct access to Aristotelian philosophy was gained in the twelfth century. . . . Christians learned to read the Bible with Greek, particularly Platonic, eyes; and Christian theology became thoroughly imbued with Greek metaphysics and cosmology. The extent of this mutual transformation was probably unrecognized by the participants, and unwanted; but unless we take cognizance of it, we cannot begin to understand the subsequent course of Western theology, philosophy, and science.

Brian Stock

The Growth of Islamic Science

Just as there has been a long but recently revised tradition of seeing science and Christianity as hostile to each other, there has been a long tradition of seeing Islamic science as important solely because it kept classical Greek knowledge alive. In the following article Brian Stock acknowledges the significance of the Arabic tradition of translations from Greek classics. But he also tries to go beyond the old stereotype to address both the question of why Islam was supportive of scientific activity and the question of what was original about Islamic science.

The first modifications of Greek science into the form in which it passed to the West were undertaken in the Hellenistic period [323–30 B.C., following the death of Alexander the Great, during which Greek culture spread throughout the ancient world and was finally absorbed into the Roman Empire]. The later stages were enacted in Islam.

In 622, a religious reformer called Muhammad, then in his forties, after attempting to propagate his ideas unsuccessfully for several years in Mecca, became "chief magistrate" in Medina, thereby gaining control of an urban community for the first time in his career. His popular acclaim coincided with the maturing of supra-tribal sentiment among the Bedouins. The result was a new, dynamic religious force, Islam. The spread of Islam began during the prophet's lifetime and gained enormous speed and momentum after his death, gradually extending Arab domination over a huge land mass stretching from contemporary Pakistan to Spain. By 650 it had overthrown the enfeebled Persian Empire and taken Syria and Egypt from Byzantium. The conquest of North Africa was completed by 698. Within a few generations the scientific traditions of the Greek East began to pass slowly into Arab hands, where they remained, as far as the West was concerned, until [following the Christian reconquest of such centers of learning as Toledo and Cordoba] Islamic scientific texts began to find their way northward, where they intermingled with translations directly from the Greek. . . .

Where urban culture shifted, science was sure to follow. Alexandria was taken during the Egyptian campaign of 642, but the tradition of its great library was not extinguished. Similar research facilities were created in Baghdad between 780 and 820. Even before the *hegira* the persecutions of the Byzantine Empire, which affected pagan philosophers as well as minor Christian sects, forced a number of scientific figures eastward. Far from disrupting the growth of interest in natural philosophy, Islam brought into being an eastern Mediterranean focus which had quietly been taking shape for some time. Situated at the intersection of the trade routes that linked East and West, Islam was the meeting point for Greek, Egyptian, Indian, and Persian traditions of thought, as well as the technology of China. It was to be expected that the strengths of Greek science would be affirmed; it would not be surprising if in some areas they were surpassed: this is precisely what happened.

Three interdependent factors stimulated the growth of an indigenous scientific tradition in Islam: translations, the cultural force of the new religion, and the successful assimilation of non-Arabs into various disciplines.

The translations were made possible by the availability of paper, a cheap, efficient medium suitable for a dry climate. Introduced into the Muslim world after the capture of Samarqand in 704, where only two years before the Chinese had set up a manufacture, it spread gradually westward, reaching Baghdad in 794. Soon a wide selection—including Egyptian wrapping tissue and a lightweight grade for pigeon post—was seen all over the Islamic world. Paper provided the means for transforming the largely oral Arab culture into a written form. It has been argued that Arabic is inherently a good language for expressing scientific constructions—a fact of considerable importance for mathematization. But what really made an Islamic scientific tradition possible was the shift from a tribal, nomadic way of life, with the clan as its center, to a more stable, geographically fixed and intensively urbanized culture, largely dependent on paper, writing, and administration. The great model for this transformation was the Koran itself. For the Muslim, Arabization was synonymous with the coming of the Book. The nomadic stage of development was looked back upon as *jahiliyya,* a time of ignorance and uncouthness. A certain amount of respect was reserved for the other "peoples of the Book," the Jews and Christians. Scribes, from the first, were a class apart. They not only transferred ideas from one tongue to another; they played a decisive role in "building up a complex and lucid Arabic philosophical vocabulary and laying the foundation of a philosophical Arabic style." With paper cheap and plentiful, the translation of Greek, Indian, and Persian scientific and philosophical texts underwent a renaissance between 800 and 1000, after which the Muslim world possessed an incomparably finer library of the ancients than was available in Latin. . . .

If one looks over the field of Muslim science as a whole, the most remarkable feature is not the frequently emphasized appearance of a handful of universal minds—al-Kindi, al-Khwarizmi, al-Razi, al-Farabi, al-Biruni, Avicenna, Alhazen, and Averroes—but, rather, that science in one or another form was the part-time or full-time occupation of so large a number of intellectuals. . . . The following may give a general idea of the place of individual sciences within Islamic culture and of their significance in influencing Western medieval development.

Mathematics

The intensity of mathematical research is clearly revealed from [the many translations and commentaries of Euclid's *Elements* mentioned] in al-Nadim's *Fihrist*. . . . The tradition outlined by al-Nadim fell into three stages: translation, exposition, and innovation. These are the categories still used by historians to describe the development of Muslim mathematics. The only aspect of the subject he omitted was the practical. He did not recall that the diffusion of Hindu-Arabic numerals and the decimal positional system was brought about by trade. Nor did he mention that Muslim mathematicians, to a much greater degree than the Greeks, interested themselves in everyday problems. Masha'allah, the noted astrologer (d. ca. 815–20), was the author of a treatise on commodity prices. Abu'l-Wafa' combined original work on Euclid and Diophantus with books bearing such titles as *What is Necessary from the Science of Arithmetic for Scribes and Businessmen* and *What is Necessary from Geometrical Construction for the Artisan*. In these works the theory was old but the examples were new. One may doubt that the most refined theory penetrated commercial circles, but commerce stimulated the theorists and oriented them toward the concrete.

Astronomy

Early historians undervalued Arab astronomy because it failed to overthrow the geocentric system, but for the practical purposes of surveying and fixing the dates of feasts Ptolemy's model was adequate. Omar Khayyam, working within the Ptolemaic framework, even succeeded in making a solar calendar more accurate than the Gregorian. A large part of Muslim astronomy dealt with civil calculations and the construction of observatories and tables. It chiefly surpassed the Greeks in application, measurement, and instrumentation. Al-Fazari (d. ca. 777), who helped to introduce Hindu notation into Islam, was one of the first to construct astrolabes. Al-'Abbas and al-Nahawandi were primarily known for having taken part in the observations at Baghdad in 829–30, at Damascus in 832–33, and at Jundishapur. A distinctive achievement of early Islamic astronomy was to separate the observatory from the temple and from astrology and to make it part of a scientific tradition. One of the first was built by Sanad ibn 'Ali for al-Ma'mun, who died in 833. From his reign onward every caliph and many minor princes felt obliged to have an observatory. . . . The observatory was one of the few institutions of Muslim science to survive the Mongol invasion and to pass directly to the modern world.

The tendency to improve the geocentric system without overthrowing it is well illustrated by the careers of al-Battani (d. 929) and al-Biruni (d. 1048). . . . Al-Battani began with mathematical considerations: the division of the heavenly sphere into signs and degrees and the principles of astronomical calculation. His improvements on Ptolemy were also calculatory. He more accurately determined the magnitude of lunar eclipses, the moon's mean longitude, and the apparent diameters of the sun and moon throughout the year. . . . Yet, if al-Battani's *De scientia stellarum* was accessible to the Latins, the works of al-Biruni were not. Thus, one of Islam's universal minds remained virtually unknown in the West until modern times. A Persian by birth, a rationalist in disposition, this contemporary of Avicenna and Alhazen not only studied history, philosophy, and geography in depth, but wrote one of the most comprehensive of Muslim astronomical treatises, the *Qanun al-Mas'udi*. In a spirited series of letters to Avicenna, his junior by seven years, al-Biruni also criticized at length Aristotle's conception of heavenly motion. Against the latter he denied the possibility of heavenly gravity, the notion that stars and planets move in circles, and the doctrine that each element has its own intrinsically suitable position. Observation and reason, he said, taught otherwise. . . .

Medicine

The oriental medical tradition began before Islam at Jundishapur, which was founded on the site of an army camp in the third century by the Sassanid king Shapur I. After 489, when the school of medicine at Edessa was closed by the Byzantine emperor, Nestorian physicians fled to the new town, bringing with them a number of Greek texts translated into Syriac. Jundishapur was further enriched after 529, when Justinian closed the pagan school at Athens, and in the seventh century, after the destruction of Alexandria and the opening of trade routes to India, along which passed new pharmacological information. Of all the sciences, medicine was held in the highest esteem by the new rulers. As a doctrine governing every aspect of life, Islam had to concern itself not only with worship but also with diet and personal hygiene. An anthology of the Prophet's writings on health is still popular in parts of the Arab world. The connection between morality and medicine was also preserved in the tradition of the hakim, who was at once a wise man and a physician. From the time of al-Mansur, who summoned Jirjis Bakhtyishu', the Nestorian doctor, to Baghdad to cure his dyspepsia in 765, Muslim capitals became centers of clinical training. Unlike the Roman *valetudinarium,* an improvised hospital which was imported from the battlefield to the city, the Muslim *bimaristan* was a highly specialized institution where, as early as the tenth century, diagnosis was an established art and candidates were examined for degrees. Also, drugs formed an important part of treatment. If Galen had brought pharmacology from an *ars conjecturalis* [an art of conjecture] to an exact discipline, in Islam the pharmacy first became institutionalized as a dispensary for drugs from doctors' prescriptions.

. . . The four greatest doctors [in the early middle ages] were Muslims. The earliest, al-Razi (d. ca. 925), was the finest clinician and a pioneer in chemical medicine. He anticipated certain modern discoveries: his treatise on natural habit looks forward to the theory of the conditioned reflex, and his monograph on children's diseases to specialized pediatrics. His famous study of measles and smallpox was printed regularly in Latin from 1489 to 1781, when it passed into several vernacular languages. Much of his enormous output was not translated directly into Latin but was absorbed into the encyclopedias of his successors. Chief among them was 'Ali ibn al-'Abbas (d. ca. 995), who was, after Avicenna, the most influential Arab physician in the West. His *Liber regius,* which was more systematic than al-Razi's *Continens* and more practical than Avicenna's *Canon,* was in common use down to the Renaissance. In surgery, the major contributor was Abu-l-Qasim (d. 1013), the Latin Albucasis, whose work had a greater vogue in the West than in the Orient, where there were more elaborate religious sanctions against the cutting of the body.

But earlier physicians were eclipsed in theory if not in practice by Ibn Sina or Avicenna (d. 1037), who was at once Islam's archetypical and most influential scientist-philosopher. Avicenna was the son of a tax collector in the eastern province of Balkh and was educated at Bukhara. Although known primarily as a philosopher and physician, he wrote on a wide variety of subjects, including natural history, physics, chemistry, astronomy, mathematics, and music. An important commentator on the Koran and Sufi doctrine (at least, so it is thought), he also composed excellent verse and, as a counselor to various princes, contributed to the solution of important economic and political problems. His two most influential works were the *Kitab al-Shifa,* the *Book of Healing* [*of the Soul*], which treated logic, physics, mathematics, and metaphysics; and the *Qanun fi'l-Tibb,* the famous *Canon of Medicine,* which was translated by Gerard of Cremona between 1150 and 1170, and remained the standard textbook of Galenism down to Harvey [in the seventeenth century]. The *Canon* is divided into five books. Book 1 deals with generalities concerning the body, sickness, and health; book 2 with the pharmacology of herbs; book 3 with pathology; books 4 and 5 with fevers, signs, symptoms, diagnostics, and prognostics. Influenced by Aristotle, Plotinus, and al-Farabi, the *Canon* was highly philosophical in language and style; yet it remained close to the real problems of health and disease and, more than Galen, returned to the pragmatism of Hippocrates. Nor was Avicenna's influence limited to medicine: in his unique codification of previous tradition one can see the roots of virtually the whole of scholastic philosophy in the West. In contrast to al-Biruni's rationalism and al-Razi's clinical empiricism, both of which illustrated the separation of science and religion, Avicenna incorporated scientific knowledge into an Aristotelian framework that was ideally suited for acceptance in Christendom. Yet within Islam he represented a turning point in another sense. So universal was his achievement that a repetition of it was unthinkable. Already overconcerned with their relation to the ancients, Muslim authors now had to contend with a philosophical classic of their own.

The Method of Science

Islam made a wide variety of contributions to scientific method. Major figures like al-Battani were also expert makers of instruments that enhanced their powers of observation and calculation. The artisan-scientist, who was considered an aberration in the ancient world, was more of a norm in Islam. The dependence on well-made apparatus meant that theory and practice were brought closer together. On at least two occasions in the *Canon,* Avicenna indicated that they were different but interdependent parts of the same discipline. Remarks of this kind were not limited to medicine, where an inevitable problem arose in relating Galenic theory to clinical practice. By introducing new disciplines like ophthalmology, or granting new status to old ones like geography, the Arabs made practice a more natural part of science itself.

Within the *quadrivium* [arithmetic, music, geometry, and astronomy, the "upper division" of the seven liberal arts taught in the middle ages] their contribution to method was threefold. They increased the range of mathematization, not only by developing algebra and trigonometry, but by attacking hitherto difficult questions through a quantitative approach. In physics al-Biruni made among the most accurate measurements of specific densities down to modern times; at a more practical level, the dispensaries of urban hospitals prescribed accurate amounts of drugs of controlled composition. Secondly, they restored Aristotle's philosophy of science to its original integrity, from which point it could stimulate the growth of new thinking in the West. This was the work of three philosophers of Muslim Spain, Avempace, Ibn Tufail, and Averroes (d. 1198). Averroes' influence on Western philosophers was so widely acknowledged that he was known simply as "the commentator" on Aristotle. His ideas turned up in Albert the Great, Thomas Aquinas, and a variety of Renaissance thinkers. A third contribution lay in the frequent resort to a prototype of the experimental method. Here perhaps the most serious single step forward after the Greeks was taken by Alhazen (d. 1039). By reorienting al-Kindi's concept of vision, Alhazen argued through a combination of geometry and physics that objects are seen by means of rays passing toward the eye, and not vice versa, as the Greeks had assumed.

Lynn White, Jr.

Natural Science and Naturalistic Art in the Middle Ages

Throughout a scholarly career that spanned fifty years, Lynn White, Jr., showed an uncanny ability to illuminate the attitudes which underlay medieval science and technology by drawing evidence from a vast range of unlikely sources. In the segments from "Natural Science and Naturalistic Art" reprinted below, he uses developments from Gothic art, from devotional literature, and from changes in sacramental practices to show that during the twelfth and thirteenth centuries there was a radical transformation in Christian attitudes toward nature—a transformation that provided a favorable climate for the revival of science.

It is strange that historians of science have paid so little attention to the development of the visual arts. Art, like science, normally deals with the objects of our physical environment, and both art and science therefore presumably reflect any modification of attitude toward that environment. In view of the fact that the first stirrings of Western scientific curiosity are detected in the twelfth century, it is surely no coincidence that historians of medieval art have discovered a related change in the modes of aesthetic expression beginning about the year 1140. The transition from Romanesque to Gothic charts the passage from an age indifferent to the investigation of nature to one deeply concerned with it.

The later Roman Empire and the early Middle Ages lived not in a world of visible facts but rather in a world of symbols. The intellectual atmosphere was so saturated with Platonic modes of thought that the first Christian millennium was scarcely more conscious of them than it was of the air it breathed. Behind every object and event lay an Idea, a spiritual entity or meaning, of which the immediate experience was merely the imperfect reflection or allegory. The world had been created by God for the spiritual edification of man, and served no other purpose. So extreme was the anthropocentrism [considering human beings and their values as the center of the universe] of late antiquity and of the early Middle Ages that even the turmoil and agony of nature was supposed to be only a shadow of Man's sinful state: when Man regained the simplicity of Eden, the harmony of the cosmos would be restored. For our regeneration God has given us two sources of spiritual knowledge: the Book of Scripture and the Book of Nature. Each is filled with hidden meanings to be searched out. In the most literal sense the men of that age found "sermons in stones and books in running brooks." They believed that the universe is a vast rebus to be solved, a cryptogram to be decoded. All that is red became to them a reminder of the blood of Christ; all that is wooden, a memento of his cross; every spring evoked a recollection of their rebirth through baptism, and its refreshing rills were the waters of the Gospel revivifying the arid world of paganism. Fishermen lowering their nets reminded them of their redemption, and hunting scenes were allegories of the Christian's struggle with the forces of sin. The crab, walking sideways, was a symbol of the fraudulent; the moon, shining by reflected light, imaged the state, which functioned at the behest of the church; the turtledove, which was supposed to refuse a second mate, was a divine rebuke to the much-married ladies of high society; the pelican, which was believed to nourish its young with its own blood, was the analogue of Christ, who feeds mankind with his blood. In such a world there was no thought of hiding behind a clump of reeds actually to observe the habits of a pelican. There would have been no point in it. Once one had grasped the spiritual meaning of the pelican, one lost interest in individual pelicans. . . .

. . . The emergence of Gothic art reflects a fundamental change in the European attitude towards the natural environment. Things ceased to be merely symbols, rebuses, *Dei vestigia,* and became objects interesting and important in themselves, quite apart from man's spiritual needs. To be sure, the concept of nature as allegory did not perish: on the contrary it continued to flourish and elaborate, particularly in mystical circles. Yet even the mystics succumbed to the new yearning for the particular and concrete: a dove might continue to represent the Holy Spirit, but it was now observed with the curiosity of an ornithologist. The meditations of the mystics took on a new factual, and therefore dramatic, quality: no longer content to contemplate Christ the Logos, the eye of the spirit now learned to follow the path of the historical Jesus step by step from Bethlehem to Calvary, reliving with him his earthly life in the most minute detail.

Indeed, at the end of the twelfth century Catholic piety suddenly concentrated itself upon an effort to bring God down to earth and to see and touch him. It was as though Europe had become populated with doubting Thomases eager to thrust their fingers into the very wounds of Christ. To an extraordinary degree the new eucharistic cult was empirical in temper, permitting the constant seeing and handling of God. The elevation of the consecrated host first appeared at Paris between 1196 and 1208; the reservation of the host for adoration became so common that the altar, hitherto conceived as a table, came to be thought of as normally supporting a tabernacle or monstrance; the dogma of transubstantiation was defined in 1215; the feast of Corpus Christi was instituted in 1264; and the first procession in honor of the host was held in 1279. Superficially the new piety might seem to be a development and expansion of the traditional sacramentalism, and as such a buttressing of the older symbolic and mediate view of nature. But, as the more conservative Eastern Church suspected, this was a sacramentalism of a new flavor, suffused with a spirit alien to that of the first Christian millennium. It seemed almost that the Latin Church, in centering its devotion upon the actual physical substance of its deity, had inadvertently deified matter. . . .

Above all, the well-known changes in the form of the crucifix illustrate the growth of the new dramatic naturalism. An object of such importance in religious practices would be subject to conservative treatment; its modification is therefore the more significant. Despite St. Bernard's ardent devotion to the sufferings of the Redeemer, until the thirteenth century crucifixes present Christ not as a bleeding victim but rather as the new Melchisedec, the priest-king, blessing with his outstretched arms. There is little sign of agony; the crown is a royal diadem, not one of thorns; the four nails are treated abstractly, and often are rendered by ornamented golden discs. Then, towards the middle of the thirteenth century, a new treatment of the crucifix emerges: Christ is contorted with pain; blood gushes from the wounds; the crown of thorns is depicted; not four nails but three appear (involving a new problem of spatial representation in the overlapping of forms) and these are long, cruel nails drawn in foreshortening. In other words, the thirteenth century crucifix reflects a new realism, both physical and psychological, a new sense of three-dimensional space, a new and vivid emotionalism, and a shift of accent in religion from divine grace to human drama.

The peril to traditional religion inherent in the new naturalism was not entirely unperceived by contemporaries. An arch-conservative Spanish ecclesiastic, Bishop Luke of Tuy, raged against the innovation of the three-nailed cross, representation of the Madonna in profile, and the depiction of God the Father or the Trinity in material form, as being the work of heretics. After seven centuries, who will say that he was wrong? The Byzantine Church, which more nearly preserved the older piety, never admitted the three-nailed crucifix to its art; for implicit in such novelties were those forces of scientific objectivity and religious subjectivism which were eventually to destroy the unity and authority of the Latin Church.

Bishop Luke should have been equally disturbed by a contemporary change in the iconography of the Annunciation. In the earlier period St. Mary is generally shown spinning, and the angel very nearly taps her on the shoulder to deliver his message. From the thirteenth century, however, the Virgin is normally depicted at prayer, either seated or kneeling, her prayerbook open before her. In other words, in primitive Christianity, when God speaks, man hears; in the new age of religious subjectivism the human mind must be prepared to receive God's message. Once religion and the realm of spirit had been objective; perception of nature, subjective. Now nature becomes objective; religion, subjective. Hitherto the natural had been merely a vehicle by which the supernatural made itself known to man; now the natural realm gains a status and importance of its own. Once natural knowledge had been entirely subordinate to sacred; now the way is prepared for that separation of secular learning from theology which was effected by the scholastics of the generation following St. Thomas, which liberated natural science and which permitted it to develop as an autonomous human endeavor.

Yet although the blackest heresy lurked behind the new attitude towards nature, the chief propagandist of the coming era was, paradoxically, the greatest saint of the Middle Ages, Francis of Assisi. Nowhere is the Gothic insistence that the eye of the spirit must be supplemented by the eye of the flesh better illustrated than in Thomas of Celano's account of that Christmas eve at Greccio when St. Francis and his friend Giovanni dramatized the scene at Bethlehem, complete with crib, stall, ox, and ass, in order to see these mysteries *corporeis oculis* [in a tangible form], and thus first popularized the créche. The Poverello and his gray-robed friars did much to spread the representational concept of art and the new emotionalism and religious subjectivism which accompanied it.

It is among the Franciscans as well that the scientific expression of the new attitude towards nature is most clearly seen. As has been remarked, the older view of the natural creation as symbol was completely anthropocentric: everything existed solely for man's spiritual benefit, and for nothing else. Against this human egotism the humility of St. Francis rebelled. To him the things of nature were indeed symbols, but they were more than that: they were fellow creatures placed on earth for God's inscrutable purposes, praising him in their proper ways as we do in ours. Such an attitude is, of course, implicit in the *Benedicite* and Psalm 148, but never before had it become explicit to such an extent within the Christian tradition. So extreme, indeed, was the reaction of St. Francis against a man-centered universe that he fell spontaneously into a cardinal deviation from orthodoxy which the church has chosen both to forgive and to minimize: he preached to birds and flowers, thus imputing moral personality to them. Earlier legends of the relations of living things to saints had emphasized that a true saint restores the harmony of Eden in which all beasts are subject to man; but St. Francis brought the wolf of Gubbio to repentance.

It may be said without exaggeration that St. Francis first taught Europe that nature is interesting and important in and of itself. No longer were flames merely the symbol of the soul's aspiration: they were Brother Fire. The ant was not simply a homily to sluggards, the worm not solely a sermon on humility: now both were autonomous entities. St. Francis was the greatest revolutionary in history: he forced man to abdicate his monarchy over the creation, and instituted a democracy of all of God's creatures. Man was no longer the focus of the visible universe. In this sense Copernicus is a corollary of St. Francis.

For an understanding of the history of science the central fact about the Poverello of Assisi is that his attitude towards his fellow creatures provided an adequate, and hitherto lacking, emotional basis for the objective investigation of nature. Indeed, is it fantastic to suggest that the saint's unwillingness to regard one creature as more important than another unconsciously assisted transition from the hierarchical-qualitative science of tradition to the egalitarian-quantitative science of modern times? At any rate it was no accident that his Order attracted men who flung themselves into furthering the new natural science and who became its leading exponents in the thirteenth century. Such activity is found throughout the Order, but centered particularly in the Franciscan school at Oxford, which became the most vital intellectual influence of the later thirteenth and early fourteenth centuries. From it emerged the two dominant figures of late scholasticism, John Duns Scotus and William of Ockham, who raised the scientific attitude of their fellow Friars Minor to the level of philosophical formulation, severed natural science from theology and paved the way for nominalism [the doctrine that only particular, individual objects or events have reality and that abstractions, or universals, are merely names with no real existence], the basic philosophy of modern times.

Joseph Ben-David

The Emergence of the Professional University Teacher in the Medieval University

The historical sociologist Joseph Ben-David is interested primarily in how institutional developments encourage or discourage the growth of science. In the following selection from The Scientist's Role in Society, *he argues that the intellectual division of labor that occurred as medieval universities became large institutions serving diverse constituencies was a key factor in the revival of science in thirteenth-century Europe because it produced a group of people whose primary professional activity was the teaching of science.*

In traditional societies the typical form of higher education was found in a master surrounded by disciples. Some of the disciples might become quite famous scholars in the lifetime of the masters, but only one could inherit the master's position. The other disciples might establish schools of their own elsewhere to carry on their masters' traditions, or they might inherit the leadership of an existing school whose master did not leave a disciple worthy or capable of inheriting from him. Rulers, rich individuals, or a community usually supported such a school by granting privileges to the scholars and maintaining hostels and teaching halls, paying the master a salary or gifts, or establishing an endowment.

Masters and scholars might be motivated by a genuine desire to understand sacred truths, to gain honor, or by anything else. The legitimation of learning, however, was that it was "practical"; it prepared the pupil for "practice." This is obvious in law and medicine but, in a sense, it is true of purely religious learning too. The study of sacred texts was regarded as desirable only if the person embodied the wisdom gained from them in his everyday life. Furthermore, the man made wise through study was expected to take a position of authority in his society. Thus the great scholars were expected to become leaders of the community, high civil servants, church dignitaries, judges. The humble scholar who lived for his learning was an exception, but even he was not honored merely for his scholarship. If he was not at the same time a saintly person whose private life was exemplary, he would not be honored. In his case, too, learning was a means to a practical end: the realization of the sacred way of life. Learning as such was not an end in itself. The conduct of learning was, therefore, an *amateur* rather than a professional activity; amateur teaching had higher prestige than the professional one. The transmission of religious and socially vital learning was not to be treated like a commodity that could be bought and sold on the market. This concept, of course, did not apply to the study of medicine or to secular studies like oratory or law, where payment was legitimate. But even in these areas, the brilliant practitioner-master who trained a few chosen disciple-apprentices, not the professional teacher, was the ideal.

This pattern explains the rudimentary organization of learning in traditional societies. Where the teacher was first and foremost a practical person or a professional practitioner, he could become only marginally involved in a complex educational organization (as is the case even today in the clinical teaching of medicine). This was a reasonable and efficient organization for legal-moral, religious, or even, up to a very developed point, medical and technological instruction. People who were capable of practical activity were not usually willing to give up their practice for teaching. Consequently, those preparing for a practical career or for a life of wisdom and saintliness preferred to become apprentices of first-class master-practitioners rather than to study with second-class masters who specialized in teaching. The exceptional creative teacher or thinker could assert himself anyway.

This state of affairs prevented any marked specialization and, in particular, specialization in theoretical studies. Specialization could not develop where a single master had to provide a comprehensive view of a whole field of learning and practice with only slight emphasis on his own preferred subject. Moreover, as long as the most respected teachers were practitioners, and the professional teachers were of low status, the practical applied approach prevailed over the theoretical one. The theorist, and therewith theory, like the person who was primarily a scholar, occupied a marginal position. Thus, natural science, mathematics, and even philosophy were marginal subjects. Even in ancient Greece and the Hellenistic world where philosophy attained a higher status and greater autonomy than elsewhere, its principal aim was still practical and moral.

Continuity of study had not been ensured under such conditions. Since the organization of teaching and study was quite informal, even famous seats of learning could decline very rapidly or even disappear without much struggle or conflict.

The European university was originally not different from the arrangements for higher learning of other traditional societies such as ancient India, China, or Islam. Students came from afar to Bologna, Paris, Montpellier, or Oxford in order to study law, theology, or medicine with famous masters, just as they would have gone to a famous master in India or Egypt. In Europe, however, the conditions were different in one important respect. The towns where the famous teachers resided were autonomous corporations, and the foreign pupils were not under the protection of the king. The townspeople and scholars in Europe were often at odds. Violence was always close to the surface, and the history of the early universities up to the fourteenth century is full of accounts of unscholarly fights, murders, disorders and drunkenness. In addition to ineffective law enforcement, there was the problem of separation of church and state, with the church claiming responsibility and authority over all spiritual matters, including education, and denying the authority of secular government over schools and scholars. This schism left the university scholars without firm regulation. To create order among the turbulent crowds of scholars and to regulate their relationship with the environing society, corporations were established. Students and scholars were formed into corporations authorized by the church and recognized by the secular ruler. The relationships of their corporation with that of the townspeople, with the local ecclesiastical officials, and with the king were carefully laid down and safeguarded by solemn oaths.

The important result of this corporate device—which was not entirely unique to Europe but which attained a much greater importance there than elsewhere—was that advanced studies ceased to be conducted in isolated circles of masters and students. Masters and/or students came to form a collective body. The European student of the thirteenth century no longer went to study with a particular master but at a particular university. A university consisted of several thousands of students (6,000 in Paris in 1300) and, at times, hundreds of masters living in an autonomous, intellectual community, relatively well endowed and privileged. This intellectual community was much more independent of the pressure of society as a whole than single intellectuals serving the state or the church (or individual scholars working as teachers) could ever be. A man commissioned by several fathers or a religious community to teach a few students could rarely command much respect. But the same person in a university community of several thousands engaged entirely in the teaching of students could become, if successful, a very popular and admired man in that community. If the university was large, rich, powerful, and famous, his status in the larger society was very high too. Thus the specialized role of the university teacher emerged; it was a role enjoying high status. Furthermore, the dependence of status on intellectual and pedagogical accomplishment within the internal system of the university, rather than on practical services rendered to laymen, permitted specialization to a considerable extent in subjects of interest only to scholars.

Of course, in order to maintain its status, the university as a whole had to emphasize subjects that were important for society (such as law, theology, and medicine). But once they had become associated with a single institution, these subjects were placed in a new perspective.

The process occurred somewhat like this. Originally, the universities concentrated on one branch of study: law in Bologna, theology and philosophy in Paris and Oxford, medicine in Salerno, and law and medicine in Montpellier. As long as the main institutional pattern of advanced learning was that of master and disciple, a famous doctor might have lived in one place and a great lawyer in another. Even if they had lived in the same place (e.g., in a capital city), they would not necessarily have had much to do with each other. However, once the universities became established, it became common for them to include all scholarly and professional subjects. It was easier to add another faculty to an existing corporation, and it was more worthwhile to do it in a place where there were undergraduates ready to enter the special faculties. These circumstances gave institutional support to the philosophical view of the existence of a coherent and comprehensive body of knowledge rendered meaningful by abstract theory. Thus, while law, medicine and, above all, theology continued to be the most important fields of learning, in the estimation of the layman within the university community itself, philosophy became the central subject. It was the basis of the intellectual culture common to all university-trained professionals.

As a result, undergraduate study in the arts faculty—in effect almost exclusively scholastic philosophy—became in many ways the most important part of the university. This situation was true at least in Paris and Oxford, even though officially the arts faculty was only preparatory to higher studies in theology, law, and medicine. At the same time, philosophy became an increasingly specialized field in its own right and a highly respected one. In the twelfth and early thirteenth centuries it was still closely tied to theology, but soon schools of philosophy arose that became more difficult to accommodate within the traditional theology. The controversies concerning the variant interpretations of Aristotle are a good example. Their potential inconsistency with religious tenets was recognized early in the twelfth century and aroused theological opposition. The conflict was eventually resolved in the synthesis of St. Thomas Aquinas, but it reappeared again in a much more violent form when Averroist influences began to penetrate the universities. The doubt these influences cast on the immortality of the soul (and some other doctrines) led to harsh and extreme denunciations and interdictions of these theories by Bishop Tempier of Paris (1270 and 1277) and by Archbishop Kilvardby of Oxford (1277). Their reactions were the same as those the Averroist doctrine provoked in Islam and among Jews. But whereas theological reaction in these latter two instances attained its end of suppressing autonomous philosophy, this did not occur in Christian Europe. The universities were strong, and the philosophical faculty was the largest and most popular. As communities of professional experts, the universities could resist the authority of other professional specialists. Eventually Siger of Brabant and William of Ockham rationalized and legitimated the differentiation of philosophic from religious thought: philosophy had its own logic leading to necessary conclusions. The contradictions of religious revelation were not to be regarded as a refutation of philosophical arguments; they showed only the existence of a higher truth beyond human reason.

This well-known story shows the importance of the differentiation and specialization which was taking place within the university. Philosophy became independent not because Ockham found the formula that made its independence somehow consistent with a "totalistic" religious outlook, but because philosophers had become a distinct and self-conscious group. That group had become large and respected enough to defend itself, which made it necessary to find a compromise formula. And they were a large and respected group because the universities were important and powerful centers of intellectual activity whose internal scale of values could not be easily disregarded.

The importance of this development lies precisely in a circumstance that is usually least appreciated: this development was originally a purely intellectual revolution. Unlike the rise of the philosophical schools in Greece and China, this philosophy did not compete with, nor did it ever replace, traditional religious doctrine. It only established a limited measure of professional freedom and equality for a new group of university intellectuals interested in a study that was neither practical nor religiously sanctioned. From the point of view of the fight for freedom of thought, as well as the intrinsic value of the philosophy thus emancipated, the achievement may appear petty. But from the point of view of the development of science and, in fact, from that of the development of institutional conditions for freedom of thought, it is exactly this apparent "pettiness" which is important. Complete victory would have led to the emergence of a new totalistic world view, competing for worldly power and influence with the religious one. The establishment, as well as the eventual overthrowing of such a philosophy, would have necessitated bloody revolutions. Since this did not happen, a first step was taken toward the separation of intellectual from political-religious revolutions. This separation was a necessary precondition of the emergence of science as an independent intellectual field.

The Peripherality of Science in the Medieval University

The intellectual division of labor arising from having different kinds of studies within one corporate organization also stimulated the further internal differentiation that gave the natural sciences their place at the universities. They were not a required part of the curriculum, and any academic degree could be acquired without knowledge of them. But inevitably among so many masters and scholars studying a considerable variety of subjects, there were some who were interested in scientific problems. Logicians took up mathematical and physical problems, and physicians considered a variety of biological problems. Informal groups emerged and were given facilities to pursue these studies outside the regular curriculum or during vacations. Even though these activities were not institutionalized, the mere size and internal differentiation of the universities permitted enough interested persons to find each other. In such a large academic "market," there was enough "demand" to maintain even a marginal intellectual field. In a small circle, by contrast, the probability of finding anyone interested in it would have been less and the stimulus to curiosity and persistent interest would have been correspondingly small.

Decentralization played its part in all these processes. The corporate autonomy of the university in any single place would not have been sufficient to withstand the onslaught of church authorities against the philosophers. Nor could, or did, scientific activity survive the disruptions caused by war, plague, and political strife in any single place. But during the thirteenth century when the opportunity to move diminished because of increased royal power (especially in Paris), it was still possible for individuals to go to a different country, as did the English and the Germans, and even some other scholars (e.g., Marsilius of Padua and John of Jandun) who left Paris for English or German universities. The decline of philosophy in France and England made Italy the new center of its study in the fifteenth and sixteenth centuries. The intellectual differentiation continued, and the fifteenth century Italian humanists or the sixteenth century neo-Aristotelians were completely secularized and specialized philosophers.

Because geometry and dynamics were mainly up to the sixteenth century cultivated by philosophers, the fate of these studies was bound to that of philosophical studies in general. The tradition of medieval natural science was started at Oxford by masters of Merton College. From there it spread to Paris, which had the closest intellectual commerce with Oxford. When the tradition declined in both places during the fourteenth century, as did philosophy, the center shifted to Italy, mainly to Padua, and to the new German, Dutch, and other universities. Thus, when special university chairs were established, in the fourteenth and fifteenth centuries this tradition, influenced probably by internal developments within the medical faculty, also led to the establishment of professorships in mathematics, astronomy, and a variety of subjects, such as natural philosophy, Aristotelian physics, and so forth, first in Italy and later everywhere in Europe. These scientific chairs were of subordinate importance; it was an advancement for their incumbents if they could be appointed as professors of philosophy, or even better, of theology, law, or medicine. In any case, it was necessary to have a degree in these latter subjects in order to be appointed to a chair. Nevertheless, by this time the natural sciences were more or less regularly taught—on however modest a level—by professors who were paid for teaching them.

A. C. Crombie

The Reception of Greco-Arabic Science in Western Christendom

In the following selection from Medieval and Early Modern Science, *A. C. Crombie summarizes the key ways in which Aristotelian philosophy, as it was interpreted by the great Islamic commentator, Averroes, seemed to conflict with traditional Christian doctrine. He identifies three general approaches to this problem which emerged in the late thirteenth century, and he argues that Islamic Aristotelianism was generally rejected in favor of positions which denied any real conflict between philosophy and theology and which assigned to each its own autonomous domain of experience.*

Most influential of all the contributions of Greco-Arabic learning to Western Christendom was the fact that the works of Aristotle, Ptolemy and Galen constituted a complete rational system explaining the universe as a whole in terms of natural causes. Aristotle's system included more than natural science as it is understood in the 20th century. It was a complete philosophy embracing all existence from 'first matter' to God. But just because of its completeness the Aristotelian system aroused much opposition in Western Christendom where scholars already had an equally comprehensive system based on the facts revealed in the Christian religion.

Moreover, some of Aristotle's theories were themselves directly contrary to Christian teaching. For instance, he held that the world was eternal and this obviously conflicted with the Christian conception of God as creator. His opinions were doubly suspect because they reached the West accompanied by Arab commentaries which stressed their absolutely determinist character. The Arab interpretation of Aristotle was strongly coloured by the Neoplatonic conception of the chain of being stretching from first matter through inanimate and animate nature, man, the angels and Intelligences to God as the origin of all. When such commentators as Alkindi, Alfarabi, Avicenna and particularly Averroës (1126–98) introduced from the Mohammedan religion into the Aristotelian system the idea of creation, they interpreted this in such a way as to deny free will not only to man but even to God himself. According to them the world had been created not directly by God but by a hierarchy of necessary causes starting with God and descending through the various Intelligences which moved the celestial spheres, until the Intelligence moving the moon's sphere caused the existence of a separate Active Intellect which was common to all men and the sole cause of their knowledge. The form of the human soul already existed in this Active Intellect before the creation of man, and after death each human soul merged again into it. At the centre of the universe within the sphere of the moon, that is, in the sublunary region, were generated a common fundamental matter, *materia prima,* and then the four elements. From the four elements were produced, under the influence of the celestial spheres, plants, animals and man himself.

Several points in this system were entirely unacceptable to the philosophers of Western Christendom in the 13th century. It denied the immortality of the individual human soul. It denied human free will and gave scope for the interpretation of all human behaviour in terms of astrology. It was rigidly determinist, denying that God could have acted in any way except that indicated by Aristotle. This determinism was made even more repulsive to Christian thinkers by the attitude of the Arab commentators and especially of Averroës, who declared:

> Aristotle's doctrine is the sum of truth because his was the summit of all human intelligence. It is therefore well said that he was created and given us by Divine Providence, so that we should know what it is possible to know.

Some allowance may be made here for oriental exaggeration, but this point of view came to be characteristic of the Latin Averroïsts. For them the world emanated from God as Aristotle had described it, and no other system of explanation was possible. Nor indeed did the extreme theological rationalism of this interpretation do violence to Aristotle's own thought. Aristotle had based his whole approach to natural science and metaphysics on the claim that it was possible to discover by reason the essence of things and of God, causing the regularities observed in the world. Plato's approach was the same, although differing over both the processes of reason involved and the nature of the essences discovered. In the brilliant *tour de force* in book 2, chapter 3 of *De Caelo,* Aristotle gave every support to the Averroïst interpretation of his cosmology. He set out to prove that his system was not only in fact true, but was necessarily true, for it alone followed from God's discovered essence and perfection. All things, he argued, existed for the ends they served and the perfection to which they tended. God's activity was eternal, and so therefore must be the motion of the heaven, which was a divine body. 'For that reason the heaven is given a circular body whose nature it is to move always in a circle . . . ; and earth is needed because eternal movement in one body necessitates eternal rest in another.' Similarly he argued that the whole actual world was necessarily as he described and explained it, and could not in the nature of things be otherwise. . . .

It was the search for means of finding an accommodation between Aristotelian philosophy and Christian theology that gave rise to the most interesting and critical developments in philosophy and in the conception of science in the 13th and 14th centuries. After some initial hesitation and embarrassment, three main general lines of policy began to clarify. The first was that of the Latin Averroïsts who took their stand on the irrefutable rational truth of Aristotelian philosophy and accepted the consequence that Christian theology was irrational or even untrue. There seems little doubt that such a man as Jean de Jandun (d. 1328) was in fact an unbeliever, but that he lightly concealed his unbelief with an irony akin to Voltaire's. A threat equally to Christian theology and to empirical science, it was the Christian doctrine of the absolute freedom of God's will that formed the basis of the critique of Averroïst rationalism, although the criticism was pressed by means of logical arguments about the possibility of there being any necessary rational truths about the world. A moderate policy, for example that of Thomas Aquinas, was to accept the rationality of science but to deny that any necessity could be discovered in God. The extreme policy of the defenders of faith appeared in the 14th century, when for example William of Ockham eliminated the threat from reason by denying the rationality of the world altogether, and reduced its order to a dependence of fact on God's inscrutable will.

The 13th century saw first the categorical condemnation of Aristotle, but by the middle of the century he had been accepted as the most important of the philosophers. In 1210 in Paris, which by the end of the 12th century had already replaced Chartres as the greatest centre of learning in France, the provincial ecclesiastical council prohibited the teaching of Aristotle's views on natural philosophy or of commentaries on them. In 1215 a similar decree was issued against reading his metaphysical and natural works, although this did not forbid private study and applied only to Paris; in fact, lectures on these works were announced in the University of Toulouse. Other prohibitions were issued subsequently, but it was not possible to enforce them. In 1231 Pope Gregory IX appointed a commission to revise some of the natural works and in 1260 William of Moerbeke began his translation from the Greek. Eminent teachers like Albertus Magnus (1193/1206–80) and his pupil Thomas Aquinas (1225–74) expounded the works of Aristotle, and by 1255 his most important metaphysical and natural works were set by the faculty of arts in Paris as a subject for examination. At Oxford the 'new Aristotle' made its first entry without attracting official opposition. Lecturing on the new logical and physical treatises had begun by the first decade of the 13th century, but it was the influence of an inspiring philosopher and teacher, Robert Grosseteste (*c.* 1168–1253), that really established medieval Oxford's enduring interest in the new science, mathematics and logic, as well as in languages and Biblical scholarship. As *Magister Scholarum* or Chancellor of the University in 1214, perhaps the first to hold this office, as lecturer to the Franciscan house in Oxford, and from 1235 as Bishop of Lincoln, in which diocese Oxford lay, Grosseteste remained the chief ornament and guide of the University's early years. . . .

Throughout the Middle Ages there were various schools of thought about the Aristotelian system of the universe. In the 13th century in Oxford the Franciscan friars, who tended to remain loyal to the main features of Augustinianism, such as the theory of knowledge and of universals, accepted some important Aristotelian additions in the explanation of such natural phenomena as the movements of the heavenly bodies, but were often hostile to Aristotle's influence as a whole. At the same time in Oxford there was an interest characteristic of another aspect of Franciscan thought, exemplified by Roger Bacon, who was keenly alive to the mathematical, physical, astronomical and medical learning of Aristotle and the Arabs and less concerned with their metaphysical views. In the University of Paris black-habited Dominicans, such as Albertus Magnus and Thomas Aquinas, accepted the main principles of Aristotle's physics and philosophy of nature . . . but rejected his absolute determinism. A fourth school of thought, represented by Siger of Brabant, who was thorough-going Averroïst, accepted an entirely determinist interpretation of the universe. Yet a fifth group was in the Italian universities of Salerno, Padua and Bologna where theological matters counted for less than in England or France and where Aristotle and the Arabs were studied principally for their medical learning.

Those mainly responsible for making Aristotle acceptable to the Christian West were, besides Grosseteste, Albertus Magnus and Thomas Aquinas. The main problem confronting them was the relation between faith and reason. In his attempt to resolve this difficulty Albertus based himself, like St. Augustine, on two certainties: the realities of revealed religion and the facts that had come within his own personal experience. Albertus and St. Thomas did not regard Aristotle as an absolute authority as Averroës had done, but simply as a guide to reason. Where Aristotle, either explicitly or as interpreted by Arab commentators, conflicted with the facts either of revelation or of observation he must be wrong: that is, the world could not be eternal, the individual human soul must be immortal, both God and man must enjoy the exercise of free will. Albertus also corrected him on a number of points of zoology. . . . But Albertus and more definitely St. Thomas realised, as Adelard of Bath had done a century earlier, that theology and natural science often spoke of the same thing from a different point of view, that something could be both the work of Divine Providence and the result of a natural cause. In this way they established a distinction between theology and philosophy which assigned to each its appropriate methods and guaranteed to each its own sphere of action. There could be no real contradiction between truth as revealed by religion and truth as revealed by reason. Albertus said that it was better to follow the apostles and fathers rather than the philosophers in what concerned faith and morals. But in medical questions he would rather believe Hippocrates or Galen, and in physics Aristotle, for they knew more about nature.

The determinist interpretation of Aristotle's teaching associated with the commentaries of Averroës was condemned by the Bishop of Paris, Êtienne Tempier, in 1277, and his example was followed in the same year by the Archbishop of Canterbury, John Pecham. In so far as this affected science it meant that in northern Christendom the Averroïst interpretation of Aristotle was banished. The Averroïsts retired to Padua where their views gave rise to the doctrine of the double truth, one for faith and another, perhaps contradictory, for reason. This condemnation of determinism has been taken by some modern scholars, notably by Duhem, as marking the beginning of modern science. The teaching of Aristotle was to dominate the thought of the later Middle Ages, but with the condemnation of the Averroïst view that Aristotle had said the last word on metaphysics and natural science, the bishops in 1277 left the way open for criticism which would, in turn, undermine his system. Not only had natural philosophers now through Aristotle a rational philosophy of nature, but because of the attitude of Christian theologians they were made free to form hypotheses regardless of Aristotle's authority, to develop the empirical habit of mind working within a rational framework, and to extend scientific discovery.

CHAPTER 3

Scientific Imagination in the Renaissance

Introduction by Richard Olson

We saw in the last chapter that Islamic science had a more applied emphasis than ancient Greek science and that European interest in Islamic science was probably initiated because the superiority of Islamic medicine, business mathematics, and navigational mathematics caught the attention of Europeans living in Moslem Spain and Sicily. But as the Aristotelian element of the new tradition was brought to the forefront, and as scientific learning found its principal home within the liberal arts faculty of the medieval university, European science returned to an almost exclusively theoretical and contemplative emphasis. A handful of medieval scientists, such as Roger Bacon (c. 1220–1292), who is usually credited with inventing eyeglasses in connection with his study of Islamic optics, retained a strong interest in applications. And a few teachers, such as Hugh of St. Victor (?–1141) in Paris, tried to get the useful "mechanical arts"—fabric making, armaments, commerce, agriculture, hunting, medicine, and theatrics—included among subjects taught at an elementary level in the cathedral schools of Europe. But the efforts of men like Hugh and Roger Bacon were largely isolated and unsuccessful.

During the early Renaissance, however, two developments associated with the humanistic movement in Northern Italy created important scientific interests outside the university structure and within the dynamic urban culture of the Renaissance city-state. Both developments involved a revived central emphasis on Neoplatonic philosophy, with its focus on mathematics. And both movements, in spite of their Platonic orientation, emphasized the *active* role of man in the universe and the search for knowledge as a foundation for *intervention* in the world and not merely for its own sake.

To a late twentieth-century reader, the Renaissance emergence of a powerful new emphasis on the visual and plastic arts (painting, sculpture, and architecture) and the Renaissance revival of interest in the occult (magic, astrology, and number mysticism) are almost certain to seem strange developments within which to find critical sources of modern science. We have been taught to think of the artistic temperament as the very antithesis of the scientific temperament; for contemporary art is intensely subjective and finds its central source of inspiration in the artist's emotional response to the world rather than in the phenomenal world itself. Similarly, we have become convinced that the critical and hardheaded rationality associated with science is the great enemy of the blind credulity and superstitious roots of magic and the occult. It is certainly true that by the end of the seventeenth century the visual arts, the occult, and science had largely parted company. Yet for a time, between about 1400 and 1600, there can be no doubt that art and magic played key roles in transforming the nature of European science.

The artistic and magical movements of the Renaissance started very differently; but each in its own way moved toward a common understanding of the relation between theory and practice which completely overcame the separation that characterized ancient Greek and medieval Christian science. In 1585 Giordano Bruno expressed this new common understanding with great eloquence in *The Expulsion of the Triumphant Beast*:

> The Gods have given man intelligence and hands, and have made him in their image, endowing him with a capacity superior to other animals. This capacity consists not only in the power to work in accordance with nature and the usual course of things, but beyond that and outside her laws, to the end that by fashioning, or having the power to fashion, other natures, other orders by the means of his intelligence, with that freedom without which his resemblance to the deity would not exist, he might in the end make himself god of the earth. That faculty, to be sure, when it is unemployed, will turn into something frustrate and vain, as useless as an eye which does not see or a hand which does not grasp. For which reason, providence has decreed that man should be occupied in action by the hands and in contemplation by the intellect, but in such a way that he may not contemplate without action or work without contemplation. Thus, in the Golden age, [i.e., classical Athens], men, through idleness, were worth not much more than dumb beasts still are today, and were perhaps more stupid than many of them. But, when difficulties beset them or necessities reappeared, then through emulation of the actions of God and under the direction of spiritual impulses, they sharpened their wits, invented industries, and discovered arts. And always, from day to day, by force of necessity, from the depth of the human mind rose new and wonderful inventions. By this means, separating themselves more and more from their animal natures by their busy and zealous employments, they climbed nearer to the divine being.

One of the most dramatic episodes in the history of Renaissance art and science occurred around 1420 when Filippo Brunelleschi (1377–1446), a Florentine goldsmith, sculptor, and would-be architect, began studying mathematical optics and mechanics with the young mathematician Paolo Toscanelli. Brunelleschi's practical concern with producing accurate two-dimensional representations of buildings was fused with Toscanelli's knowledge of the mathematical treatment of the appearance of three-dimensional objects, which had been initiated by Euclid in Greek antiquity. From this fusion there emerged a new artistic and architectural tradition obsessed with the problem of perspective and dedicated to the proposition that every practicing artist and architect needs to become familiar with formal mathematics.

Giorgio de Santillana's "The Role of Art in the Scientific Renaissance" focuses on the role of Brunelleschi in the transformation of Renaissance art and architecture from traditional medieval crafts to professions that drew inspiration from the learned tradition of mathematics. Brunelleschi was typical of Renaissance artists in combining an interest in painting with other interests that now belong to engineering and architecture. (If one needs other examples, Leonardo da Vinci, Michelangelo, and Brunelleschi's literary voice, Leon Battista Alberti, come immediately to mind.) So when a mathematical dimension entered painting through perspective it was rapidly extended to the more complex problems of building and military engineering. As Santillana suggests, mathematics entered Renaissance art and architecture within the context of a Neoplatonic philosophy that shared the ancient Pythagorean belief in the basically mathematical structure of the entire universe. So once artists and architect-engineers became convinced of the utility of mathematics, they almost universally sought to extend their mathematical knowledge. It is for this reason that Italian Renaissance architect-engineers were central in both the recovery and translation of the most sophisticated mathematical works of antiquity—that is, those of Archimedes, Hero of Alexandria, and Pappus—and in the emergence of a vital new tradition of mathematics, which included the solution of certain classes of cubic equations by the engineer Niccolo Tartaglia in about 1535.

The new artistic tradition of the Renaissance was an artisan-based movement which began among a group of goldsmiths, sculptors, and painters in Florence, but which led to the emergence of a new professional cadre of architects, urban planners, and military engineers who began to view the theoretical knowledge of mathematics, mechanics, optics, and so on as a necessary prerequisite for their productive activities. The magical movement, on the other hand, began within the intellectual elite among humanist scholars—a group of men symbolized by Marcilio Ficino and Pico della Mirandola—who were seeking to discover the most esoteric secrets of ancient Greek thought to pass them on to the Latin-speaking world. It moved from a strange form of mystical Christianity, through such occult sciences as astrology, cabalistic (quasireligious, esoteric, secret, and mysterious) magic, and Islamic alchemy, increasingly toward a dual insistence that human beings were granted the ability to know so that they might play a central role in completing God's unfinished work of creation and human knowledge of the universe could only come from a detailed mathematical and empirical study of the world of nature itself.

Although the magical tradition drew from many sources, it received its single greatest impetus around 1460 when the literary agents of Cosimo de Medici brought a Greek manuscript purporting to be the writings of Hermes Trismegistus to Florence. In 1463 Ficino, who was busy translating Plato's writings into Latin for Cosimo, was told to interrupt his work and translate the Hermetic writing first, because there was a long tradition, beginning with St. Augustine, which identified Hermes as an ancient Egyptian wise man, or magus, predating Moses by over 1,000 years. In fact, the Hermetic writings were a collection of texts from the second century A.D. which mixed Christian, Neoplatonist, and Gnostic (pertaining to pre-Christian cults that held knowledge to be the key to salvation) elements; but for our purposes it is critical to recognize that well into the seventeenth century the Hermetic writings were dated from the second millennium B.C.

Given its presumed early dating, the Hermetic Corpus was truly amazing; for it clearly seemed to anticipate Mosaic, Pythagorean, Platonic, and even Christian doctrines and to justify a belief that there was a single ancient theological and philosophical tradition, originating with Hermes and culminating in Plato and Christ.

From the present perspective, the most important feature of the Hermetic writings lay in the notion that man was created by God with an intellect so that he could know and appreciate God's works *and* with a body so that he could use his spark of divine knowledge "*to regulate all things . . . and take care of terrestrial things and govern them.*" Every man is potentially a magus, capable of both recognizing and directing the forces inherent—but sometimes hidden—in nature, in order to fashion the world according to his own designs. In the selection on "Cornelius Agrippa's Survey of Renaissance Magic," Dame Frances Yates characterizes the world of the Renaissance magus, pointing out that the search for ways of acting to govern the world provided a direct stimulus to studies of medicine, natural philosophy, astronomy, and above all, mathematics, as well as "talismanic" magic and "ceremonial" magic.

By the middle of the sixteenth century, several of the scholarly traditions that had been stimulated by both the artistic-architectural tradition flowing from Brunelleschi and the magical tradition flowing from the Hermetic Corpus, had taken on lives of their own. This was certainly true, for example, of alchemy, which became a serious science in its own right through the work of Paracelsus and his followers; and it was also true of mathematics, which spread its influence into almost every aspect of scholarly and commercial life. In his selection on Paracelsus and the Paracelsians, Allen Debus discusses the virulent anti-Aristotelian and strongly empirical thrust of the Christian-Neoplatonic alchemical movement which built upon Hermetic magic. And in her essay on "The Uses of Mathematics," Marie Boas explores the sixteenth-century explosion of applied mathematics, especially in the field of mapmaking and navigation.

Just a few years after Giordano Bruno's death at the hands of the Roman Inquisition, one of his English friends, Francis Bacon, chided those who remained in the old tradition of theoretical science:

> In natural philosophy practical results are not only the means to improve human well-being, they are also the guarantee of truth. There is a true rule in religion that a man must show his faith by his *works*. The same rule holds good in natural philosophy. Science, too, must be known by its works. It is by the witness of *works* rather than by logic or even by observation that truth is revealed and established. It follows from this that the improvement of man's lot and man's mind are one and the same thing.

A selection from Bacon's *New Atlantis*, one of the chief manifestos of a new European vision of experimental science as the great engine of material progress, concludes the set of readings for this chapter, for it is in an important sense the direct outcome of the forces loosed within the Renaissance artistic and magical traditions. At the same time it is a prophetic progenitor of the pervasive modern attitude so well expressed in the long-time corporate motto of E. I. du Pont, which exhorted us to seek "better living through chemistry."

Giorgio de Santillana

The Role of Art in the Scientific Renaissance

A strong case can be made that both the study of anatomy and the study of mathematics in the Renaissance were greatly encouraged by a group of Florentine artists that included Leonardo da Vinci and Michelangelo. Giorgio de Santillana focuses in the following selection on the figure of Filippo Brunelleschi, who was the chief architect responsible for constructing the great dome of the cathedral in Florence. Brunelleschi, he argues, was the key figure in establishing the centrality of mathematical learning for the artists, architects, and engineers of the Italian Renaissance, just as he was the key figure in raising the status of the architect-engineer from that of traditional craftsman to learned professional.

Erwin Panofsky has analyzed with great learning and insight the contribution of art to science in many fields. He has rightly insisted that above all the discovery of perspective, and the related methods of drawing three-dimensional objects to scale, were as necessary for the development of the "descriptive" sciences in a pre-Galilean period as were the telescope and the microscope in the next centuries, and as is photography today. This was particularly the case, in the Renaissance, with anatomy. In this field, it was really the painters, beginning with Pollaiolo, and not the doctors, who practiced the thing in person and for purposes of exploration rather than demonstration. In all this I cannot but accept Panofsky's verdict: Art, from a certain point on, provides the means of transmitting observations which no amount of learned words could achieve in many fields. . . .

What I intend to do, in this essay, is to concentrate on the early period, which centers around Filippo Brunelleschi (1377–1446). . . . Brunelleschi around 1400 should be considered the most creative scientist as well as the most creative artist of his time, since there was nothing much else then that could go by the name of creative science. . . .

Let us then try to make a landfall at a point where art and science, undeniably, join. Brunelleschi created his theory of perspective by experimental means. He built the earliest optical instrument after the eyeglasses. We have Manetti's description of the device, a wooden tablet of about half an ell, in which he had painted "with such diligence and excellence and care of color, that it seemed the work of a miniaturist," the square of the Cathedral in Florence, seen from a point three feet inside the main door of the cathedral. What there was of open sky within the painting he had filled in with a plate of burnished silver, "so that the air and sky should be reflected in it as they are, and so the clouds, which are seen moving on that silver as they are borne by the winds." In the front, at the point where the perpendicular of vision met the portrayed scene, he had bored a hole not much bigger than the pupil of the eye, which funneled out to the other side. Opposite the tablet, at arm's length, he had mounted a mirror. If you looked then, through the hole from the back of the tablet at its reflection in the mounted mirror, you saw the painting exactly from its perspective point, "so that you thought you saw the proper truth and not an image."

The next step, as we see it now, is to invert the device and let in the light through the pinhole, to portray by itself the exterior scene on an oiled paper screen. This is the *camera obscura*, . . . and it took time to be properly understood. But the whole train of ideas originated with Brunelleschi, between 1390 and 1420.

We have thus not *one* device, but a set of

experimental devices of enormous import, comparable in importance to that next device which came two centuries later, namely, Galileo's telescope. . . .

Surely, here we have the beginnings of a science— . . . so scientific that there is even an apparatus designed for it. But what is it for? To help us portray rightly what we see around us—essentially, to give the illusion of it. Illusionism is a strong motive, inasmuch as the most direct application is to scenographic design. This in itself is no mean thing, nor merely a way to amuse the rich. It has been proved that those monumental town perspectives—like that of the main square of Urbino by a pupil of Piero—are no mere exercises in drawing, they are actually projects for an architecture that is not yet there, and the first sketches of town planning. This is in the true line of development, for Brunelleschi himself had devised his instrument as an aid to architectural planning. . . .

Here we have, then, for the first time the Master Engineer of a new type, backed by the prestige of mathematics and of the "recondite secrets of perspective," . . . the man whose capacity is not supposed to depend only on long experience and trade secrets, but on strength of intellect and theoretical boldness, . . . who can speak his mind in the councils of the city and is granted patents for his engineering devices. . . . He is, in fact, the first professional engineer as opposed to the old and tradition-bound figure of the "master builder": He is the first man to be consulted by the Signoria as a professional military engineer and to design the fortifications of several towns: . . . he is acknowledged to the end of his life as the great designer and artist; not only that, but as the man who masters the philosophical implications of what he is doing. . . . Brunelleschi . . . stands as an intellectual. . . .

Now, as to Brunelleschi; how much was he, whose business after all it was to make things, aware of the intellectual implications? We know that he was very keen on theory. . . . We would have to know more about his background to have an exact answer. The real books of his early education, we are told, were Dante and the Bible. . . . Manetti and Toscanelli cannot be properly called his teachers, since they are younger than he. But when he decided to move from sculpture into architecture, they became his advisers and confidantes. There is some documentary evidence of their influence in his planning and thinking. . . . What these men imparted was not so much the elements of geometry which were needed for perspective and graphical statics; it was a certain frame of thought connected with the subject, which had come down to them from the world of scholars. . . . What did those men, then, actually transmit? The feeling of the high dignity of mathematics, for one, as vouched for by Platonic wisdom; the hope in mathematics, as revived in the preceding century by famous scholars like Bacon and Grosseteste; most of all, the beliefs and wondrous intimations to which Christian intellectual mysticism had clung through the centuries, and which expressed themselves in the "metaphysics of light."

With this body of doctrine we are fully acquainted through many expositions. It starts from the Platonic analogy of God with the sun, that is set forth in the myth of the Cave. It pursues the analogy to suggest that, as God is the life of the soul, so the physical world is held together and animated by the force of light and heat, which should turn out to be, as it were, the ultimate constituent of reality. . . . They were disembodied and yet very pregnant ideas. They had superadded themselves to the old traditional wonderings about the cosmic role of the five geometrical solids. They are the inspiration that teachers, from Toscanelli to Pacioli, communicated with passion. At this point, in these years, they hit fertile ground, with the invention of perspective by Brunelleschi and Uccello; . . . we may understand better all that Brunelleschi would see in his invention—if he was the intellectual that history shows him to be. We have not yet investigated Toscanelli as an individual; but we have characterized enough of his ideas to provide an adequate first answer. . . .

The documents will then tell us that Toscanelli was an intensely ascetic and spiritual man, "a physician, philosopher, and astrologer of most holy life". . .; a man who shunned meat and held it was wrong to take the life of animals.

When we see this unusual doctrine so strongly emphasized by his successor Leonardo, with an explicit and equally unusual Pythagorean connotation, we may conclude that we know something of Toscanelli's line of thought. . . .

. . . At a time when what *we* mean by science was still beyond the horizon, when the *name* of science was monopolized by scholastic officials, who officially denied to mathematics any link with physical reality, these men had conceived of an original prototype of science based on mathematics, which was to provide them with a creative knowledge of reality, repeat—creative, and could claim the name of true knowledge in that it dealt with first and last things. . . .

It is this intellectual element, overarching that whole early Renaissance, which explains many things in the rise of the new architecture. . . . It will drive Uccello and Piero deep into geometrical speculation, it will eventually find its concluding manifesto in Pacioli's *Divina Proportione,* which is expressly dedicated to Piero della Francesca, "the most worthy monarch of the art of our time." In this kind of theory, elementary abstraction has hardened and simplified much of what had been profound and original creation. . . .

Whither does it lead us? To what Alberti, as essentially a man of letters, could not see, nor for that matter Pacioli (who was barely above a teacher of the abacus), but seems to be present as a deep intuition in the creative masters. We mean, to an impressive generalization of the Pythagorean system. In the original doctrine, there were only a few entities which embodied geometry, but they were felt to be enough in that they acted by participation. The circle, the cube, the Harmonic Fifth, existed as absolutes, and caused things to behave "in imitation" of them. Of such manifest imitation the cases could not but be very few, and so science was limited to the charismatic regions of music and astronomy. But if we begin to think in projective terms, then we begin to see circularity inherent in ellipses, squareness in rhomboidal shapes, and so on. The "imitation" of the Ideas turns up everywhere, even if the original forms be no longer obvious: The square becomes one of its own perspectives, the circle a conic section. In fact, what used to be "form" becomes a collection of very abstract relations whose mathematical treatment is surely not elementary—and may have been above anyone's resources at the time. . . . This is, then, a vision of the Pythagorean system become truly universal and permeating all of reality.

Frances A. Yates

Cornelius Agrippa's Survey of Renaissance Magic

Frances Yates's Giordano Bruno and the Hermetic Tradition *has probably done more to influence scholarship on Renaissance science than any other work written during the past quarter century. Through this work she called our attention to the central role of Hermetic magic in the intellectual life of the Renaissance. Many of her detailed claims—those regarding the impact of Hermeticism on the emergence and acceptance of Copernican astronomy, for example—have been challenged by subsequent scholars. But the more general point that Renaissance magical attitudes and aspirations served as a significant stimulus to early modern science is widely accepted. In the following selection she summarizes the major doctrines of a "typical" Renaissance magus, Cornelius Agrippa (1486–1535), and emphasizes the way in which they focused on the active intervention of man in the natural world.*

Henry Cornelius Agrippa of Nettesheim is by no means the most important of the magicians of the Renaissance, nor is his *De occulta philosophia* really a text-book of magic, as it has sometimes been called. It does not fully give the technical procedures, nor is it a profound philosophical work, as its title implies, and Cardanus, a really deep magician, despised it as a trivial affair. Nevertheless the *De occulta philosophia* provided for the first time a useful and—so far as the abstruseness of the subject permitted—a clear survey of the whole field of Renaissance magic. . . .

The universe is to be divided, says Agrippa in the first two chapters of his first book, into three worlds, the elemental world, the celestial world, the intellectual world. Each world receives influences from the one above it, so that the virtue of the Creator descends through the angels in the intellectual world, to the stars in the celestial world, and thence to the elements and to all things composed of them in the elemental world, animals, plants, metals, stones, and so on. Magicians think that we can make the same progress upwards, and draw the virtues of the upper world down to us by manipulating the lower ones. They try to discover the virtues of the elemental world by medicine and natural philosophy; the virtues of the celestial world by astrology and mathematics; and in regard to the intellectual world, they study the holy ceremonies of religions. Agrippa's work is divided into three books; the first book is about natural magic, or magic in the elemental world; the second is about celestial magic; the third is about ceremonial magic. These three divisions correspond to the divisions of philosophy into physics, mathematics, and theology. Magic alone includes all three. Eminent magicians of the past have been Mercurius Trismegistus, Zoroaster, Orpheus, Pythagoras, Porphyry, Iamblichus, Plotinus, Proclus, Plato.

Book I or Natural Magic

After chapters on the theory of the four elements, he comes to the occult virtues in things and how these are infused "by the Ideas through the World Soul and the rays of the stars." This is based on the first chapter of Ficino's *De vita coelitus comparanda,* which is quoted verbally, and Agrippa has understood that Ficino is there talking about the star images as the medium through which the Ideas descend. "Thus all the virtues of inferior things depend on the stars and their images . . . and each species has a celestial image which corresponds to it." In a later chapter on "The Spirit of the World as the Link between Occult Virtues" he is again quoting Ficino and reproducing his *spiritus* theory. Then follow chapters on the plants, animals, stones, and so on belonging to each planet, and to the signs of the zodiac, and on how the "character" of the star is imprinted in the object belonging to it, so that if you cut across the bone of a solar animal or the root or stem of a solar plant, you will see the character of the sun stamped upon it. Then come instructions on how to do natural magic by manipulations of the natural sympathies in things and thus through arrangements and correct uses of the lower things to draw down the powers of the higher things.

So far, what Agrippa has been talking about is Ficino's natural magic as done in the elemental world that is through occult stellar virtues in natural objects. But, . . . Agrippa does not follow Ficino in taking care to avoid the demonic side of this magic by aiming only at attracting stellar influences and not the influences of spiritual forces beyond the stars. For you can draw down in this way, says Agrippa, not only celestial and vital benefits (that is benefits from the middle or celestial world) but also intellectual and divine gifts (that is benefits from the intellectual world). "Mercurius Trismegistus writes that a demon immediately animates a figure or statue well composed of certain things which suit that demon; Augustine also mentions this in the eighth book of his *City of God.*" Agrippa fails to add that Augustine mentions this with strong disapproval. "For such is the concordance of the world that celestial things draw supercelestial things, and natural things, supernatural things, through the virtue running through all and the participation in it of all species." Hence it was that ancient priests were able to make statues and images which foretold the future. Agrippa is aiming at the full demonic magic of the *Asclepius,* going far beyond the mild Neoplatonised magic of Ficino which he has been describing in the earlier chapters. He knows that there is an evil kind of this magic, practised by "gnostic magicians" and possibly by the Templars, but adds that everyone knows that a pure spirit with mystical prayers and pious mortifications can attract the angels of heaven, and therefore it cannot be doubted that certain terrestrial substances used in a good way can attract the divinities.

There follow chapters on fascination, poisons, fumigations (perfumes sympathetic to the planets and how to make them), unguents and philtres, rings, and an interesting chapter on light. Light descends from the Father to the Son and the Holy Spirit, thence to the angels, the celestial bodies, to fire, to man in the light of reason and knowledge of divine things, to the fantasy, and it communicates itself to luminous bodies as colour, after which follows the list of the colours of the planets. Then we have gestures related to the planets, divinations, geomancy, hydromancy, aeromancy, pyromancy, *furor* and the power of the melancholy humour. There is then a section on psychology followed by discussion of the passions, their power to change the body, and how by cultivating the passions or emotions belonging to a star (as love belonging to Venus) we can attract the influence of that star, and how the operations of the magician use a strong emotional force.

The power of words and names is discussed in the later chapters of the book, the virtue of proper names, how to compose an incantation using all the names and virtues of a star or of a divinity. The final chapter is on the relation of the letters of the Hebrew alphabet to the signs of the zodiac, planets, and elements which give that language a strong magical power. Other alphabets also have these meanings but less intensely than the Hebrew.

Book II. Celestial Magic

Mathematics are most necessary in magic, for everything which is done through natural virtue is governed by number, weight, and measure. By mathematics one can produce without any natural virtue, operations which seem natural, statues and figures which move and speak. (That is, mathematical magic can produce the living statues with the same powers as those made by using occult natural virtues, as described in the *Asclepius* which Agrippa has quoted on such statues.) When a magician follows natural philosophy and mathematics and knows the middle sciences which come from them—arithmetic, music, geometry, optics, astronomy, mechanics—he can do marvellous things. We see to-day remains of ancient works, columns, pyramids, huge artificial mounds. Such things were done by mathematical magic. As one acquires natural virtue by natural things, so by abstract things—mathematical and celestial things—one acquires celestial virtue, and images can be made which foretell the future, as that head of brass, formed at the rising of Saturn.

Pythagoras said that numbers have more reality than natural things, hence the superiority of mathematical magic to natural magic.

There follow chapters on the virtues of numbers and number groupings, beginning with One which is the principle and end of all things, which belongs to the supreme God. There is one sun. Mankind arose from one Adam and is redeemed in one Christ. Then come chapters on two to twelve, with their meanings and groupings, as Three for the Trinity, three theological virtues; three Graces; three decans in each sign; three powers of the soul; number, measure, and weight. The letters of the Hebrew alphabet have numerical values and these are most potent for number magic. Then follows an exposition of magic squares, that is numbers arranged in a square (either the actual numbers or their Hebrew letter equivalents) which are in accordance with planetary numbers and have power to draw down the influence of the planet to which they are related.

Then comes a treatment of harmony and its relation to the stars, harmony in the soul of man, the effects of music rightly composed in accordance with universal harmony in harmonising the soul.

After the long discussion of number in celestial magic, we have a very long discussion of images in celestial magic, with long lists of such images, images for the planets, images for the signs, nor does Agrippa fear actually to *print* the images of the thirty-six decan demons. . . .

The philosophy of magic in this book is important. Some of it is the usual material about the soul of the world, . . . but Agrippa is also using material from the *Corpus Hermeticum* from which he constantly quotes (of course in the form of opinions or sayings of Hermes Trismegistus). In relation to the world soul, he quotes from "Mercurius' Treatise *De communi*", one of the Hermetic treatises . . . with its optimist gnosis of the divinity of the world and its animation, exemplified from the continual movement of the earth as things grow and diminish, which movement shows that the earth is alive. Agrippa was thus not only using the *Asclepius* and its magic, but other treatises of the *Corpus Hermeticum* the philosophy of which he incorporated into his magical philosophy. . . .

[I have] only hinted in a partial and fragmentary way, and with but a few examples, at a theme which I believe may be of absolutely basic importance for the history of thought—namely, Renaissance magic as a factor in bringing about fundamental changes in the human outlook.

The Greeks with their first class mathematical and scientific brains made many discoveries in mechanics and other applied sciences but they never took whole-heartedly, with all their powers, the momentous step which western man took at the beginning of the modern period of crossing the bridge between the theoretical and the practical, of going all out to apply knowledge to produce operations. Why was this? It was basically a matter of the will. Fundamentally, the Greeks did not *want* to operate. They regarded operations as base and mechanical, a degeneration from the only occupation worthy of the dignity of man, pure rational and philosophical speculation. The Middle Ages carried on this attitude in the form that theology is the crown of philosophy and the true end of man is contemplation; any wish to operate can only be inspired by the devil. Quite apart from the question of whether Renaissance magic could, or could not, lead on to genuinely scientific procedures, the real function of the Renaissance Magus in relation to the modern period (or so I see it) is that he changed the will. It was now dignified and important for man to operate; it was also religious and not contrary to the will of God that man, the great miracle, should exert his powers. It was this basic psychological reorientation towards a direction of the will which was neither Greek nor mediaeval in spirit, which made all the difference.

What were the emotional sources of the new attitude? They lie, it may be suggested, in the religious excitement caused by the rediscovery of the *Hermetica,* and their attendant Magia; in the overwhelming emotions aroused by Cabala and its magico-religious techniques. It is magic as an aid to gnosis which begins to turn the will in the new direction.

And even the impulse towards the breaking down of the old cosmology with heliocentricity may have as the emotional impulse towards the new vision of the sun the Hermetic impulse towards the world, interpreted first as magic by Ficino, emerging as science in Copernicus, reverting to gnostic religiosity in Bruno. . . . Bruno's further leap out from his Copernicanism into an infinite universe peopled with innumerable worlds certainly had behind it, as its emotional driving power, the Hermetic impulse.

Thus "Hermes Trismegistus" and the Neoplatonism and Cabalism associated with him, may have played during his period of glorious ascendance over the mind of western man a strangely important role in the shaping of human destiny.

Allen G. Debus

The Paracelsian Chemical Philosophy

Allen G. Debus has done as much as any scholar to examine the centrality of alchemy and "the chemical philosophy" within early modern science. In this selection he summarizes the character and major empirical thrust of Paracelsian alchemy.

Born near Zurich in the small town of Einsiedeln in 1493, Philippus Aureolus Theophrastus Bombastus von Hohenheim was only later referred to as "Paracelsus," or "greater than Celsus." As a child he was exposed to a heady mixture of Renaissance thought. His father was a country physician who dabbled in alchemy, and the son was never to lose his interest in either medicine or the chemical laboratory. Young Paracelsus was to study under the famed abbot and alchemist, Johannes Trithemius (1462–1516), and was to learn the lore of mines when he worked as an apprentice in the Fugger mines in Villach when his father moved there in 1500. This experience was to bear fruit later in his speculations on the growth of metals and his book on the diseases of miners, the first book ever written on an occupational health problem.

At the age of fourteen Paracelsus left home to study, and over a period of more than two decades he traveled widely. He visited many universities and may have received a medical degree at Ferrara, but if so, he was willing to work in the far less prestigious position of surgeon with the armies that were ever on the move throughout Europe. By the third decade of the century his travels become easier to follow. Now in his thirties, he confined his travels to Central Europe, where he moved constantly from town to town both writing and offering his services as a physician. There were occasional moments of glory such as his appointment as municipal physician at Basel in 1527, but these were always short due to his rash temper. He made no effort to disguise his contempt for the universities and their academic circles. As for the physicians, they need hardly be considered:

"I need not don a coat of mail or a buckler against you, for you are not learned or experienced enough to refute even one word of mine . . . you defend your kingdom with belly-crawling and flattery. How long do you think this will last? . . . Let me tell you this: every little hair on my neck knows more than you and all your scribes, and my shoe-buckles are more learned than your Galen and Avicenna, and my beard has more experience than all your high colleges."

Such outbursts were to lose him position after position, for they offended even those who most wanted to help him. As a result he was constantly on the move; he died in Salzburg in 1541, where he had only recently been called by the bishop suffragan Ernest of Wittelsbach. . . .

At the death of Paracelsus there was little to indicate that his work would become the focal point for debate among scholars for more than a century. True, he had been a controversial figure during his lifetime, but relatively few of his voluminous writings had been published while he had been alive. The flood of Paracelsian texts began to issue from the presses only later. The legend of the man's near-miraculous cures began in the years after 1550 and soon there was a widespread search for his manuscripts, which were often published with notes and commentaries. Toward the end of the century vast collected editions were printed, and a whole school of Paracelsians battled with Aristotelians and Galenists over the course of natural philosophy and medicine alike.

Because of the late publication of the texts it is as proper to speak of the philosophy of the Paracelsians as it is that of Paracelsus. But even if we make this allowance the chemical philosophy is difficult to reconstruct, partly because no simple textbooks were published and partly because the views of these men are alien to those of the twenthieth-century scientists.

Actually there is much in the work of the Paracelsians that is reminiscent of other Renaissance natural philosophers. Above all, they sought to overturn the traditional, dominant Aristotelianism of the universities. For them Aristotle was a heathen author whose philosophy and system of nature was inconsistent with Christianity, a point of considerable concern during the Reformation. They stated that his influence on medicine had been catastrophic because Galen had uncritically accepted his work and the Aristotelian-Galenic system had subsequently become the basis of medical training throughout Europe. For them the universities were hopelessly moribund and unyielding in their adherence to antiquity.

The Paracelsians hoped to replace all this with a Christian neo-Platonic and Hermetic philosophy, one that would account for all natural phenomena. They argued that the true physician might find truth in the two divine books: the book of divine revelation—Scripture—and the book of divine Creation—nature. Thus, the Paracelsians applied themselves on the one hand to a form of biblical exegesis, and on the other to the call for a new philosophy of nature based on fresh observation and experiment. An excellent example of this may be found in the work of the important early systematizer of the Paracelsian corpus, Peter Severinus (1540–1602), physician to the king of Denmark, who told his readers that they must sell their possessions, burn their books, and begin to travel so that they might make and collect observations on plants, animals, and minerals. After their *Wanderjahren* [years of wandering] they must "purchase coal, build furnaces, watch and operate with the fire without wearying. In this way and no other you will arrive at a knowledge of things and their properties."

One senses a strong reliance on observation and experiment in the work of these men even though their concept of what an experiment is and its purpose was often quite different from our own. At the same time one notes an underlying distrust of the use of mathematics in the study of nature. They might well, as Platonists, speak of the divine mathematical harmonies of the universe. Paracelsus, in addition, spoke firmly of true mathematics as the true natural magic. But it was more customary for the Paracelsians to react with distaste to the logical, "geometrical," method of argument employed by the Aristotelians and Galenists. They condemned this "mathematical method" along with the traditional scholastic emphasis on geometry and they very specifically attacked mathematical abstraction in the study of natural phenomena—particularly the study of local motion. Their reason for this was primarily religious, and they were particularly incensed by the *Physics* of Aristotle. There—through the study of motion—it was argued that the Creator God must be immobile. The Paracelsian chemists of the period of the Reformation stated firmly that any argument imposing such a restriction on the omnipotent Deity could not be accepted—and that for this reason alone the texts of the ancients were sacrilegious and must be discarded. The chemical philosophy was to be a new science based firmly on observation and religion. Those who turned to quantification might recall that God had created "all things in number, weight and measure." This was interpreted as a mandate for the physician, the chemist, and the pharmacist—men who weighed and measured regularly in the course of their work. . . .

If the Paracelsians rejected what they called the "logico-mathematical" method of the schools, they turned to chemistry with the conviction that this science was the basis for a new understanding of nature. It was an observational science, and its scope was universal. These claims were to be found in the traditional chemical texts. For Paracelsus alchemy had offered an "adequate explanation of all the four elements," and this meant literally that alchemy and chemistry might be used as keys to the cosmos either through direct experiment or through analogy. Paracelsus explained the Creation itself as a chemical unfolding of nature. The later Paracelsians agreed and amplified this theme. Gerhard Dorn (fl. 1565–1585) gave a detailed description of the first two chapters of Genesis in terms of the new chemical physics, and Thomas Tymme argued that the

Creation had been nothing but an "Halchymicall Extraction, Separation, Sublimation, and Conjunction."

The chemical interpretation of Genesis helped to focus attention on the problem of the elements as the required first fruit of the Creation. Although the Paracelsian *tria prima* (salt, sulfur, and mercury) was a modification of both the earlier sulfur-mercury theory of the metals and other elemental triads, it has a special significance in the rise of modern science. The Aristotelian elements (earth, water, air, and fire) served as the basis of the accepted cosmological system. They were used by the alchemists as a means of explaining the composition of matter, by the physicians (through the humors) as a system for the interpretation of disease, and by the physicists as the basis for the proper understanding of natural motion. The introduction of a new elemental system thus ran the risk of calling into question the whole framework of ancient medicine and natural philosophy.

Marie Boas

The Uses of Mathematics

During the sixteenth century there was a rapid growth in mathematical studies and the application of mathematics to a wide variety of practical problems. Within the universities, chairs of mathematics and mixed mathematics were established; new mathematical classes and books aimed at accountants, artisans, engineers, and navigators proliferated; and new applications of mathematics and mathematical instruments in such fields as cartography were developed. This boom in mathematical awareness and sophistication, which set the stage for rapid developments in the physical sciences during the seventeenth century, is the subject of Marie Boas's article.

Though geometry was the branch of mathematics which had been most esteemed by the Greeks, they had not neglected other branches. The Pythagoreans had judged mathematics to consist of four divisions: geometry, arithmetic (number theory), astronomy and music; for they regarded astronomy as being applied geometry, and music as applied arithmetic. This classification had persisted, to reappear in the quadrivium of the mediaeval university. Plato, influenced by the Pythagoreans, had emphasised the role of mathematics in science as well as in philosophy. Pure mathematics, in Platonic doctrine, because it dealt with the world of perfect, unchanging, abstract ideas, was the best possible training for the philosopher who wished to study the nature of ideas, forms and essences. Mathematics reflected the unchanging reality behind the flux and uncertainty of the world of the senses; hence for the Platonist to study nature was to search for the mathematical laws which govern the world. Though Aristotle had protested that magnitude and body were different things, and natural philosophy and mathematics could not be the same, the Platonic tradition continued to appeal to many minds. The fifteenth century's intensification of interest in Platonic and neo-Platonic doctrine helped to encourage the view that mathematics was not only the key to science, but included within its competence the greater part of what the seventeenth century was to call natural philosophy. . . .

The Platonic tradition was of enormous consequence for Renaissance mathematics. Most obviously, it encouraged the study of pure mathematics and the search for previously neglected Greek mathematical texts. It stimulated the founding of chairs of mathematics in the new humanist schools, like the Collège Royale in France, though these were intended as linguistic centres. It helped the revival of professorships of mathematics in the established universities, though it did not raise the professors' salaries. It suggested that mathematics was better training for the mind than dialectic. It offered a number of useful varieties of mathematics, suitable for non-academic education: fortification for the gentleman-soldier, surveying for the landed proprietor, practical astronomy and some knowledge of the use of maps for all. On a less rational plane, Platonism and neo-Platonism encouraged so much number mysticism and astrology that to the layman "mathematicus" and "astrologer" were identical.

. . . Mathematicians were eager to exploit the host of newly discovered ways in which they could aid the unlearned, from teaching the merchant how to reckon his profits to showing the instrument-maker how to draw the scales on the brass plates of his wares. So great was the demand that there sprang up a new profession of semi-learned mathematical practitioners, men skilled in the practical aspects of mathematics, who knew how to apply geometry and trigonometry to the problems of scientific measuring devices. Many of these gave mathematical lectures in the vernacular, a practice especially common in London in the second half of the sixteenth century, and wrote books of elementary instruction in plain, simple and easy language.

A fair example is *A Booke Named Tectonicon* by Leonard Digges, published in 1556 and often reprinted. Digges said he had planned a "volume, containing the flowers of the Sciences Mathematicall, largely applied to our outward practise, profitably pleasant to all manner men in this Realme"; while waiting to complete it he produced this smaller work, whose subtitle declares it to be a book

> briefly shewing the exact measuring, and speedie reckoning all manner of Land, Squares, Timber, Stone, Steeples, Pillers, Globes, &c. Further declaring the perfect making and large use of the Carpenters Ruler, containing a Quadrant Geometricall. Comprehending also the rare use of the Square. And in the end a little Treatise adjoyning, opening the composition and appliancy of an Instrument called the Profitable Staffe. With other things pleasant and necessary, most conducible for surveyers, Land-meaters, Joyners, Carpenters and Masons.

Truly an indispensable mathematical handbook, suitable for learned and unlearned alike. . . .

Navigational problems such as fifteenth-century astronomers had tried to solve were still in the domain of the applied mathematician. . . .

Whether by dead-reckoning or astronomical methods (increasingly sophisticated now, as more and more mathematicians compiled tables, developed simplified methods and published books) all navigation involved the use of maps and charts. By the beginning of the sixteenth century nearly all land maps were based on some form of projection, but the "plane chart" still held supremacy at sea. In the plane chart distances between meridians were the same at all latitudes, whether near the equator or the poles, and large errors were thereby introduced at high latitudes. The Portuguese mathematician Pedro Nuñez (1502–1578) . . . tried to analyse the problem mathematically in his *Tracts* (1537); his analysis became better known when a Latin version appeared in 1566 under the title *On the Art of Sailing*. Nuñez discovered that on a sphere a rhumb line or loxodrome (a line of constant compass heading) is not a straight line, as it is on a plane, but a spiral terminating at the pole. He also noted that since the meridians on a globe converge, a true sea chart should not have its meridians everywhere equally spaced. Nuñez designed a quadrant which would enable one to find the number of leagues in a degree along each parallel, but he was unable to solve the much more important mathematical problem of finding a projection which would give the required convergence and make rhumbs straight lines.

Many references to the problem are to be found in subsequent books on mathematical navigation; the next real step towards its solution was made by Gerard Mercator (1512–94). Mercator studied mathematics under Gemma Frisius and lectured at Louvain until his Protestant faith made it necessary for him to leave the Low Countries for Germany. There he became a mathematical instrument-maker and a globe and map designer and publisher. His globes reflect both his mathematical ingenuity and his knowledge of the work of Nuñez, whose loxodromic spiral he engraved on some of them. . . . His world map of 1569, not a true sea chart though ostensibly "for the use of mariners," further utilised the notions of Nuñez: here Mercator spaced out the meridians towards the poles, apparently by guess-work, though he may have used trigonometric methods. . . .

. . . Edward Wright's *Certaine Errors in Navigation, Arising either of the ordinarie erroneous making of the Sea Chart, Compasse, Crosse staff, and Tables of declination of the Sunne, and fixed Starres detected and corrected* (1599). . . circulated for some time in manuscript before it was published; he claimed at last to make it public only to forestall a pirated edition under another's name. . . .

Wright's intention was to analyse all the errors commonly associated with the usual methods of dead-reckoning: in particular he treated the errors inherent in the use of the plane chart, showing their geometrical and physical sources and the ways of avoiding them. Wright supplied tables of rhumbs, showed how to use these tables and the new charts based upon them; how to find the distance from one place to another on the new charts, given latitude and longitude, and how best to plot a course. In fact everything the practical man needed to know, and with the tediums of calculation and computation removed as far as possible. . . .

. . . It was necessary to work out tables to permit the construction of maps upon [Mercator's] projection. Wright did both; and after the publication of his work any map-maker could draw a map on the now familiar Mercator projection, so particularly suited to the sea chart, since now a rhumb line is a straight line, and a constant compass course can be laid out with a ruler. That a great circle route is not so simple was still, obviously, of no concern to seamen not interested in finding the shortest distance between two points, since wind and current would never permit them to sail it even if it had been easily plotted.

The new projection did not become instantly popular, though it was fairly common within a generation.

Francis Bacon

Salomon's House

Francis Bacon (1561–1626), Lord Chancellor of England, humanistic scholar, and magnificent prose stylist, has long been seen as one of the great spokespersons for the experimental method in science and for the utilitarian value of scientific knowledge. Though his ideas were less unique than scholars once believed, his efforts as a promoter and propagandist on behalf of social support for science were important well into the nineteenth century. The description of an imaginary state-supported research institution—Salomon's House—from his utopian novella, The New Atlantis *(1627), which is presented here, seems to have played a significant role in promoting the establishment of both the Royal Society of London in 1662 and the Académie des Sciences at Paris in 1666.*

The Father of Salomon's House . . . caused me to sit down beside him, and spake to me thus . . .

"God bless thee, my son; I will give thee the greatest jewel I have. For I will impart unto thee, for the love of God and men, a relation of the true state of Salomon's House. Son, to make you know the true state of Salomon's House, I will keep this order. First, I will set forth unto you the end of our foundation. Secondly, the preparations and instruments we have for our works. Thirdly, the several employments and functions whereto our fellows are assigned. And fourthly, the ordinances and rites which we observe.

"The End of our Foundation is the knowledge of Causes and secret motions of things, and the enlarging of the bounds of Human Empire, to the effecting of all things possible.

"The Preparations and Instruments are these. We have large and deep caves of several depths; the deepest are sunk six hundred fathom, and some of them are digged and made under great hills and mountains, so that if you reckon together the depth of the hill and the depth of the cave, they are (some of them) above three miles deep. For we find that the depth of a hill and the depth of a cave from the flat is the same thing, both remote alike from the sun and heaven's beams and from the open air. These caves we call the Lower Region. And we use them for all coagulations, indurations, refrigerations, and conservations of bodies. We use them likewise for the imitation of natural mines and the producing also of new artificial metals by compositions and materials which we use, and lay there for many years. We use them also sometimes (which may seem strange) for curing of some diseases and for prolongation of life in some hermits that choose to live there, well accommodated of all things necessary; and indeed live very long, by whom also we learn many things.

"We have burials in several earths, where we put divers cements, as the Chineses do their porcelain. But we have them in greater variety, and some of them more fine. We have also great variety of composts and soils for the making of the earth fruitful.

"We have high towers, the highest about half a mile in height, and some of them likewise set upon high mountains, so that the vantage of the hill with the tower is in the highest of them three miles at least. And these places we call the Upper Region, accounting the air between the high places and the low as a Middle Region. We use these towers, according to their several heights and situations, for insolation, refrigeration, conservation, and for the view of [meteorological phenomena] as winds, rain, snow, hail, and some of the fiery meteors also. And upon them, in some places, are dwellings of hermits, whom we visit sometimes, and instruct what to observe. . . .

"We have also large and various orchards and gardens, wherein we do not so much respect beauty as variety of ground and soil, proper for divers trees and herbs, and some very spacious, where trees and berries are set whereof we make divers kinds of drinks, besides the vineyards. In these we practise likewise all [experiments] of grafting and inoculating, as well of wild trees as fruit trees, which produceth many effects. And we make (by art) in the same orchards and gardens trees and flowers to come earlier or later than their seasons, and to come up and bear more speedily than by their natural course they do. We make them also by art greater much than their nature, and their fruit greater and sweeter and of differing taste, smell, colour, and figure, from their nature. And many of them we so order as they become of medicinal use. . . .

"I will not hold you long with recounting of our brew-houses, bakehouses, and kitchens, where are made divers drinks, breads, and meats, rare and of special effects. Wines we have of grapes, and drinks of other juice of fruits, of grains, and of roots, and of mixtures with honey, sugar, manna, and fruits dried and decocted. Also of the tears or woundings of trees and of the pulp of canes. And these drinks are of several ages, some to the age . . . of forty years. We have drinks also brewed with several herbs and roots and spices, yea with several fleshes and white meats, whereof some of the drinks are such as they are in effect meat and drink both, so that divers, especially in age, do desire to live [on] them, with little or no meat or bread. And above all, we strive to have drinks of extreme thin parts to insinuate into the body, and yet without all biting, sharpness, or fretting, insomuch as some of them put upon the back of your hand will, with a little stay, pass through to the palm, and yet taste mild to the mouth. We have also waters which we ripen in that fashion, as they become nourishing, so that they are indeed excellent drink, and many will use no other. Breads we have of several grains, roots, and kernels; yea and some of flesh and fish dried, with divers kinds of leavenings and seasonings, so that some do extremely move appetites; some do nourish so, as divers do live of them without any other meat, who live very long. So for meats: we have some of them so beaten and made tender and mortified, yet without all corrupting, as a weak heat of the stomach will turn them into good [chyle], as well as a strong heat would meat otherwise prepared. We have some meats also and breads and drinks, which taken by men enable them to fast long after, and some other that, used, make the very flesh of men's bodies sensibly more hard and tough, and their strength far greater than otherwise it would be.

"We have dispensatories, or shops of medicines. Wherein you may easily think, if we have such variety of plants and living creatures more than you have in Europe (for we know what you have), the simples, drugs, and ingredients of medicines must likewise be in so much the greater variety. We have them likewise of divers ages and long fermentations. And for their preparations, we have not only all manner of exquisite distillations and separations, and especially by gentle heats and percolations through divers strainers, yea and substances, but also exact forms of composition, whereby they incorporate almost, as they were natural simples.

"We have also divers mechanical arts, which you have not, and stuffs made by them, as papers, linen, silks, tissues, dainty works of feathers of wonderful lustre, excellent dyes, and many others; and shops likewise as well for such as are not brought into vulgar use amongst us as for those that are. For you must know that of the things before recited, many of them are grown into use throughout the kingdom, but yet if they did flow from our invention, we have of them also for patterns. . . .

"We have also furnaces of great diversities and that keep great diversity of heats: fierce and quick; strong and constant; soft and mild; blown, quiet; dry, moist; and the like. But above all, we have heats in imitation of the sun's and heavenly bodies' heats, . . . whereby we produce admirable effects. Besides, we have heats of dungs, and of bellies and maws of living creatures, and of their bloods and bodies, and of hays and herbs laid up moist, of lime [unslaked], and such like. Instruments also which generate heat only by motion. And farther, places for strong insolations; and again, places under the earth, which by nature or art yield heat. These divers heats we use, as the nature of the operation which we intend requireth.

"We have also perspective-houses, where we make demonstrations of all lights and radiations, and of all colours; and out of things uncoloured and transparent we can represent unto you all several colours, not in rainbows, as it is in gems and prisms, but of themselves single. We represent also all multiplications of light, which we carry to great distance, and make so sharp as to discern small points and lines; also all colorations of light, all delusions and deceits of the sight in figures, magnitudes, motions, colours, all demonstrations of shadows. We find also divers means, yet unknown to you, of producing of light originally from divers bodies. We procure means of seeing objects afar off, as in the heaven and remote places, and represent things near as afar off and things afar off as near, making feigned distances. We have also helps for the sight, far above spectacles and glasses in use. We have also glasses and means to see small and minute bodies perfectly and distinctly, as the shapes and colours of small flies and worms, grains and flaws in gems, which cannot otherwise be seen, observations in urine and blood, not otherwise to be seen. We make artificial rainbows, haloes, and circles about light. We represent also all manner of reflexions, refractions, and multiplications of visual beams of objects. . . .

"We have also engine-houses, where are prepared engines and instruments for all sorts of motions. There we imitate and practise to make swifter motions than any you have, either out of your muskets or any engine that you have; and to make them and multiply them more easily and with small force by wheels and other means, and to make them stronger and more violent than yours are, exceeding your greatest cannons. . . . We represent also ordnance and instruments of war, and engines of all kinds, and likewise new mixtures and compositions of gunpowder, wildfires burning in water and unquenchable. Also fireworks of all variety both for pleasure and use. We imitate also flights of birds; we have some degrees of flying in the air; we have ships and boats for going under water, and brooking of seas, also swimming-girdles and supporters. We have divers curious clocks . . . and some perpetual motions. We imitate also motions of living creatures by images of men, beasts, birds, fishes, and serpents. We have also a great number of other various motions, strange for equality, fineness, and subtilty.

"We have also a mathematical house, where are represented all instruments, as well of geometry as astronomy, exquisitely made.

"We have also houses of deceits of the senses, where we represent all manner of feats of juggling, false apparitions, impostures, and illusions, and their fallacies. And surely you will easily believe that we that have so many things truly natural which induce admiration could in a world of particulars deceive the senses, if we would disguise those things and labour to make them seem more miraculous. But we do hate all impostures and lies; insomuch as we have severely forbidden it to all our fellows, under pain of ignominy and fines, that they do not show any natural work or thing adorned or swelling, but only pure as it is, and without all affectation of strangeness.

"These are, my son, the riches of Salomon's House. . . .

". . . And this we do also: we have consultations, which of the inventions and experiences which we have discovered shall be published, and which not; and take all an oath of secrecy for the concealing of those which we think fit to keep secret, though some of those we do reveal sometimes to the state, and some not.

"For our ordinances and rites, we have two very long and fair galleries: in one of these we place patterns and samples of all manner of the more rare and excellent inventions, in the other we place the statuas of all principal inventors. . . .

"Lastly, we have circuits or visits of divers principal cities of the kingdom, where, as it cometh to pass, we do publish such new profitable inventions as we think good. And we do also [make scientific forecasts] of diseases, plagues, swarms of hurtful creatures, scarcity, tempests, earthquakes, great inundations, comets, temperature of the year, and divers other things; and we give counsel thereupon what the people shall do for the prevention and remedy of them."

And when he had said this, he stood up, and I, as I had been taught, kneeled down, and he laid his right hand upon my head, and said, "God bless thee, my son, and God bless this relation which I have made. I give thee leave to publish it for the good of other nations, for we here are in God's bosom, a land unknown."

CHAPTER 4

Patronage and Printing

Introduction by Robert S. Westman

Before we examine the cultural forms of sixteenth- and seventeenth-century science, we must carefully avoid historical anachronism by putting quotation marks around the word "scientific"; this shows that it is our word and not a term that early modern thinkers used in a modern sense. Phrases like "natural knowledge" and "natural philosophy" better express the scientific language of the age; they referred to topics such as the motion of bodies on earth or in the heavens but might also include subjects like vision, sound, memory, and the intellect. The group of subjects that early moderns considered part of the proper study of natural philosophy have since been divided into separate disciplines such as physics and psychology.

If early moderns classified knowledge somewhat differently than we do, it should not surprise us too much to learn that the major cultural forms of scientific practice also differed from ours. Even a brief listing of institutions will convey a proper sense of diversity.

(1) In 1662, the English king, newly restored to the throne after more than a decade of civil war, gave formal institutional recognition to men who had been meeting informally in the homes of noblemen, in university lodgings, or coffee houses: Charles II (ruled 1660–1685) provided a charter that founded the Royal Society of London. Four years later, Louis XIV, under the guidance of his powerful minister Colbert, provided a special place in his library together with the adjoining apartment, where learned men met regularly to conduct experiments and to discuss scientific subjects. From these gatherings, the Academy of Sciences of Paris was formed. Yet, such well-known institutions hardly cover the full range of cultural formations.

(2) Men of scientific bent had already been meeting without royal patronage in the polite and fashionable Paris salons, often organized by a high-placed woman of leisure. And the coffee houses of late seventeenth-century London served as significant environments where natural knowledge was exchanged.

(3) Italy, too, had its learned societies. These numerous voluntary organizations emerged especially during the sixteenth and seventeenth centuries, created by Italian princes as an alternative political culture to the universities. In these environments men

of different social ranks could mingle, enjoying the common medium of erudite conversation lubricated by good food and drink. Among such societies, the Accademia dei Segreti formed around Giambattista della Porta in Naples, Prince Federigo Cesi's Accademia dei Lincei and Prince Leopold's Accademia del Cimento are the best known for their explicit goals of natural investigation; but one should be careful not to separate them too sharply from the academies whose stated aims were literary, such as the Accademia della Crusca.

(4) The print shops, as the historian Elizabeth Eisenstein has recently reminded us, were important loci of intellectual enterprise where author and publisher sometimes lived together and discoursed about the texts that they produced.

Finally, (5) we should not forget the prominence of the Catholic orders and the various Protestant ecclesiastical preferments. The former spawned such varied scientific types as Giordano Bruno and Tomasso Campanella (Dominicans), Marin Mersenne (Minim), Galileo's disciples Benedetto Castelli (Benedictine), Bonaventura Cavalieri (Jesuate) and Paolo Foscarini (Carmelite) and a train of important Jesuits too lengthy to list here in full, but including Galileo's friend, Christopher Clavius, who successfully inaugurated a powerful tradition of mathematical teaching in the Jesuit colleges. Among Protestants who held degrees of divinity or were practicing ministers, one can cite the examples of Kepler's teacher, Michael Maestlin (Lutheran pastor in Wuerttemberg) and Newton's teacher, Isaac Barrow (Master of Trinity College, Cambridge).

The great range of forms of cultural life surveyed above would seem to make it difficult at best to detect any simple unifying pattern in the social landscape of science. Yet, two important centers of scientific cultural activity continually stand out: the universities and the courts of kings, princes, and emperors.

As we saw in Chapter 2, medieval and early modern universities were quite different from our own institutions of higher learning. To begin with, they were *religious* foundations. They were established initially in the twelfth century by the legal acts of popes and kings who issued them their charters. Members of the medieval university came under ecclesiastical law. This meant that academics had the right to be tried in the more lenient church courts rather than in royal courts—an advantage for students who brawled with townspeople. Since faculty were *legally* members of the clergy, they were also required to take vows of celibacy, a practice that continued in England at Oxford and Cambridge even after the Protestant Reformation of the sixteenth century. While medieval professors more resembled monks, Protestant professors after the Reformation, especially in Germany, were allowed to marry. Martin Luther himself nicely illustrates this transition: Luther was a member of the Augustinian order and professor of theology at the University of Wittenberg. After his break with the Catholic Church, he took a wife. Within a decade or so, men often began academic careers first by living in the home of a professor and then, if all went according to schedule, by marrying the professor's daughter. Professorial chairs sometimes stayed in families for a century or more, as in the case of the Bartholin family at Copenhagen who dominated chairs of natural philosophy, medicine, and mathematics throughout the seventeenth century. In the Catholic church, teaching posts and other offices would be kept in the family by passing them down from uncle to nephew. As an example, the astronomer Copernicus received through his uncle, the Bishop of Warmia, a church "scholastry" or teaching position at a church in Wroclaw; he never taught a class there but had no trouble in collecting the monies that accrued through his proxies.

It may seem strange to us that professorships could be acquired through family connections, but such personal circumstances were the rule rather than the exception.

Because jobs and positions were valuable assets controlled and allocated by powerful secular nobles and church elite, one had to curry favor in order to gain rewards or protection. Universities provided educated men for a variety of social tasks. They prepared men for the bureaucracies of church and state, trained schoolmasters and priests, advised rulers on legal, diplomatic, and medical affairs. Yet, men moved into these positions, both inside and outside the university's walls, by dedicating their poems, plays, histories, planetary tables, weather forecasts, and works of art to a patron. In short, European society organized and distributed its resources according to rules of patronage.

In the first selection that follows, Arthur Marotti succinctly characterizes the English patronage system in the early seventeenth century. Marotti's description might easily apply to the situation faced by almost all contributors to the cultural life of the English elite, be they artists, historians, astrologers, or painters. It suggests to us that the cultural products of this period were inevitably tied to the conditions of patron-client relations.

Renaissance Italian painting is no exception to our theme. As Michael Baxandall shows eloquently, a painting can be viewed profitably as the product of a negotiation between a man who solicits the work and a man who makes it within a given set of conventions. Like a poem, a painting could also bear the residues of the patron-client relationship.

Let us turn now to the culture of the European courts. The courts were, of course, the households of the rulers and the centers of government. But they also served other purposes. As the focus of a realm, they symbolized through their monumental buildings the power and stability of the ruling class. Their brilliant pageantry and art works were designed to have a deep impact on visitors, celebrating the ruler and his power. From one court to the next, there were often substantial differences in style, but whatever their nature one can still point to certain consistent structural features of the court system. The historian Werner Gundersheimer has drawn interesting parallels between patronage systems in the courts of preindustrial Europe and tribal societies of New Guinea studied by anthropologists. In this regard, he cites the observation of the anthropologist Mary Douglas about practices of tribal domination. One has the dominance of "a leader who will gather his own network of allegiances powerfully around himself and create a centre of force for the rest of society. . . . There are few overriding community interests to check the leader's impetus. The greater his influence, the more support he attracts."[1] Gundersheimer goes on to extend Douglas's point:

> [The leader] may have rivals, but this does not threaten the integrity of the system itself. In the anthropologist's vocabulary, societies exhibiting this structure are called 'Big Man' systems. The presence of such Big Man systems all over the world—in Melanesia, among the Indians of the Pacific Northwest, in the Philippines, and elsewhere—may alert us to their possible impact in pre-industrial Europe. . . . The same elements are there: the growing accrual of power in the hands of Big Men; the evolution of patterns of deference and patronage; the competition between rivals and their client groups, both in politics and the arts. (p. 13)

Like the universities, the courts were hierarchical in their organization; yet, unlike the former, the patron's power was not mediated by the higher faculties of learning (theology, medicine, law) but flowed directly from the ruler himself. Indeed, those who created intellectual and material artifacts for the courts—art, music, poetry, mechanical devices—were immediately dependent upon the ruler's whim, and their situations could be often quite insecure. For the court environment was filled with tensions. While courtly virtue produced mannered and "civilized" behavior, it frequently created the opposite, as we learn in this early seventeenth-century criticism of courtiers:

> Base sycophants, crumb-catching parasites,
> Obsequious slaves, which bend at every nod:
> Insatiate harpies, gormandizing kites,
> Epicures, atheists, which adore no God
> But your own bellies and your private gain,
> Got by your oily tongues and bewitching train.[2]

In such a world, natural philosophers, astrologers, astronomers, alchemists, botanists, and other "scientific types" of this era were meant to entertain, praise, tutor, and sometimes provide practical advice to the Big Man.

Bruce Moran's article provides a nice glimpse into the world of German princes who were themselves involved in the very technical and mechanical practices that they patronized. He focuses particularly on the value placed on *precision*—in clockmaking and mapmaking, mining and metallurgy, weather forecasting, and so on—and, in general, on the importance of collecting and systematizing technical knowledge.

The court of Philip II, King of Spain, reveals to us another style. Authoritarian and intolerant, Philip still had a place for scientific practices at his court. In his interesting article, David Goodman describes Philip's interest in the improvement of public health, alchemical experimentation, mapmaking and surveying, architecture, and the teaching of mathematics.

Astronomers had a particularly important place among the scientific types of the early modern era. Within the universities, they were teachers of various mathematical subjects, including not only arithmetic, geometry, and trigonometry, but any disciplines that required a mathematical component, such as optics, mechanics, and music. It was not uncommon for them to hold chairs of medicine as well, since astrology was often an important discipline in the physician's tool kit. As evinced here and in the next chapter, the astronomers whose names and works are best known to us were almost all attached to court environments.

If courts and universities were significant loci of scientific work, the print shops deserve no less consideration. Printing transformed the possibilities for scientific work. Most important, publication created the notion of public priority. It is interesting to note that astronomers were prominent among the earliest publishers. And, indeed, there was a great public appetite for writings about the heavens, since people believed the motions of the planets to have a significant influence on their lives. In her stimulating book on printing, Elizabeth Eisenstein opens up new thinking about the importance of the printing press in early modern society. One might even say that without the printing press, no reform of the heavens—as discussed in the next chapter—would have been possible.

1. Mary Douglas, *Natural Symbols* (London, 1973), 89–90; cited in Gundersheimer, "Patronage in the Renaissance: An Exploratory Approach," in *Patronage in the Renaissance,* Guy Fitch Lytle and Stephen Orgel, eds. (Princeton: Princeton University Press, 1981), 13.
2. Cited in Sidney Anglo, "The Courtier: The Renaissance and Changing Ideals," in A. G. Dickens, ed. *The Courts of Europe. Politics, Patronage and Royalty, 1400–1800* (London: Thames and Hudson, 1977), 33.

Arthur F. Marotti

The Rewards of Patronage

Unlike modern industrial societies in which rewards are distributed according to systems of wages and profits and in which favoritism independent of merit is frowned upon, early modern Europe created structures whereby many of its main rewards were allocated according to personal criteria. In this brief excerpt from a longer study of the English poet and sermon writer John Donne, Arthur Marotti draws a succinct picture of the main features of the patronage system of Renaissance England, in which competition, dependency, and career ambition emerge as central themes.

In the Tudor and early Stuart period, patronage affected all aspects of English social, economic, and political life. Hence its influence on literature was inevitable. For most authors, patronage meant much more, however, than the financial support and social protection that allowed them to pursue aesthetic and intellectual enterprises. Often, as Lawrence Stone has noted, it provided "the necessary leverage to thrust them into comfortable jobs in the Church, the universities, and royal administration." Literary patronage was really inseparable from the systems of social and political patronage. Both amateur and professional, courtly and non-courtly writers, those who addressed recognized benefactors and those who communicated their work to an audience of social equals, were involved in the society's system of patronage. Their work either expresses the shared wishes for the rewards patronage could bring or was used, sometimes indirectly, as an instrument to obtain them. The term "literature of patronage" should not be limited to complimentary works or to works provided with complimentary dedications designed to get financial and social favors, for almost all English Renaissance literature is a literature of patronage. . . .

The social, political, and economic hierarchies of Renaissance England imply a functioning system of patronage. Gifts and rewards flowed not only from the monarch, but also from major and minor nobility and gentry, royal favorites, government civilian and military officers, virtually anyone who was positioned advantageously to offer, sell, or bargain over those tangible and intangible benefits ambitious men sought. The prizes of patronage included cash, titles and honors, lands, leases, grants, licenses, monopolies, pensions, educational and ecclesiastical positions, parliament seats, and places in the employ of the nobility, government officials, and the monarch. Young men who passed through the universities and the Inns of Court in the late sixteenth and early seventeenth centuries in record numbers competed heatedly for places, seeking careers and positions in which they could exercise their talents, make use of their training, and improve their social and economic lots. Gentlemen of limited means and those who, as younger brothers, had poor prospects for inherited wealth were especially eager, if not desperate, for success.

One of the effects of the Tudor centralization of power in the monarchy, a process Queen Elizabeth fostered with clear determination, is that the system of royal patronage was strengthened at the expense of the system of aristocratic patronage, making the Court more than ever the focus for men's hopes and ambitions. [The historian] Wallace MacCaffrey has observed that "the imposing stability of the Elizabethan regime depended upon a number of conditions, among which the successful distribution of patronage must be numbered. Most of the important gentlemen of England became beneficiaries of the Crown, bound to it by the interest of favors received and hoped for." MacCaffrey has estimated that in the latter part of Elizabeth's reign approximately 2,500 men were continually vying for some 1,200 places at the Queen's disposal: "This was a political society of which most of the members knew one another directly or indirectly and were almost all personally known to the leading ministers." In the small competitive world of London the educated and politically or socially active gentlemen knew one another, followed fashions, related gossip, attended major and minor state occasions, pursued their ambitions, and, occasionally, wrote poetry. For such individuals the royal Court was the acknowledged center of the realm, and such environments as the Inns of Court, the universities, London business and professional circles, the country, even the English-occupied areas of the Low Countries and Ireland, were satellites to it, dependent on the decisions of the monarch and her officers.

Michael Baxandall

Painting and Patronage in Fifteenth-Century Italy

The painter was dependent upon patrons to sustain his work. The work of visual art, like its verbal counterpart, was created not for a general market, but instead was commissioned by a particular individual or group. As the eminent art historian Michael Baxandall shows, the relationship between the artist and his patron could take different forms, but, in the end, even if the artist was salaried, the relationship was always an intensely personal one.

A fifteenth-century painting is the deposit of a social relationship. On one side there was a painter who made the picture, or at least supervised its making. On the other side there was somebody else who asked him to make it, provided funds for him to make it and, after he had made it, reckoned on using it in some way or other. Both parties worked within institutions and conventions—commercial, religious, perceptual, in the widest sense social—that were different from ours and influenced the forms of what they together made.

The man who asked for, paid for, and found a use for the painting might be called the *patron*, except that this is a term that carries many overtones from other and rather different situations. This second party is an active, determining and not necessarily benevolent agent in the transaction of which the painting is the result: we can fairly call him a *client*. The better sort of fifteenth-century painting was made on a bespoke basis, the client asking for a manufacture after his own specifications. Ready-made pictures were limited to such things as run-of-the-mill Madonnas and marriage chests painted by the less sought after artists in slack periods; the altar-pieces and frescoes that most interest us were done to order, and the client and the artist commonly entered into a legal agreement in which the latter committed himself to delivering what the former, with a greater or lesser amount of detail, had laid down.

The client paid for the work, then as now, but he allotted his funds in a fifteenth-century way and this could affect the character of the paintings. The relationship of which the painting is the deposit was among other things a commercial relationship, and some of the economic practices of the period are quite concretely embodied in the paintings. Money is very important in the history of art. It acts on painting not only in the matter of a client being willing to spend money on a work, but in the details of how he hands it over. A client like Borso d'Este, the Duke of Ferrara, who makes a point of paying for his paintings by the square foot—for the frescoes in the Palazzo Schifanoia Borso's rate was ten Bolognese *lire* for the square *pede*—will tend to get a different sort of painting from a commercially more refined man like the Florentine merchant Giovanni de' Bardi who pays the painter for his materials and his time. Fifteenth-century modes of costing manufactures, and fifteenth-century differential payments of masters and journeymen, are both deeply involved in the style of the paintings as we see them now: paintings are among other things fossils of economic life.

And again, pictures were designed for the client's use. It is not very profitable to speculate about individual clients' motives in commissioning pictures: each man's motives are mixed and the mixture is a little different in each case. One active employer of painters, the Florentine merchant Giovanni Rucellai, noted he had in his house works by Domenico Veneziano, Filippo Lippi, Verrocchio, Pollaiuolo, Andrea del Castagno and Paolo Uccello—along with those of a number of goldsmiths and sculptors—'the best masters there have been for a long time not only in Florence but in Italy.' His satisfaction about personally owning what is good is obvious. Elsewhere, speaking now more of his very large expenditure on building and decorating churches and houses, Rucellai suggests three more motives: these things give him 'the greatest contentment and the greatest pleasure because they serve the glory of God, the honour of the city, and the commemoration of myself.' In varying degrees these must have been powerful motives in many painting commissions; an altarpiece in a church or a fresco cycle in a chapel certainly served all three. And then Rucellai introduces a fifth motive: buying such things is an outlet for the pleasure and virtue of spending money well, a pleasure greater than the admittedly substantial one of making money. It is a less whimsical remark than it seems at first. For a conspicuously wealthy man, particularly someone like Rucellai who had made money by charging interest, by usury indeed, spending money on such public amenities as churches and works of art was a necessary virtue and pleasure, an expected repayment to society, something between a charitable donation and the payment of taxes or church dues. As such gestures went, one is bound to say, a painting had the advantage of being both noticeable and cheap: bells, marble paving, brocade hangings or other such gifts to a church were more expensive. Finally, there is a sixth motive which Rucellai—a man whose descriptions of things and whose record as a builder are not those of a visually insensitive person—does not mention but which one is ready to attribute to him, an element of enjoyment in looking at good paintings; in another context he might not have been shy of speaking about this.

The pleasure of possession, an active piety, civic consciousness of one or another kind, self-commemoration and perhaps self-advertisement, the rich man's necessary virtue and pleasure of reparation, a taste for pictures: in fact, the client need not analyse his own motives much because he generally worked through institutional forms—the altarpiece, the frescoed family chapel, the Madonna in the bedroom, the cultured wall-furniture in the study—which implicitly rationalized his motives for him, usually in quite flattering ways, and also went far towards briefing the painter on what was needed. And anyway for our purpose it is usually enough to know the obvious, that the primary use of the picture was for looking at: they were designed for the client and people he esteemed to look at, with a view to receiving pleasing and memorable and even profitable stimulations.

. . . The one general point to be insisted on is that in the fifteenth century painting was still too important to be left to the painters. The picture trade was a quite different thing from that in our own late romantic condition, in which painters paint what they think best and then look round for a buyer. We buy our pictures ready-made now; this need not be a matter of our having more respect for the artist's individual talent than fifteenth-century people like Giovanni Rucellai did, so much as of our living in a different sort of commercial society. The pattern of the picture trade tends to assimilate itself to that of more substantial manufactures: post-romantic is also post-Industrial Revolution, and most of us now buy our furniture ready-made too. The fifteenth century was a period of bespoke painting, however, and . . . about the customer's participation in it.

In 1457 Filippo Lippi painted a triptych for Giovanni di Cosimo de' Medici; it was intended as a gift to King Alfonso V of Naples, a minor ploy in Medici diplomacy. Filippo Lippi worked in Florence, Giovanni was sometimes out of the city, and Filippo tried to keep in touch by letter:

> I have done what you told me on the painting, and applied myself scrupulously to each thing. The figure of St. Michael is now so near finishing that, since his armour is to be of silver and gold and his other garments too, I have been to see Bartolomeo Martelli: he said he would speak with Francesco Cantansanti about the gold and what you want, and that I should do exactly what you wish. And he chided me, making out that I have wronged you.
>
> Now, Giovanni, I am altogether your servant here, and shall be so in deed. I have had fourteen florins from you, and I wrote to you that my expenses would come to thirty florins, and it comes to that much because the picture is rich in its ornament. I beg you to arrange with Martelli to be your agent in this work, and if I need something to speed the work along, I may go to him and it will be seen to. . . .
>
> If you agree . . . to give me sixty florins to include materials, gold, gilding and painting, with Bartolomeo acting as I suggest, I will for my part, so as to cause you less trouble, have the picture finished completely by 20 August, with Bartolomeo as my guarantor . . . And to keep you informed, I send a drawing of how the triptych is made of wood, and with its height and breadth. Out of friendship to you I do not want to take more than the labour costs of 100 florins for this: I ask no more. I beg you to reply, because I am languishing here and want to leave Florence when I am finished. If I have presumed too much in writing to you, forgive me. I shall always do what you want in every respect, great and small.
>
> Valete. 20 July 1457.
> Fra Filippo the painter, in Florence.

Underneath the letter Filippo Lippi provided a sketch of the triptych as planned. Left to right, he sketched a St. Bernard, an Adoration of the Child, and a St. Michael; the frame of the altarpiece, the point about which he is particularly asking approval, is drawn in a more finished way.

A distinction between 'public' and 'private' does not fit the functions of fifteenth-century painting very well. Private men's commissions often had very public roles, often in public places; an altarpiece or a fresco cycle in the side-chapel of a church is not private in any useful sense. A more relevant distinction is between commissions controlled by large corporate institutions like the offices of cathedral works and commissions from individual men or small groups of people: collective or communal undertakings on the one hand, personal initiatives on the other. The painter was typically, though not invariably, employed and controlled by an individual or small group.

It is important that this should have been so, because it means that he was usually exposed to a fairly direct relationship with a lay client—a private citizen, or the prior of a confraternity or monastery, or a prince, or a prince's officer; even in the most complex cases the painter normally worked for somebody identifiable, who had initiated the work, chosen an artist, had an end in view, and saw the picture through to completion. In this he differed from the sculptor, who often worked for large communal enterprises—as Donatello worked so long for the Wool Guild's administration of the Cathedral works in Florence—where lay control was less personal and probably very much less complete. The painter was more exposed than the sculptor, though in the nature of things clients' day-to-day interference is not usually recorded; Filippo Lippi's letter to Giovanni de' Medici is one of rather few cases where one can clearly sense the weight of the client's hand. But in what areas of the art did the client directly intervene?

There is a class of formal documents recording the bare bones of the relationship from which a painting came, written agreements about the main contractual obligations of each party. Several hundred of these survive, though the greater part refer to paintings that are now lost. Some are full-dress contracts drawn up by a notary, others are less elaborate *ricordi,* memoranda to be held by each side: the latter have less notarial rhetoric but still had some contractual weight. Both tended to the same range of clauses.

There are no completely typical contracts because there was no fixed form, even within one town. One agreement less untypical than many was between the Florentine painter, Domenico Ghirlandaio, and the Prior of the Spedale degli Innocenti at Florence; it is the contract for the *Adoration of the Magi* (1488) still at the Spedale.

> Be it known and manifest to whoever sees or reads this document that, at the request of the reverend Messer Francesco di Giovanni Tesori, presently Prior of the Spedale degli Innocenti at Florence, and of Domenico di Tomaso di Curado [Ghirlandaio], painter, I, Fra Bernardo di Francesco of Florence, Jesuate Brother, have drawn up this document with my own hand as agreement contract and commission for an altar panel to go in the church of the abovesaid Spedale degli Innocenti with the agreements and stipulations stated below, namely:
>
> That this day 23 October 1485 the said Francesco commits and entrusts to the said Domenico the painting of a panel which the said Francesco has had made and has provided; the which panel the said Domenico is to make good, that is, pay for; and he is to colour and paint the said panel all with his own hand in the manner shown in a drawing on paper with those figures and in that manner shown in it, in every particular according to what I, Fra Bernardo, think best; not departing from the manner and composition of the said drawing; and he must colour the panel at his own expense with good colours and with powdered gold on such ornaments as demand it, with any other expense incurred on the same panel, and the blue must be ultramarine of the value about four florins the ounce; and he must have made and delivered complete the said panel within thirty months from today; and he must receive as the price of the panel as here described (made at his, that is, the said Domenico's expense throughout) 115 large florins if it seems to me, the abovesaid Fra Bernardo, that it is worth it; and I can go to whoever I think best for an opinion on its value or workmanship, and if it does not seem to me worth the stated price, he shall receive as much less as I, Fra Bernardo, think right; and he must within the terms of the agreement paint the predella of the said panel as I, Fra Bernardo, think good; and he shall receive payment as follows—the said Messer Francesco must give the abovesaid Domenico three large florins every month, starting from 1 November 1485 and continuing after as is stated, every month three large florins. . . .
>
> And if Domenico has not delivered the panel within the abovesaid period of time, he will be liable to a penalty of fifteen large florins; and correspondingly if Messer Francesco does not keep to the abovesaid monthly payments he will be liable to a penalty of the whole amount, that is, once the panel is finished he will have to pay complete and in full the balance of the sum due.

Both parties sign the agreement.

This contract contains the three main themes of such agreements: (i) it specifies what the painter is to paint, in this case through his commitment to an agreed drawing; (ii) it is explicit about how and when the client is to pay, and when the painter is to deliver; (iii) it insists on the painter using a good quality of colours, specially gold and ultramarine. Details and exactness varied from contract to contract. . . .

Payment was usually in the form of one inclusive sum paid in instalments, as in Ghirlandaio's case, but sometimes the painter's expenses were distinguished from his labour. A client might provide the costlier pigments and pay the painter for his time and skill: when Filippino Lippi painted the life of St. Thomas in S. Maria sopra Minerva at Rome (1488–93) Cardinal Caraffa gave him 2,000 ducats for his personal part and paid for his assistants and the ultramarine separately. In any case the two headings of expenses and of the painter's labour were the basis for calculating payment: as Neri de Bicci noted, he was paid 'for gold and for applying it and for colours and for my workmanship.' The sum agreed in a contract was not quite inflexible, and if a painter found himself making a loss on a contract he could usually renegotiate: in the event Ghirlandaio, who had undertaken to provide a predella for the Innocenti altarpiece under the original 115 florins, got a supplementary seven florins for this. If the painter and client could not agree on the final sum, professional painters could act as arbitrators, but usually matters did not come to this point. . . .

Of course, not all artists worked within institutions of this kind; in particular, some artists worked for princes who paid them a salary.

Mantegna, who worked from 1460 until his death in 1506 for the Gonzaga Marquises of Mantua, is a well documented case and Lodovico Gonzaga's offer to him in April 1458 is very clear: 'I intend to give you fifteen ducats monthly as salary, to provide lodgings where you can live comfortably with your family, to give you enough grain each year to cover generously the feeding of six mouths, and also the firewood you need for your own use. . . .' Mantegna, after much hesitation, accepted and in return for his salary not only painted frescoes and panels for the Gonzagas, but filled other functions as well. Lodovico Gonzaga to Mantegna, 1469:

> I desire that you see to drawing two turkeys from the life, one cock and one hen, and send them to me here, since I want to have them woven by my tapesters: you can have a look at the turkeys in the garden at Mantua.

Cardinal Francesco Gonzaga to Lodovico Gonzaga, 1472:

> . . . I beg you to order Andrea Mantegna . . . to come and stay with me [at Foligno]. With him I shall entertain myself by showing him my engraved gems, figures of bronze and other fine antiques; we will study and discuss them together.

Duke of Milan to Federico Gonzaga, 1480:

> I am sending you some designs for pictures which I beg you to have painted by your Andrea Mantegna, the famous painter . . .

In practice Mantegna's position was not quite as tidy as Gonzaga's offer proposed. His salary was not always regularly paid; on the other hand, he was given occasional privileges and gifts of land or money, and fees from outside patrons. But Mantegna's position was unusual among the great Quattrocento [fifteenth-century] painters; even those who produced paintings for princes were more commonly paid for a piece of work than as permanent salaried retainers. It was the commercial practice expounded in the contracts, and seen at its clearest in Florence, that set the tone of Quattrocento patronage.

Bruce T. Moran

Princely Patronage of Science

Much of the artistic, literary, and scientific work of the Renaissance was conducted within the context of royal, imperial, or princely courts. In this selection, the historian of science Bruce T. Moran provides a more detailed examination of the kinds of scientific practices to be found at certain late sixteenth-century German courts. Precision was a characteristic feature of court science in both its earthly and heavenly concerns: land surveying and mapmaking served as tools of princely accounting systems; stone-crushing machines assisted in the extraction of precious ores; large-scale astronomical instruments pushed the capabilities of the naked eye to its maximum in recording the motions of heavenly objects. As Moran shows, court astronomers and mechanicians did not work in isolation. Remarkably good routes of communication existed both between neighboring courts and between courts and universities.

The expansion of mathematical and technical literature during the 15th and 16th centuries, often through the patronage of Italian princes and ecclesiastics, is only one aspect of a many-sided relationship between European courts, technology, and science. A deeper understanding of aristocratic involvement in technology and science may be attained by developing a picture of a specific princely type—one who not only patronizes technical and mathematical projects, but who is himself a technical and mathematical practitioner. The study of prince-practitioners in Germany will serve to illustrate the function of several European courts as institutional nodes of technical activity. I suggest that such courts, by tailoring technical and scientific roles to the special projects of princes; by emphasizing the procedural values of technical precision, exact and critical observation and the organized collection of information; and by promoting collaborative efforts through informal routes of communication added aristocratic support and authority to the development of important features of the experimental approach to the study of nature.

The influence of the court in the development of new technologies and in the creation of procedures useful to science has so far gained only peripheral attention. This discussion, while in no way attempting to limit the authority of Bacon's philosophy in the establishment of the experimental method nor favoring a particular explanation of the origin of Baconian principles, seeks at least to widen the social context of methodological discovery in the late Renaissance by defining the technical and scientific aspects of such an aristocratic milieu. Recent studies have determined the important role of Italian prince-practitioners in the founding of the first scientific society geared to experiment, the Accademia del Cimento. While the relationship of Bacon's philosophy to the members of the Accademia is anything but clear, what is certain is that aristocratic participation in scientific operations and in the construction of instruments useful to a scientific technology was not limited to the dukes of Tuscany. Their endeavors express a tradition of princely activity in scientific and technical projects, whose beginnings may be traced to the Middle Ages, which nurtured procedural features distinctive of early modern science.

Within the preface of a work concerning the design of sundials and astrolabes (dedicated to the Emperor Maximilian II) the Nürnberg mathematician Andreas Schöner (1528–90) considered the merits of several German princes who were themselves experienced in the design and manufacture of mathematical instruments:

> There are now many princes who even excel in the knowledge of the mathematical arts. Through their own effort they betake themselves to these studies as well as to a certain widening of the intellect. They design [*pingunt*] instruments and indeed I have seen many of this sort designed by princes. They observe the motion of the heavens and perform duties similar to those of their mathematicians. Undoubtedly among these are the illustrious prince August, the Elector, Duke of Saxony; Wilhelm, Landgraf of Hesse; Johann Friedrick and his brothers, dukes of Saxony; and Johann, the Elector of Brandenburg. I mention all these for the sake of [their] honor and for their immortal praise and glory. . . . [However] . . . of the heralds to Your Majesty, the greatest in these arts, both learned as well as most keen of judgment and most zealous of study, is the illustrious prince, Wilhelm, the Landgraf of Hesse.

Schöner insists that learned Protestant princes, in developing their own mathematical interests, aspired toward intellectual values which were first established at the courts of the Holy Roman Emperors. Certainly the most influential courts in northern Europe in the development of aristocratic attention to technology and science were the imperial Hapsburg courts in Vienna, and, in particular, the fabulous court of Rudolf II in Prague. For a brief period, the Prague court took shape as a center of scientific activity and mechanical artisanry. Astronomers, alchemists, naturalists, and magicians resided at Prague for various lengths of time, enjoying the patronage and protection of the emperor. Rudolf secured as well the talents of numerous clock and instrument makers, among whom Jost Bürgi (1552–1632); Erasmus Habermel (died 1606); Thomas Ruckert; and the Augsburg clockmakers Georg Roll (ca. 1546–92), Mattias Rungel (ca. 1563–1630), and Christoph Schissler (died 1609) were preeminent contributors to the Prague collections of ornate mechanical artifacts. The design of machines, both fanciful and practical, became an overall preoccupation at court and involved Rudolf himself in the construction of a self-orienting travelers chart which was operated by a concealed compass. In Hesse, the Palatinate, Württemberg, Braunschweig, and Saxony, Protestant princes adopted the values of Hapsburg patronage and extended them to accommodate princely involvement in practical mathematics, instrument making, and observation.

From his court in Kassel, the *Landgraf* of Hesse, Wilhelm IV (1532–92), organized projects of exact stellar measurement. . . . Wilhelm's participation began with observations of the fixed stars in 1566–67 and involved him in the observation of solar meridians at Kassel even after his mathematician, Christoph Rothmann (ca. 1550–ca. 1605), received overall charge of the court's measurements. Above all, Wilhelm's observations of the new star of 1572 were frequently compared by 16th-century astronomers and later gained the attention of Galileo, who viewed them as among the most accurate measurements of the star. . . .

The fascinating task of evaluating the incentives which influenced technical interest and which emphasized the value of precision at court would take us too far from the immediate purpose of this study. Pragmatic, technological, artistic, and intellectual factors can be found which, to various degrees, guided the interests of princes to the development of precision-measuring instruments and to the creation of an economically useful technology. On the other hand, designs of instruments and machines whose construction suggests little consideration for practical effect or economic worth appear frequently in the technical literature of the late 16th century—works normally intended for the aristocratic consumer. In these designs, a concern for detail and mechanical extravagance coincides with artistic values associated with the courtly style of Renaissance mannerism. At some courts, as at the imperial court of Rudolf II in Prague, mannerist forms of art evolved from the whole pattern of courtly living. Mannerist elements extended therefore as much to the court's patronage of the fine arts as to its involvement in science and technology.

Among prince-practitioners, however, interest in precision and in the production of machines and measuring instruments has a substantive basis in practical problems arising from efforts toward political consolidation and exploration as well as territorial and commercial expansion. More than artistic style, these concerns emphasized skills pertaining to surveying, cartography, mining, and fortification and attached important political and economic functions to the projects of mathematicians and artisans. At such courts, similarities of procedure, reaffirming precise description and the organized collection of information, characterized both scientific and essentially administrative programs.

Economic as well as scientific projects coincide at Hesse-Kassel with the values of systematic collection and observational accuracy inherent in both. While the *Landgraf's* attempt to construct a literal catalog of reliable stellar positions depended upon precise astronomical measurement. Wilhelm's desire to produce a sound basis for determining economic policy in Hesse resulted in the organization of detailed statistical surveys structured personally by the prince. Through individual inquiry, officials of the Kassel court determined the population, resources, and possessions, as well as property and forest rights of each village, town, and estate throughout the principality, compiling this information also into a form of catalog referred to by Wilhelm as his *Ökonomische Staat*.

One of the most common sorts of inventory aiding the consolidation of political, legal, and economic rights in the 16th century was the drawing of maps. Maps determined rights to forest lands and fields; offered proof of sovereign power and legal jurisdiction; specified taxable lands; and defined possession of economically productive regions such as mining areas, saltworks, woodlands, and transportation routes. Their construction required a general knowledge of geography as well as the development of accurate surveying techniques and instruments. In Germany, Protestant universities, linked to the needs and prestige of local sovereigns, provided necessary instruction in applied mathematics, particularly surveying and other aspects of practical geography. At the University of Wittenberg, the reforms of the Lutheran theologian Philip Melanchthon (1497–1560) made practical geography (i.e., the computation of distances of places from longitudes and latitudes) a requirement for students wishing a bachelors degree in the faculty of arts. As with many aspects of the curriculum at Wittenberg, Melanchthon's reform of mathematical studies became a model for other Protestant universities. The statutes of the University of Helmstedt for instance, founded by Duke Julius of Braunschweig (1528–89) in 1576, call for two professors of mathematics within the philosophy faculty. Besides arithmetic and geometry, the *mathematicus inferior* was to offer instruction in geography with reference to the globe, to grids, and to the longitude and latitude of places. As textbooks, students should consult the *Arithmeticae practicae methodus facilis* (Köln, 1576) of Gemma Frisius, the *Rudimenta cosmographica* (Tiguri, 1546) of the Kronstadt theologian, mapmaker, pedagogue, and publisher Johannes Honter (1498–1549), Pliny's *Historia naturalis,* and, if possible, Ptolemy's *Geographia*. The principal works of Euclid and Ptolemy were to be the required texts of the *mathematicus superior*. In astronomical instruction, more modern authors, including Peurbach, Regiomontanus, and the Copernican-based tables of Erasmus Reinhold might also be employed. Extending theory to practice, students should learn the use of several observational instruments, including the astrolabe, quadrant, torquetum, and Ptolemaic rule. This approach was to be followed also in the study of geography. Aside from preparing students to calculate planetary motions, the *mathematicus superior* was obliged "to investigate longitudes and latitudes of places by means of instruments and observations."

At court, pragmatically motivated interests in cartography increased princely involvement in new technology and directed their participation in the construction of mathematical instruments useful in surveying operations. From workshops in Nürnberg, Augsburg, and Dresden, Elector August of Saxony commissioned various odometers and pedometers. With other Protestant princes, including Christian of Denmark (1503–59) and Wilhelm IV of Hesse, August exchanged surveyors and surveying instruments and involved himself with the construction and improvement of measuring devices. To the odometers manufactured by his own mechanicians, Valentine Thau and Johannes Homelius (1518–62), August attached his own innovations. Measurements which he obtained with these instruments, and descriptions of their use, were recorded by August in manuscripts preserved at the elector's Dresden library.

Of the various practitioners of surveying, the mining surveyor held a preeminent position at princely courts. Since mining often contributed significantly to the wealth of emerging centralized principalities, it does not seem surprising to find that economic interests in mining helped to shape the technical projects of prince-practitioners. At least four compass dials were designed by Elector August for use in the Saxon mines. In Braunschweig, the interests of Duke Julius in mining and metallurgy inspired plans for the construction of several machines which the prince intended to be employed in the mining operations of the Harz. These inventions make up part of a remarkable manuscript in twenty-six chapters named by Julius, the *Instrumentenbuch* [the Book of Instruments]. The manuscript, which according to the subtitle was "in part conceived by Julius and drawn and painted by his own hand, contains numerous illustrations of tools and machines supposedly useful for the extraction, lifting, loading, and transportation of stones. One of Julius's designs is a machine used for crushing and mixing lime. When set in motion, the device employs sets of wheels to produce a simultaneous movement in its crushing elements and mixing arms. The instrument's elaborate design would have delighted a noble dilettante, yet Julius's intentions in creating the machine (one among several in the manuscript) were entirely practical and economic. The instrument's greatest advantage, he claims, lies in its ability to decrease the expenditure of physical labor, since by use of the machine only two men are required to do the work which, without its assistance, would occupy fifty-five laborers.

Many princely courts not only provided an environment for the development of technical proficiency and innovation but also, through long-standing personal, familial, and political lines of communication, initiated informal networks of scientific and technical correspondence. In the early 17th century an important network of communication based upon the meteorological interests of the Tuscan prince Ferdinand II (1610–70) arose as part of the work of the Accademia del Cimento. Having first commissioned the construction of several atmospheric instruments, Ferdinand entrusted thermometers, barometers, and hygrometers to select observers throughout Italy and later to observers in Paris, Innsbruck, and Warsaw. The information collected from individual stations was thereafter recorded and sent back to the academy for comparison. . . .

The diffusion of ideas and information through the replication of correspondence, or through the transference of *marginalia* [marginal notes] among copies of the same scientific treatise, forms an important aspect of the exchange of technical and scientific knowledge throughout the Renaissance. Like printed material, such information could be dispersed among a large number of interested persons who held no direct contact with the information source. In her recent major study of the transition from scribal to print culture, Elizabeth Eisenstein delineates various features of printing in the early modern era which altered the manner of collecting, systematizing, and disseminating scientific and technical knowledge. The argument, however, that the advent of print culture constituted a "communications revolution" (which itself played a key role in the establishment of the scientific revolution) should be evaluated against a backdrop of persisting, pretypographical conditions which affected the acquisition and transmission of information. In this regard, while the invention of movable type rendered an unprecedented advantage to the dissemination of *formalized* ideas, one must be cautious of overestimating the significance of printing in the kinds of information exchange which take place during the actual process of developing ideas and the initial sharing of discoveries. Although the ultimate aim may have been to record and circulate information through the printed word, the initial transmission of recent discoveries and new information among individuals actively involved at shared "research fronts" continued to take place throughout the shift from scribal to print culture, to a large extent, through informal interaction. . . .

[In sum,] the courts of princely patrons and practitioners (1) provided an alternative institutional setting and proposed new problem situations which influenced the development of scientific and technical skills; (2) created projects which emphasized the values of operational precision, critical observation, and systematic collection of information; and (3) established a means of collaboration through informal networks of communication based upon courtly correspondence.

Scholars beyond the German states called attention in the 16th century to the cultivation of mathematical studies at the universities and courts of German princes. The operational values which accompanied the technical and scientific projects of Hapsburg emperors and prominent Lutheran princes must also have been highly visible.

In England, the articulation of the empirical and utilitarian components of Baconian philosophy reflected both the development of the mechanical arts generally and the extension of technical literature in the Renaissance and coincided with a narrow consequence of a more far-reaching aristocratic involvement in technology (i.e., the increased interest in the practical use of mathematics and mathematical instruments among the English gentry).

Apart from the dependency of trade upon the availability and improvement of navigational instruments, the dissolution and redistribution of monastic lands in England prompted the extensive development of practitioner surveying and instrumentation throughout the mid- and later 16th century. The valuation of land, and the process of reassembling or parceling out larger estates, rested upon accurate land surveys which required the use of instruments and at least a rudimentary knowledge of practical geometry. An aid to such operations appeared in 1556 in Leonard Digges's *A Booke Named Tectonicon* . . . . which provided the surveyor and general artisan with a discussion of helpful mathematical principles as well as a description of measuring instruments and their use. The work underwent numerous printings but was initially published by one of England's first really talented instrument makers, Thomas Gemini (fl. 1524–62). Gemini emigrated from the Low Countries in the early 1540s, transferring to London the skills of engraving and instrument making which he had acquired at the workshop of Gemma Frisius in Louvain. Illustrations from copperplate (the first truly fine examples in England) which he produced in an imitative copy of Vesalius's *Corporis Fabrica* (1545) won him a royal pension, and soon thereafter he began the construction of astrolabes, in the Louvain style, for Queen Elizabeth and other members of the English aristocracy. A more famous instrument maker, Humphrey Cole (ca. 1530–91) most likely began his own career as an associate or apprentice in Gemini's workshop, where the instruments described and recommended by Digges in the *Tectonicon* could be had by special order. Cole was himself an expert in mining and metallurgy and practiced as well the skills of the engraver, diesinker, and instrument maker. His nautical instruments accompanied the Frobisher expedition to the northwest in 1576, and many other devices, dating from 1568 to 1586, display a skill in precision engraving unequaled by any English contemporary. Through his influence the first important circle of English instrument makers, which further promoted the involvement of commercial craftsmen and gentlemen alike, in projects of practical mathematics and instrument making, was founded.

As patrons and practitioners, princes, both within and outside Germany, fused together courtly and technical roles which both bridged the gulf between scholar and craftsman and helped to narrow the distance between the activities of the social elite and scientific and technical operations traditionally reserved for the artisan. That the privileged aristocracy would willingly collaborate with academic and artisan practitioners became a basic feature of utilitarian science as outlined in Bacon's "instauration" [reform] of learning. Whether or not Bacon found a model for such collaboration in the courts of Renaissance prince-practitioners, aristocratic participation in technical and scientific projects accentuated the development of many of the procedural values characteristic of 17th-century experimental science.

David Goodman

Patronage of Science at the Court of Philip II

Each court, depending upon the character and interests of its ruler, generated its own intellectual style. Mid-sixteenth-century Spain affords an interesting example of a powerful, intolerant ruler who clamped down sharply on freedom of expression and travel outside the Iberian peninsula. Yet, as the historian of science David Goodman shows in this excerpt, Philip was well read in the sciences and there was a considerable degree of scientific activity both in the Spanish universities and at his court.

Philip II a patron of the sciences? This aspect of his turbulent reign, like many others, has brought conflicting assessments. He has been praised for his enterprise and blamed for isolating Spain from the scientific revolution. More information has now become available as a result of research on related themes, and it seems opportune to reconsider Philip's relations with the sciences. This has not attracted much attention outside of Spain because of the general neglect of the history of Spanish science. Yet Spain was no intellectual backwater—it had a rich scientific culture which was still alive in the sixteenth century. And the question of Philip's patronage is important since Castile had become the heart of the most powerful empire in the West at a time when Europe was experiencing a new phase of the Reformation and the beginnings of scientific revolution.

There was little sign of Philip's patronage of science or scholarship in the first years of his reign. Those years, the 1550s, were a terrible period of religious intolerance, when the earlier hopes of a religious settlement between Catholics and the reformed religions had vanished. Servetus [the Spanish theologian and physician] had been executed in Calvin's Geneva; Protestants perished in the fires at Smithfield; and in Rome the new pope Paul IV had begun a relentless persecution of heretics and issued the first Roman index of prohibited books.

In the same repressive spirit Philip introduced stern measures which restricted intellectual freedom in his realms. In September 1558 a pragmatic issued at Valladolid announced his attention to strengthen the censorship of books to protect his subjects from novelties and heresies which threatened the Catholic religion. From now on no bookseller would be allowed to import, hold or sell any book prohibited by the Holy Office of the Inquisition; the Holy Office would prepare an index for public display in bookshops. Books could not be published until every page had been scrutinized by members of the royal council (grammars and children's books were exempt, but even these had to go before the local authority). The penalties for unauthorized printing or sale of prohibited books were made more severe. Under the Catholic Monarchs there had been fines and confiscation of prohibited books; now the death penalty was introduced.

Philip ordered books to be inspected in shops, monasteries and universities. He instructed the rector and masters of the University of Salamanca to search the library, and also to report to the Inquisition members of the university who had been found to possess suspect works, or who taught or held Lutheran doctrines.

The promised Index was published in 1559. Prepared by Fernando Valdés, archbishop of Seville and Inquisitor General, the list concentrated on theological works of the Reformation and vernacular translations of the bible. Some works on natural philosophy were also prohibited: Cardan's *De subtilitate rerum* and *De varietate rerum;* Gesner's *Historia animalium,* because the author was a Zwinglian; Leonhart Fuchs' *Institutiones medicinae, Paradoxorum medicorum libri tres* and all of his other books, because he was a Lutheran; and the complete works of Brunfels on account of his conversion to Lutheranism. . . .

What were the effects of these new regulations on Spanish science? One unfortunate result was the check given to the dissemination within Spain of the botanical studies of Brunfels and Fuchs; though some contact with the new botany was maintained by Clusius' travels in the peninsula. More generally, recent research has found that after the 1550s there was a sharp decline in the publication of scientific books in Spain. . . .

All of the obstructions which Philip II had put in the way of the sciences had come from his uncompromising opposition to the Reformation and not from any hostility towards the sciences themselves. In fact he was keenly interested in the mathematical sciences. His education had been put in the hands of Martinez Siliceo who had studied mathematics at Paris and published arithmetical works. When he was a young man of eighteen Philip bought a large number of books at Salamanca and Medina del Campo. These included Vitruvius' *De architectura;* Peter Apian's *Cosmographia;* Archimedes in Greek and Latin; an alchemical work by 'Geber'; Reisch's *Margarita philosophica;* Ptolemy's *Almagest;* and, particularly noteworthy, Copernicus' *De revolutionibus* just two years after its publication. He later acquired scientific works by Hippocrates, Aristotle, Dioscorides, Galen, Pliny and Hermes Trismegistus; Regiomontanus' *De triangulis* and Agricola's *De re metallica.* All of them eventually went to form part of the great library of Philip's Escorial [royal residence and monastery].

Philip gathered around him mathematicians, engineers and physicians. He was sympathetic to their plans, and promoted them by his royal power. He was particularly interested in improving medical care. When the Cortes of Castile assembled in Madrid in 1563 he was petitioned to improve hospital facilities by amalgamating several hospitals; Philip agreed to the project and by the end of the reign the general hospital of Madrid was established (1596). And when in 1566 the city of Salamanca complained to Philip of the shortage of skilled surgeons, he wrote to the rector of the university requesting the foundation of a chair of surgery. Andrés Alcázar was appointed in 1567 and soon did notable work on trepanning. Petitions from the Cortes continued to ask for university-trained surgeons, and in December 1593, perhaps persuaded by his physician Luis Mercado, Philip ordered the creation of more chairs in surgery. Within a few weeks of his order Francisco Ruiz was appointed at Valladolid and Luis de Victoria at Alcalá. The occupants of the chairs were to have the title 'surgeon royal' because Philip saw them as potentially in his service. The new surgery courses would be taken not only by students wishing to become surgeons; they were made compulsory for medical students as well. And licenses would be given to surgeons from these universities only if they had also taken some arts courses, a medical course and practised with an experienced surgeon.

Whatever the success of these plans to produce well-trained surgeons, the enterprising spirit is clear. Apart from Padua, Bologna and Valencia universities elsewhere in Europe found no place for surgery, a craft that was generally disparaged in the academic world. . . .

At Philip's court encouragement was given to the alchemical arts of distillation. One of the royal gardeners, Gillis Holbeeck, a Fleming, was appointed 'distiller of waters and essences'. And when Philip called Nardo Recchi from the kingdom of Naples to become a physician of the royal household, his duties included cultivating medicinal herbs in the royal gardens, supervising work on distillation and teaching the art of distillation to other court physicians. In his Escorial, at once a royal palace, monastery and mausoleum, Philip made provision for an alchemical laboratory. It was equipped with alembics [apparatus for distillation], retorts, and a giant 'philosophical tower', made of brass, and hollow so that steam could enter and fill the numerous alembics which it housed. In several rooms distillers prepared essences from herbs, spices, and metals. During the winter of 1567 Philip waited expectantly for the results of experiments, conducted day and night, to convert mixtures of copper, silver and lead into gold. To his secretary Pedro de Hoyo, who witnessed the experiments and for a while believed in their success, Philip wrote: 'although I am incredulous of these things, I am less so about this'. But he soon wrote: 'these transformations do not satisfy me'. Philip did not have the wholehearted enthusiasm for alchemy displayed by his nephew the Emperor Rudolf II at the Hradschin in Prague.

Philip's patronage of the mathematical sciences was more enthusiastic. Around 1566 he asked Pedro Esquivel, professor of mathematics at the University of Alcalá, to carry out an exact survey of Spain. Esquivel designed his own instruments for the task. Although he died before he was able to complete the work, he achieved one of the earliest geodetic triangulations; his accurate maps are preserved in the Escorial. Closely connected with the geodetic survey, and reflecting Philip's interests in geography and effective government, were the *relaciones* which Philip commissioned in 1575 and 1578. These were systematic attempts to collect detailed statistical information on the localities and population of peninsular Spain. Over six hundred accounts of towns and villages, mostly in New Castile, are the surviving records of Philip's ambitious project; they too are in the Escorial.

Philip's mathematical interests were strengthened through his close association with Juan de Herrera, first the assistant architect and then, from 1567, architect in charge of the construction of the Escorial. For that enormous undertaking Herrera designed special giant rotating cranes to lift large stones from quarries and position them in the rising buildings. He provided Philip with an Archimedean explanation of the mechanism of the cranes, based on the principle of the lever.

When Herrera advised the creation of a teaching institution for mathematical sciences, Philip welcomed the proposal and soon took the necessary steps. Writing on Christmas day 1582 from Lisbon, the capital of his recently conquered Kingdom of Portugal, he announced that 'for the benefit of our subjects and the provision in our realm of men proficient in mathematics, architecture and other related sciences and skills, we have appointed to our service Juan Bautista de Labaña to organize in our court matters relating to cosmography, geography and topography, and to teach mathematics in a form and place which we will order'. . . .

Patron and society—in the case of Philip the two are inseparable, because he changed the character of Spanish society by reducing its intellectual freedom. Unfortunately this was one of the most durable of his legacies; the force of his pragmatics was felt throughout the seventeenth century. His overriding interest was to protect the Catholic faith from the Reformation, and to achieve this he was willing to create conditions which were not at all favourable to independent scientific thought. And favourable conditions did not return until the middle of Charles II's reign (1665–1700) when Philip's rigorous censorial controls began to be relaxed. It was then that the European scientific revolution, so far largely resisted, began to penetrate the peninsula.

Robert S. Westman

The Astronomer's Role in the Sixteenth Century

The discipline of astronomy was practiced, for the most part, in two, quite different, social environments in the early modern period. Unlike heavenly observers in modern research universities, astronomers in early modern Europe did not engage in anything like "research." They were essentially teachers of mathematics. The primary purpose for studying the heavens was to cast horoscopes and to try to make predictions of political and religious significance. Gradually, in the course of the sixteenth century, the astronomer's role changed as astronomers began to challenge the structure of reality that was presented in the writings of natural philosophers and theologians. This shift in the astronomer's role was led by a group of writers who lived and worked in court contexts, freed from the pedagogical values and demands of the universities and concerned, instead, to glorify their patrons.

The sixteenth century professor's institutional role had little resemblance to his counterpart in a bureaucratic university of the post-industrial revolution age, which one historian has called a "factory system"—"the student . . . a 'pair of hands' working for the greater glory of his supervisor, the department as a conveyor belt for the production of Ph.D.s, the publication of paper as a sort of dividend". The sixteenth century academic had more in common with the textbook-writing pedagogue of the seventeenth and eighteenth century German universities. Not unlike the high school or secondary modern teacher with pedagogical competences in many subjects, the early modern academic could often teach in several different disciplines whilst respecting and never challenging the lines separating them. But however the dual goals of teaching and research were weighted, it would be a mistake not to recognize in the sixteenth century academic a certain kind of disciplinary consciousness. An instructor of astronomy—the focus of much of this paper—could turn, among other things, to a common historiography of astronomical authorities from the Greeks, the Arabs, and the Medieval Latins, a shared written language (Latin), a set of traditional problems (calculating planetary positions, predicting eclipses, and constructing observational instruments) and common rhetorical modes (the commentary, epitome, theorica, etc.). This disciplinary sense was strongly assisted by the eagerness of printers to publish a vast spectrum of literature on the heavens, which easily crossed national boundaries and divisions of confessional loyalty.

In the tradition-bound world of the preindustrial university, . . . the appearance of a fundamental scientific novelty explicitly raises several interesting questions about disciplinary communities and their relations. First, what were the conventional expectations and obligations of someone who considered himself to be the member of a discipline? Secondly, what kinds of social controls were exerted on those who attempted, in putatively illegitimate ways, to cross the boundary lines separating disciplinary domains? Thirdly, what kinds of strategies were employed by 'discipline bridgers' and their followers? The intent of these questions is to illuminate the role of the astronomer in the sixteenth century, not as a fixed set of static relations and structures, but as a dynamic, evolving, 'negotiated' process in which analogies to diplomatic exchange, military encounter, legal negotiation and competition for social status are not out of place. . . .

In this section, we examine those features of the early modern university which sustained conservative role alignments. We shall do this by briefly examining the career of the mathematics professor. This is an appropriate approach to the problem since the overwhelming number of those who lectured in astronomy held chairs or other institutional positions in mathematics. A fully detailed account cannot be offered here but some preliminary sketches can be made.

. . . The higher faculties of sixteenth century universities were oriented towards three professions: law, theology and medicine. In Northern Europe, the emphasis was on preparation for the clergy, as it had been during the Middle Ages, and such training was often followed by entry into some post in governmental administration. Consequently, training in canon law was very important, and law and theology faculties were correspondingly powerful. Beginning in the mid-sixteenth century, Spanish universities began to give enormous emphasis to juridical studies. By the end of the seventeenth century, students of law at the important universities of Salamanca and Valladolid outnumbered theologians by twenty to one. In the Italian universities, the emphasis was on law and medicine with theology, again, the less populous faculty. To understand the typical career of the mathematics professor, we need to realize that at none of these institutions was there provision for the doctorate in mathematics or astronomy, no licensing, that is, which recognized that symbol of full disciplinary autonomy. If one wished to make an academic career as an astronomer, therefore, one had a choice of seeking a degree in one of the faculties previously mentioned. It is for this reason that the figures, both major and minor, who populate the scientific scene in this period all received higher degrees either in law, theology or medicine. Copernicus, for example, studied law at Bologna and medicine at Padua in the course of which he received private instruction in mathematics from the professor in whose house he lived in Bologna, Domenico Maria de Novara (1454–1504). Michael Maestlin and his pupil Kepler exemplify the common Tübingen pattern of taking higher degrees in theology, and Maestlin, for a time, was a practising pastor. These are typical examples of how a later career in one of the liberal arts was initially tied to a profession with extramural status. By and large the most striking and rarely noticed feature of sixteenth century academic practitioners of the mathematical arts is that their tenure in that position occurred either while studying for a higher degree or while practising or while holding a joint chair in *medicine*. This seems to have been a phenomenon not known on a European-wide scale but it seems to have been most prevalent at the German Protestant universities and it was already a marked pattern at medieval Bologna and Padua.

There are at least two reasons for this situation and for both there is earlier precedent.

First, there is the cognitive link between medicine and mathematics occasioned by a common connection with astrology. Belief that the visible heavens control the human body by means of invisible forces was, of course, common in the ancient world and was the basis for such texts as Ptolemy's *Tetrabiblos* and the Hermetic writings. The rise of the status of the physician in the Middle Ages, especially Arabic and Jewish physicians, brought with it a desire to articulate more firmly the cosmological grounding of medicine. The most important Islamic philosophers were trained in, and often practised, medicine and the great twelfth century translations of scientific works were initiated often by Jewish physicians and philosophers. One of the most important functions of the doctor, prognosticating the course of an illness and its outcome, was tied to the skill of forecasting the positions of the planets. The practising physician, then, needed to know some planetary theory although use of tables of mean motions for calculation could be done without any expertise in model building. Moreover, predicting the position of a planet could be separated entirely from the question of how the planet acted physically on the human body across space. But the nature of that force was obviously something that demanded a theory at least among more philosophically inclined physicians. The revival of Hermetic, Neoplatonic and Stoic texts by fifteenth century Florentine humanist philosophers and physicians, like Marsilio Ficino and Pico della Mirandola, gave new legitimacy to old conceptual alternatives. Celestial–terrestrial forces other than moving material spheres were now wedded more firmly to astrological medical theory. The Stoic *pneuma,* a breathlike substance with both celestial and terrestrial properties, light, music and magnetism all became candidates for interdepartmental forces between higher and lower regions of the universe. But while astrology combined the predictive function of the astronomer with the explanatory role of the natural philosopher in the person of the academic physician, the alliance was still essentially a conservative one. The astrologer who calculated horoscopes or sought improved tables of mean motions never tried to challenge the explanatory fundamentals of his discipline. He was, in short, parasitic on astronomy for his calculational tools and on natural philosophy for his concepts of force and cosmological order. Small wonder, then, that the professors of mathematics and medicine who were largely responsible for producing the popular and numerous almanacs, prognostications and iatromathematical [medical-mathematical] treatises of the sixteenth century were neither receptive to cosmological innovation nor to the habit which produced it.

A second reason for the connection that we have been exploring is economic and institutional. The medical chair was a higher position; it was not only more prestigious but it was also more lucrative. At Wittenberg and Marburg, for example, a professor of medicine could earn twice as much as a professor of mathematics. In other words, there were incentives for a man, who was professor of mathematics already, to continue studying medicine in the hopes of gaining the more desirable chair. Once achieved, such an individual could also supplement his income through practice outside the university. The mathematics–medicine 'career track' meant that there were certain built-in incentives for social mobility between disciplines, with mathematics usually occupying a *propaedeutic* status, an occupation through which a man passed on his way to becoming an academic or practising physician, a theologian or a linguist. . . .

The effect of the 'career structure' that I have been outlining suggests that mathematics was frequently in the position of a *service role* to other areas of knowledge, such as medicine and theology. What counted as an 'improvement' in astronomy, then, usually meant some practical contribution: progress in its ability to make better predictions for the astrological physician, adjustments in the principles of the calendar, improved books on how to construct and operate well known instruments of observation, treatises on the computation of latitude and longitude at sea, or rationalization of astronomical pedagogy through better organized and more clearly illustrated textbooks. While the academic *mathematicus* possessed a certain consciousness of his discipline as an area with well-defined problems and techniques, his social position in the university was such as to dampen incentives toward making the mathematics professorship into a research role in problems of cosmology and theoretical astronomy; it was largely a pedagogical position for an undergraduate subject. . . .

The academic astronomer was called upon mainly to produce knowledge for certain . . . groups, namely, the undergraduate student and his lecturer, the physician and the general public who purchased almanacs. There existed another constituency, however, which represented a growing demand for astronomical and astrological knowledge, the landed nobility and the royal courts. There is, of course, a long tradition of kings and princes supporting astronomical activity; university patronage was a comparatively late development. In the sixteenth century, Hapsburg Imperial patronage, especially during the Rudolfine era (1576–1612), was perhaps the outstanding source of support for astronomical work in Europe. While financial inducements were not always stable, as Kepler sadly learned when he tried to collect his salary, there existed an atmosphere of what R. J. W. Evans calls "cosmopolitan freedom" among the intellectual elite who populated the Hradčany Palace in Prague. No other royal court of this period seems to have sustained such a diversity and number of intellectual types. Important as such associations were, however, the rise in status of the astronomer is connected even more significantly with a new development: the noble patron himself becomes involved in the very activity which he patronizes; the consumer becomes a producer. While there is good precedent for this in the arts—Henry VIII composed music, James I wrote poetry, etc.—this development is a new one in astronomy. The outstanding examples are not in the royal and imperial courts but in the smaller princely ones of Germany and Denmark. The central cases are Tycho Brahe in Hveen and Wilhelm IV in nearby Hesse-Cassel.

Court humanism, as opposed to the pedagogical orientation of university humanism, was directed toward the goal of glorifying the ruler. One of the ways in which the demands of aristocratic egos were gratified and inter-aristocratic status competition intensified, was in the building of a concrete display of scientific wealth—accurate and beautiful observing instruments, a coterie of the best practical mechanicians and mathematically trained men who, in sharing their learning with the ruler, gave prestige to him as he, in turn, rendered higher status to their occupation. The social organization of the court made possible a new kind of role for the academic mathematics professor *away from pedagogical constraints and expectations*. Correspondence between universities and courts became more frequent, and professors of medicine and mathematics served princes in the roles of physicians, mechanicians and mathematical tutors or assistants. Virtually absent from the courts of the North was the academic philosopher.

The example of the famous Danish astronomer, Tycho Brahe, provides one of the best cases for inspecting the shift in role norms associated with the court. Tycho himself deviated from the norms of his noble family by choosing to study a subject which they thought to be beneath his status and by rejecting the training in law and rhetoric that was urged upon him. Tycho's attitude toward the university is therefore of great interest. He never obtained a formal degree but he used the universities of Germany and Switzerland as a training resource, travelling from one to the other, befriending the professors and later corresponding with and employing them. From various pieces of evidence, we can infer some of Tycho's motivations for choosing not to make the university his own place of work. First, there is his important 1574 *Oratio de disciplinis mathematicis* delivered at the University of Copenhagen at the urging of the Danish King. The *Oratio* is a typical piece of mathematical-humanist propaganda exhorting the university to grant high rank to mathematical studies. . . . Stressing the now familiar themes of the certitude and utility of mathematics (especially astronomy) and the value of astronomy for medicine, Tycho argues for the primacy of mathematics in the curriculum: "I believe that the ancient Philosophers mounted to such heights of erudition because they taught Geometry from early childhood while most of us waste the best years of adolescence in the study of Grammar and languages.'' Much of the *Oratio* seeks to refute the objections of theologians, philosophers, doctors and lawyers to the claims of astrology. In May 1577, Tycho was offered the rectorship of the university, a position usually reserved for one of the professors, but even this honour he turned down on the grounds that he was too busy with affairs on newly occupied Hveen. While formal association with the university might have been resisted by Tycho's family and peers, Tycho evidently perceived the atmosphere of the university as simply unsuitable to the sort of endeavours that he was to pursue on his island. Some of his misgivings were expressed in a letter, no longer extant, to his friend Henricus Brucaeus, who held chairs of mathematics and medicine at Rostock. Brucaeus's reply to Tycho defends the academy and in so doing mentions some of Tycho's characterizations of the professors: ''shadow chasers'', and ''those engaged in idle formalities and empty processions of words''. . . .

Tycho Brahe's social position was not unlike that of a feudal lord. Consider the following: on 23 May 1576, King Frederick II of Denmark granted Tycho a small island ''to have, enjoy, use and hold quit and free'', says the document in its unmistakably feudal legal language, ''without any rent, all the days of his life, and as long as he lives and likes to continue and follow his *studia mathematices* . . .''. The legal position of Tycho was such that his only obligations were to his liege-lord, the King of Denmark, and *not* to the rector and senate of a university. His social position was such that he could take the economic and status privileges of nobility and confer them upon the activity of astronomical investigation. The result was a new role model and new prestige for astronomical activity. Tycho, the noble lord and vassal to the Danish king, surrounded himself with an entourage—not an army of knights and foot soldiers such as his family would have preferred, but of skilled mathematicians, observers and instrument builders. His weapons were not spears and arrows but his giant observing instruments and, from 1584 onward, his printing press. Tycho could publish what he liked, when he liked (except when there were paper shortages) and, more importantly, *for whom* he liked. The first publication from his presses appeared in 1588, containing his account of the comet of 1577 and his new cosmology—and it was *not for sale*. As a feudal lord, Tycho could avoid the free capitalist market and send copies of the book to those whom he thought would praise and understand it. Whilst Tycho's arguments against the existence of solid spheres have often been celebrated as a great victory for empiricism, we must not forget that the actual rejection of the spheres was a social act which involved the invasion of one discipline into the assumed domain of another. When Tycho marched his army of mathematicians into the fields of natural philosophy, the way had already been prepared for his attack on the authority of Aristotle by his abandonment of the university and its entrenched role divisions. And the new cosmology which he proudly waved as his leading banner became a symbol of his rights to theoretical property.

Elizabeth L. Eisenstein

Science and the Printing Press

In her long, stimulating study of the "communications revolution" of the early modern era, Elizabeth Eisenstein focuses our attention on the print shop as a new social context for books and authors. In contrast to the environment of the medieval scribes, who worked in isolation from one another, the print shop created a new work space in which the author, illuminators, goldsmiths and leatherworkers, and the printer-entrepreneur were brought together in a new kind of enterprise. As an early capitalist, the printer raised venture capital, hired workers, and took risks in marketing his product in places far removed from his shop. The title page, an innovation of the typographical age, fixed both the date and place of the product and thereby converted ideas into a kind of "intellectual property." Among the earliest printers were Johannes Müller (Regiomontanus) of Nuremberg, Peter Apianus of Ingolstadt, and Tycho Brahe of Hveen. All had their own printing presses and all were practicing astronomers who published their own works.

Social historians . . . need to be alerted to the new interplay between diverse occupational groups which occurred within the new workshops that were set up by early printers. The preparation of copy and illustrative material for printed editions led to a rearrangement of all bookmaking arts and routines. Not only did new skills, such as typefounding and presswork, involve veritable occupational mutations; but the production of printed books also gathered together in one place more traditional variegated skills. In the age of scribes, bookmaking had occurred under the diverse auspices represented by stationers and lay copyists in university towns; illuminators and miniaturists trained in special ateliers; goldsmiths and leather workers belonging to special guilds; monks and lay brothers gathered in scriptoria; royal clerks and papal secretaries working in chanceries and courts; preachers compiling books of sermons on their own; humanist poets serving as their own scribes. The advent of printing led to the creation of a new kind of shop structure; to a regrouping which entailed closer contacts among diversely skilled workers and encouraged new forms of cross-cultural interchange.

Thus it is not uncommon to find former priests among early printers or former abbots serving as editors and correctors. University professors also often served in similar capacities and thus came into closer contact with metal workers and mechanics. Other fruitful forms of collaboration brought astronomers and engravers, physicians and painters together, dissolving older divisions of intellectual labor and encouraging new ways of coordinating the work of brains, eyes and hands. Problems of financing the publication of the large Latin volumes that were used by late medieval faculties of theology, law, and medicine also led to the formation of partnerships that brought rich merchants and local scholars into closer contact. The new financial syndicates that were formed to provide master printers with needed labor and supplies brought together representatives of town and gown. As the key figure around whom all arrangements revolved, the master printer himself bridged many worlds. He was responsible for obtaining money, supplies and labor, while developing complex production schedules, coping with strikes, trying to estimate book markets and lining up learned assistants. He had to keep on good terms with officials who provided protection and lucrative jobs, while cultivating and promoting talented authors and artists who might bring his firm profits or prestige. In those places where his enterprise prospered and he achieved a position of influence with fellow townsmen, his workshop became a veritable cultural center attracting local literati and celebrated foreigners; providing both a meeting place and message center for an expanding cosmopolitan Commonwealth of Learning. . . .

As the prototype of the early capitalist . . . the printer embraced an even wider repertoire of roles. [The printer] Aldus' household in Venice, which contained some thirty members, has recently been described as an 'almost incredible mixture of the sweat shop, the boarding house and the research institute.' A most interesting study might be devoted to a comparison of the talents mobilized by early printers with those previously employed by stationers or manuscript bookdealers. Of equal interest would be a comparison of the occupational culture of Peter Schoeffer, printer, with that of Peter Schoeffer, scribe. The two seem to work in contrasting milieux, subject to different pressures and aiming at different goals. Unlike the shift from stationer to publisher, the shift from scribe to printer represented a genuine occupational mutation. Although Schoeffer was the first to make the leap, many others took the same route before the century's end.

As self-serving publicists, early printers issued book lists, circulars and broadsides. They put their firm's name, emblem and shop address on the front page of their books. Indeed, their use of title pages entailed a significant reversal of scribal procedures; they put themselves first. Scribal colophons had come last. They also extended their new promotional techniques to the authors and artists whose work they published, thus contributing to the celebration of lay culture-heroes and to their achievement of personal celebrity and eponymous fame. Reckon masters and instrument makers along with professors and preachers also profited from book advertisemments that spread their fame beyond shops and lecture halls. Studies concerned with the rise of a lay intelligentsia, with the new dignity assigned to artisan crafts or with the heightened visibility achieved by the 'capitalist spirit' might well devote more attention to these early practitioners of the advertising arts.

Their control of a new publicity apparatus, moreover, placed early printers in an exceptional position with regard to other enterprises. They not only sought ever larger markets for their own products; but they also contributed to, and profited from, the expansion of other commercial enterprises. What effects did the appearance of new advertising techniques have on sixteenth-century commerce and industry? Possibly some answers to this question are known. Probably others can still be found. Many other aspects of job printing and the changes it entailed clearly need further study. The printed calendars and indulgences that were first issued from the Mainz workshops of Gutenberg and Fust, for example, warrant at least as much attention as the more celebrated Bibles. Indeed the mass production of indulgences illustrates rather neatly the sort of change that often goes overlooked. . . .

One must distinguish . . . between literacy and habitual book reading. By no means all who mastered the written word have, down to the present, become members of a book-reading public. Learning *to read* is different, moreover, from learning by reading. Reliance on apprenticeship training, oral communication and special mnemonic devices had gone together with mastering letters in the age of scribes. After the advent of printing however, the transmission of written information became much more efficient. It was not only the craftsman outside universities who profited from the new opportunities to teach himself. Of equal importance was the chance extended to bright undergraduates to reach beyond their teachers' grasp. Gifted students no longer needed to sit at the feet of a given master in order to learn a language or academic skill. Instead they could swiftly achieve mastery on their own, even by sneaking books past their tutors—as did the young would-be astronomer, Tycho Brahe. 'Why should old men be preferred to their juniors now that it is possible for the young by diligent study to acquire the same knowledge?' asked the author of a fifteenth-century outline of history.

As learning by reading took on new importance, the role played by mnemonic aids was diminished. Rhyme and cadence were no longer required to preserve certain formulas and recipes. The nature of the collective memory was transformed. . . .

A useful service might be performed by discriminating more clearly between the new functions performed by the first astronomer-printer and those previously performed when astronomers had had to serve as scribes. Until 1470, Regiomontanus pursued a course similar to that followed by his immediate predecessors such as Toscanelli or Peurbach who had also been wandering scholars and versatile servants of cardinals and kings. Insofar as he combined librarianship with astronomy, he followed a pattern that had been characteristic of Alexandrians and Arabs alike. When he left the library in Buda in 1470 to set up his Nuremberg press, however, Regiomontanus crossed an historical great divide. He had, up to then, served colleagues and patrons in a traditional fashion. By gaining the backing of a wealthy Nuremberg citizen to set up a press and observatory, by turning out duplicate tables of sines and tangents, series of Ephemerides and advertisements for instruments, by training apprentices to carry on with the printing of technical treatises and by issuing his own advance publication list, he served later generations in new ways. He died prematurely in 1476. Some of his manuscripts did not get printed until more than fifty years after his death. Others were lost so completely that his ultimate plans for reforming astronomy will never be known. Moreover the great library at Buda that he had culled had its contents dispersed and was sacked by the Turks in 1527. Similar catastrophes had set astronomy back in the past. The course pursued by Regiomontanus gained momentum instead. The serial publications he launched never stopped. Ephemerides and trigonometry tables flowed in an uninterrupted stream which seems to have no stopping point even now. His efforts to train assistants in scientific printing and instrument-making also produced increasingly useful results. Insofar as he embodied the roles of printer and scientist 'in one and the same person,' Regiomontanus was the first but not the last of a new breed. Others carried on where he left off – as did those who prepared *De Revolutionibus* for a Nuremberg press more than sixty years after his death, and as Erhard Ratdolt and his co-workers did in Venice right away.

Innovations associated with Ratdolt's press: the first 'modern' title page, the first use of Arabic numerals for dating a book, the earliest extant type-specimen sheet, the first list of errata, the first three-color printed illustrations, the first engraved diagrams in a printed geometry book – are often noted in books aimed at bibliophiles. But their broader significance for the history of science remains to be explored. . . .

The *editio princeps* (which Ratdolt published in Venice in 1482 and Copernicus consulted as a young man) not only made the works of Euclid more available, it also arrested corruption of the text and introduced sharp-edged visual aids. The six-hundred-odd diagrams, which were ingeniously devised for the *editio princeps,* illustrated Euclidean proofs somewhat less 'dimly' than had been done in many hand-copied books. Printed diagrams endowed the *Elements* with a clarity and uniformity that they had not possessed before. 'Is it too fanciful to suggest that Euclid was associated, thereafter, less with Latin verbiage and more with triangles, circles and squares?

What Ratdolt's diagrams did for plane geometry, other early publication programs (beginning perhaps with Luca Pacioli) did for solid geometry – thereby making new use of the artistic invention of focused perspective developed in quattrocento Italy. The sharp and clear perspective renderings of three-dimensional geometric forms that illustrate Pacioli's *Divina Proportione* and other later sixteenth-century books seem to have supplied the models Kepler had in mind when he envisaged the planetary orbits as a series of nesting 'perfect solids.' It has been noted that *De Revolutionibus* also reflects a new concern with diagrammatic unity and visually symmetrical models. Is it not possible that such concerns owed something to the new forms of book illustration that were ushered in by printing?

CHAPTER 5

The Reform of the Heavens

Introduction by Robert S. Westman

When Nicolaus Copernicus (1473–1543) entered the University of Cracow in 1491, he listened to lectures on a variety of subjects from masters who commented on medieval or Greek texts. Among these texts were Aristotle's treatise *On the Heavens* and a short medieval treatise, *The Sphere*. The master of philosophy described a physical universe of concentric spheres, made up of an invisible, impermeable, unchanging *aether* that moved the planets and stars uniformly (i.e., equal angles in equal times) in circles around the universe's center. These spheres surrounded an inner core of "elements" beneath the moon—earth, air, water, and fire—identifiable by certain pairs of sensible qualities, such as heat and dryness, cold and moisture, and also by the directions in which those elements moved by virtue of their "nature," that is, up (away) from, or down to (toward), the center of the universe.

The master of mathematics accepted the physics of the Aristotelian spheres but, for purposes of calculating planetary positions on the sphere, certain difficulties imposed themselves. To an observer on a presumably stationary earth, the planets do not *appear* to move uniformly in circles: apparent variations occur in their velocities; brightness varies, as measured by the apparent size of their diameters; the planets' directions appear to change as the brightness increases; and, Mercury and Venus alone appear to remain tied to the direction in which the sun lies, shuttling back and forth across the sun at predictable intervals. The complexity of these "phenomena" or appearances defied the master of philosophy's simple, though appealing, physical model. As a result, a series of ingenious off-center ("ex-centric") geometrical devices were proposed as early as the first century B.C. in order to "save the phenomena": the epicycle, on which the planet rides; the deferent, a large circle whose center is the eccentric point and whose circumference defines the path of the epicycle's center; and the equant point. The center of the epicycle revolves *nonuniformly* as viewed both from the eccentric point and from the earth, but uniformly as computed from the noncentral equant point (situated as far from the center on one side as the earth is on the other). Put otherwise, the center of the sphere's uniform rotation is not an axis passing through its own center, but an off-center axis. Imagine the wobbling motion that would result from a grapefruit made to rotate around a pencil that does not pass through the center of the fruit!

Such worries about celestial wobbling formed the starting point for Copernicus's reform of ancient astronomy: a proper astronomy must be one in which the physics and the mathematics of the heavens were mutually coherent. Sometime between 1508 and 1515, Copernicus composed a short treatise, the *Commentariolus,* in which he laid out his ideas. This treatise (excerpted for the first reading in this chapter) circulated privately among Copernicus's friends for some years but was not published. Copernicus knew that his theory would be disturbing not only to the followers of Aristotle but also to some theologians who might find the theory in conflict with a literal reading of the Bible. In fact, Copernicus might never have published his book were it not for a visit from Georg Joachim Rheticus (1514–1574), a young mathematics professor from Wittenberg. After Rheticus had stayed two or three months, Copernicus allowed a short version of his major work to appear under Rheticus's name (*Narratio Prima,* 1540). Even more important, Copernicus allowed Rheticus to take the manuscript of his work with him in order to have it published.

The preface to Copernicus's *De revolutionibus orbium coelestium* (On the Revolutions of the Heavenly Spheres) is a most significant document. It is addressed not to "mathematicians" or to a secular patron, but to Pope Paul III. Copernicus describes how he long withheld publication of his work and has finally allowed it to be published only at the urging of his friends, all of whom were, like him, members of the Church. He then characterizes the astronomical tradition as filled with confusion and uncertainty. It is like a monster, the parts of whose body are disconnected and out of order. By contrast, the theory that he proposes will restore *symmetria* to the universe; it is like a body whose parts fit together into a well-proportioned whole. The logic of this claim bears an interesting relation to the *Commentariolus.* What Copernicus now embodied in *De revolutionibus* (excerpted below) was the full machinery of planetary calculation. If postulates 3 and 6 of the *Commentariolus* are granted to him, then it follows that the relative distances of the planets can be computed with respect to a common earth-sun distance. Only those with mathematical training will understand such a claim, he thinks; and he appeals to the pope and several bishops as those within the Church who possess that sort of understanding. He also hopes that the pope will protect him from those who would attack his new theory.

Reaction to Copernicus's theory in the sixteenth century cannot be divided simply into "for" and "against." Of course, there was negative response; there were those who belittled the theory on the grounds that if the earth moved, buildings would collapse and birds would be plastered into the west by a continuous east wind. Yet, it is an interesting feature of this theory that parts of it could be accepted while other sections were rejected or ignored. In particular, the mathematical models were easily adapted to the old premise that the earth rests at the center of a finite sphere. This meant that, in practice, Copernicus's calculational mechanisms were actively studied and used while his more radical proposals were ignored or rejected.

From the 1570s onward, a network of mathematical astronomers became intimately familiar with Copernican astronomy. They used the wide margins of Copernicus's book to keep detailed annotations on their reading. Many of these notes then passed into the copies of other owners so that gradually a common reading of certain problems emerged. Among those who were "plugged in" to this network of Copernican annotators was the Danish nobleman, Tycho Brahe (1546–1601). Although it was not considered fitting for nobility to engage in the practice of stargazing, Brahe defied his family and became an astronomer. In his case, this meant that he rejected the universities and became a princely astronomer, with his own castle, assistants, printing press, and large-scale observing instruments. In the short selection from Brahe's book on the comet of 1577, we see that he has inserted a chapter on what he called the "world system." The complex story of how he came upon this system is not told; what emerges clearly, however, is Brahe's discovery of a middle ground between the traditional Aristotelian-Ptolemaic cosmology and the new planetary arrangement of Copernicus. And most remarkably, Brahe rejects the existence of hard, impervious, solid spheres on the grounds that there would be no place for those recent comets, calculated to exist in the region above the moon.

Over more than two decades, Brahe made new and better observations of the planetary motions. If the heavens were to be reformed so that one could accurately predict important astrological events, such as the conjunction of three planets, then one needed accuracy. One way in which Brahe created a reputation was by sending copies of his books, printed at his own press, to various astronomers all over Europe. It was in this way that Michael Maestlin (1550–1631), a professor of mathematics at the University of Tubingen, received a copy of Brahe's treatise on the comet, and he was kind enough to lend it to his best student, Johannes Kepler (1571–1630). Young Kepler was impressed, but he had his own ideas. Kepler had also been allowed to borrow Maestlin's heavily annotated copy of *De revolutionibus* and, like his teacher, he was enraptured by Copernicus's theory. Thus, when in 1596 Kepler wrote his first book, entitled *The Cosmographic Mystery,* it was a Copernican world system that he described—and that could hardly have been to Brahe's liking when he thumbed through the copy that the young man sent to him. Yet Brahe was always looking for new disciples and, in Kepler, he saw great potential; for his part, Kepler was eager to see Brahe's treasure of observations, since he hoped that more accurate data might confirm his own theory explaining the spacing of the Copernican orbs.

Owen Gingerich deftly describes the encounter between Kepler and Brahe and Kepler's eventual creation of a new kind of astronomy. Kepler's *New Astronomy* incorporated certain insights of Copernicus and Brahe—the ordering of the planets and the destruction of the solid spheres—but it also went much further in challenging the ancient axiom of astronomy that Copernicus and Brahe still accepted: that all celestial motions are uniform and circular or composed of uniform and circular motions. Kepler built his three so-called laws of planetary motion on that new foundation (described in the reading by Gingerich).

The heavenly reforms introduced by the Catholic Copernicus and the Lutherans Brahe and Kepler culminated with the dramatic scientific contributions of an Italian natural philosopher and mathematician named Galileo Galilei (1564–1642). Whereas there had been earlier resistance to Copernicus from both Catholic and Protestant theologians who claimed conflict with the literal meaning of Holy Scripture, Galileo's defense led to the first sustained difficulties with conservatives within the Church. However, the Church was no more monolithic in the early seventeenth century than it is now; there were voices of reform from among certain members of the minor religious orders. Among these was the Florentine Carmelite Paolo Foscarini. In 1615 Foscarini wrote a small treatise seeking to reconcile the Bible with Copernicus's theory. He argued that the Copernican theory was "probably true" and that, in any event, the Bible was not meant to be taken literally, for its language had been accommodated to the understanding of the common people. In the excerpts below we see first the reply to Foscarini written in April 1615 by the powerful, yet moderate, Jesuit Cardinal Robert Bellarmine. Bellarmine was willing to accept the need to reexamine interpretations of the Bible, but only if a "true demonstration" of the physical world could be provided. Copernicus's "postulates" would not be sufficient to cause the Church to change official interpretations of the Bible. In his private notes on Bellarmine's critique, Galileo replies that he never questions the truth of the Bible but only the competence of theologians, untrained in astronomy, to make reasonable judgments on those portions of Scripture pertaining to the physical world. Furthermore, like

Foscarini, Galileo maintains that the Bible is not a science textbook but a moral document about man's salvation.

On 22 June 1633 Galileo was taken to a hall in the convent of Santa Maria sopra Minerva, made to kneel while a sentence was read, and ordered to abjure "the false opinion that the sun is the center of the universe and immovable, and that the earth is not the center of the same." He was told that he was "vehemently suspected of heresy." Much has been made of this incident, but the relative moderation of the sentence and the charge are sometimes overlooked. Galileo was not charged with heresy but only with the "vehement suspicion" of it. He was not tortured, although he was shown the instruments of torture—a standard practice. He was not imprisoned in the dungeons of the Inquisition but allowed to return to his country estate near Florence, where he remained under house arrest for the rest of his life. During this time he completed a major work, *Dialogue on the Two New Sciences*. Nonetheless, the trial had a cautionary effect in some parts of Europe.

Among those affected was the philosopher René Descartes (1596–1650). When Descartes heard of Galileo's condemnation, he was living in the politically secure city of Amsterdam; yet he remarked to a friend: "It is not my temperament to set sail against the wind." Descartes had just completed a treatise around 1633 that he entitled *The World, or Treatise on Light;* because of its Copernican assumptions, however, he chose to withhold publication. Nonetheless, Descartes's plans were ambitious. Trained at a Jesuit college, he had imbibed the Jesuit passion for systematization. But even more important, he had also come to the conclusion that the entire structure of natural philosophy needed to be reformed. "Instead of explaining only one phenomenon, I have resolved to explain all the phenomena of nature, that is to say, all of physics," Descartes wrote to a friend in 1629. Descartes's object was to replace the physics of Aristotle with a completely new vision of the world. The elements of that universe were first worked out in *The World* but not set forth until his *Principles of Philosophy* appeared in 1644. A small excerpt from that large treatise is reproduced here in order to complete our discussion of the reform of the heavens. In the selection provided, Descartes tries to reconcile his physics with an astronomy that preserves the best features of Copernicus and Tycho Brahe. His main assumption, highly questionable from the viewpoint of Galilean relativity, is that the earth is at rest with respect to the boundaries of its celestial whirlpool or vortex; the whirlpool itself, however, moves with respect to the other vortices in the universe. In this way, Descartes's earth both moves and rests; he has his cake and eats it, too. When the young Isaac Newton (1642–1727) entered Cambridge University in 1661, he kept a notebook of his readings and among the most extensive notes were those on Descartes's *Principles of Philosophy;* and this was the start of Newton's own scientific work.

Nicolaus Copernicus

Commentariolus

Sometime between 1508 and 1515, a young church administrator named Nicolaus Copernicus (1473–1543), from a region of East Prussia that is now part of Poland, composed a short treatise laying out a new series of models for calculating the heavenly motions. In this treatise, the author made several assumptions, one of which was that the earth revolves about the sun and that the sun rests near the center of the universe. The manuscript of this work circulated among Copernicus's friends in his own lifetime but was not actually published until the nineteenth century. In this excerpt, Copernicus succinctly lists the seven postulates of his new cosmology. A postulate, according to Aristotle, is any provable proposition that is assumed and used without being proved. Copernicus evidently assumed postulates 3 and 6 with 2, 4, 5, and 7 as consequences. Postulate 1 stands by itself as the main principle of planetary motion, excluding the possibility of a set of purely concentric spheres without epicycles.

The following list provides a brief explanation of the meaning of each postulate:

1. *Rejects the principle underlying the concentric cosmology. An "orb" is any individual spherical shell. The composite spherical shell containing all the partial orbs is a "complete orb" or "total sphere."*
2. *Rejects the fundamental physical proposition of Aristotelian-Ptolemaic astronomy that "down" is a point, the center of the universe. As it stands, this postulate is unproved; however, it follows necessarily from the next postulate and from number 6.*
3. *The center of the earth's orbit is the common center of the planetary spheres. That center is not the sun itself but a point* near *the sun. Strictly speaking, Copernicus's universe is heliostatic rather than heliocentric. Since there is no necessity binding us to this proposition, it too must be regarded as a postulate.*
4. *Here Copernicus anticipates an objection. If the earth revolves around the sun, then it follows that we ought to observe an annual shift in the apparent position of a star. However, because the stars are so distant from us, no such shift can be observed with the naked eye. It was not until 1838 that the so-called annual parallax was measured by the astronomer Bessel.*
5. *Copernicus explains here the principle of relative motion. In modern terms: think of the experience of sitting in a railroad car at rest in a station next to another train. As the other car pulls out of the station, we have the illusion that it is we who are moving and not the other train. Copernicus claims that, although the stars fixed in their giant celestial vault appear to move around the earth, it is really the earth itself, together with the other elements, that turns daily on its own axis.*
6. *Like postulate 3, this is another fundamental assumption. Here, he assumes that the sun's two observed motions—its daily rising and setting from east to west* with respect to *the horizon and its annual slow 365+ day drift from west to east* with respect to *the sphere of the fixed stars—can be explained by the earth's dual motions.*
7. *This follows from postulates 3 and 6.* If the earth spins on its axis daily and revolves annually around the sun, *then all observable heavenly phenomena can be accounted for. In this postulate, Copernicus mentions only one of those phenomena, the "retrograde" motion or change in direction of planet as seen from the earth.*

First Postulate

There is no one center of all the celestial [orbs] (*orbium*) or spheres (*sphaerarum*).

Second Postulate

The center of the earth is not the center of the universe, but only the center towards which heavy things move and the center of the lunar sphere.

Third Postulate

All spheres surround the sun as though it were in the middle of all of them, and therefore the center of the universe is near the sun.

Fourth Postulate

The ratio of the distance between the sun and earth to the height of the sphere of the fixed stars is so much smaller than the ratio of the semidiameter of the earth to the distance of the sun that the distance between the sun and earth is imperceptible compared to the great height of the sphere of the fixed stars.

Fifth Postulate

Whatever motion appears in the sphere of the fixed stars belongs not to it but to the earth. Thus the entire earth along with the nearby elements rotates with a daily motion on its fixed poles while the sphere of the fixed stars remains immovable and the outermost heaven.

Sixth Postulate

Whatever motions appear to us to belong to the sun are not due to [motion] of the sun but [to the motion] of the earth and our sphere with which we revolve around the sun just as any other planet. And thus the earth is carried by more than one motion.

Seventh Postulate

The retrograde and direct motion that appears in the planets belongs not to them but to the [motion] of the earth. Thus, the motion of the earth by itself accounts for a considerable number of apparently irregular motions in the heavens.

Nicolaus Copernicus

On the Revolutions of the Heavenly Spheres

Copernicus received a published copy of his main work just as his life expired. He had turned over responsibility for publishing the manuscript to Georg Joachim Rheticus (1514–1574), a young Lutheran mathematician from Wittenberg. Rheticus, who had to return to his teaching duties, left the manuscript in the hands of Andreas Osiander. Without seeking permission from either Rheticus or Copernicus, Osiander inserted an anonymous "Letter to the Reader" (reprinted below) in which he declared that the discipline of astronomy can never hope to offer true explanations for the heavenly motions but must content itself with angular predictions. Appropriately, as a life-long member of the clergy, Copernicus dedicated his book to Pope Paul III and to an audience of mathematically learned ecclesiastics. This preface, together with a brief section of Book I, Chapter 10, contains the kernel of his main strategy of proof. The essence of his claim is that his theory provides an absolute criterion called symmetria *whereby the parts of the universe may be brought into common agreement; the other astronomical theories of his time failed to satisfy this standard.*

Anonymous Foreword [by Andreas Osiander]

To the Reader Concerning the Hypotheses of this Work

There have already been widespread reports about the novel hypotheses of this work, which declares that the earth moves whereas the sun is at rest in the center of the universe. Hence certain scholars, I have no doubt, are deeply offended and believe that the liberal arts, which were established long ago on a sound basis, should not be thrown into confusion. But if these men are willing to examine the matter closely, they will find that the author of this work has done nothing blameworthy. For it is the duty of an astronomer to compose the history of the celestial motions through careful and expert study. Then he must conceive and devise the causes of these motions or hypotheses about them. Since he cannot in any way attain to the true causes, he will adopt whatever suppositions enable the motions to be computed correctly from the principles of geometry for the future as well as for the past. The present author has performed both these duties excellently. For these hypotheses need not be true nor even probable. On the contrary, if they provide a calculus consistent with the observations, that alone is enough. Perhaps there is someone who is so ignorant of geometry and optics that he regards the epicycle of Venus as probable, or thinks that it is the reason why Venus sometimes precedes and sometimes follows the sun by forty degrees and even more. Is there anyone who is not aware that from this assumption it necessarily follows that the diameter of the planet at perigee should appear more than four times, and the body of the planet more than sixteen times, as great as at apogee? Yet this variation is refuted by the experience of every age. In this science there are some other no less important absurdities, which need not be set forth at the moment. For this art, it is quite clear, is completely and absolutely ignorant of the causes of the apparent nonuniform motions. And if any causes are devised by the imagination, as indeed very many are, they are not put forward to convince anyone that they are true, but merely to provide a reliable basis for computation. However, since different hypotheses are sometimes offered for one and the same motion (for example, eccentricity and an epicycle for the sun's motion), the astronomer will take as his first choice that hypothesis which is the easiest to grasp. The philosopher will perhaps rather seek the semblance of the truth. But neither of them will understand or state anything certain, unless it has been divinely revealed to him.

Therefore alongside the ancient hypotheses, which are no more probable, let us permit these new hypotheses also to become known, especially since they are admirable as well as simple and bring with them a huge treasure of very skillful observations. So far as hypotheses are concerned, let no one expect anything certain from astronomy, which cannot furnish it, lest he accept as the truth ideas conceived for another purpose, and depart from this study a greater fool than when he entered it. Farewell.

To His Holiness, Pope Paul III, Nicholas Copernicus' Preface to His Books on the Revolutions

I can readily imagine, Holy Father, that as soon as some people hear that in this volume, which I have written about the revolutions of the spheres of the universe, I ascribe certain motions to the terrestrial globe, they will shout that I must be immediately repudiated together with this belief. For I am not so enamored of my own opinions that I disregard what others may think of them. I am aware that a philosopher's ideas are not subject to the judgement of ordinary persons, because it is his endeavor to seek the truth in all things, to the extent permitted to human reason by God. Yet I hold that completely erroneous views should be shunned. Those who know that the consensus of many centuries has sanctioned the conception that the earth remains at rest in the middle of the heaven as its center would, I reflected, regard it as an insane pronouncement if I made the opposite assertion that the earth moves. Therefore I debated with myself for a long time whether to publish the volume which I wrote to prove the earth's motion or rather to follow the example of the Pythagoreans and certain others, who used to transmit philosophy's secrets only to kinsmen and friends, not in writing but by word of mouth, as is shown by Lysis' letter to Hipparchus. And they did so, it seems to me, not, as some suppose, because they were in some way jealous about their teachings, which would be spread around; on the contrary, they wanted the very beautiful thoughts attained by great men of deep devotion not to be ridiculed by those who are reluctant to exert themselves vigorously in any literary pursuit unless it is lucrative; or if they are stimulated by the non-acquisitive study of philosophy by the exhortation and example of others, yet because of their dullness of mind they play the same part among philosophers as drones among bees. When I weighed these considerations, the scorn which I had reason to fear on account of the novelty and unconventionality of my opinion almost induced me to abandon completely the work which I had undertaken.

But while I hesitated for a long time and even resisted, my friends drew me back. Foremost among them was the cardinal of Capua, Nicholas Schönberg, renowned in every field of learning. Next to him was a man who loves me dearly, Tiedemann Giese, bishop of Chelmno, a close student of sacred letters as well as of all good literature. For he repeatedly encouraged me and, sometimes adding reproaches, urgently requested me to publish this volume and finally permit it to appear after being buried among my papers and lying concealed not merely until the ninth year but by now the fourth period of nine years. The same conduct was recommended to me by not a few other very eminent scholars. They exhorted me no longer to refuse, on account of the fear which I felt, to make my work available for the general use of students of [mathematics]. The crazier my doctrine of the earth's motion now appeared to most people, the argument ran, so much the more admiration and thanks would it gain after they saw the publication of my writings dispel the fog of absurdity by most luminous proofs. Influenced therefore by these persuasive men and by this hope, in the end I allowed my friends to bring out an edition of the volume, as they had long besought me to do.

However, Your Holiness will perhaps not be greatly surprised that I have dared to publish my studies after devoting so much effort to working them out that I did not hesitate to put down my thoughts about the earth's motion in written form too. But you are rather waiting to hear from me how it occurred to me to venture to conceive any motion of the earth, against the traditional opinion of [mathematicians] and almost against common sense. I have accordingly no desire to conceal from Your Holiness that I was impelled to consider a different system of deducing the motions of the universe's spheres for no other reason than the realization that [mathematicians] do not agree among themselves in their investigations of this subject. For, in the first place, they are so uncertain about the motion of the sun and moon that they cannot establish and observe a constant length even for the tropical year. Secondly, in determining the motions not only of these bodies but also of the other five planets, they do not use the same principles, assumptions, and explanations of the apparent revolutions and motions. For while some employ only homocentrics, others utilize eccentrics and epicycles, and yet they do not quite reach their goal. For although those who put their faith in homocentrics showed that some nonuniform motions could be compounded in this way, nevertheless by this means they were unable to obtain any incontrovertible result in absolute agreement with the phenomena. On the other hand, those who devised the eccentrics seem thereby in large measure to have solved the problem of the apparent motions with appropriate calculations. But meanwhile they introduced a good many ideas which apparently contradict the first principles of uniform motion. Nor could they elicit or deduce from the eccentrics the principal consideration, that is, the structure of the universe and the [sure] symmetry of its parts. On the contrary, [with them it is just as though someone were to join together hands, feet, a head, and other members from different places, each part well drawn, but not proportioned to one and the same body, and not in the least matching each other, so that from these (fragments) a monster rather than a man would be put together.] Hence in the process of demonstration or "method", as it is called, those who employed eccentrics are found either to have omitted something essential or to have admitted something extraneous and wholly irrelevant. This would not have happened to them, had they followed sound principles. For if the hypotheses assumed by them were not false, everything which follows from their hypotheses would be confirmed beyond any doubt. Even though what I am now saying may be obscure, it will nevertheless become clearer in the proper place.

For a long time, then, I reflected on this confusion in the [mathematical] traditions concerning the derivation of the motions of the universe's spheres. I began to be annoyed that the movements of the world machine, created for our sake by the best and most systematic Artisan of all, were not understood with greater certainty by the philosophers, who otherwise examined so precisely the most insignificant trifles of this world. For this reason I undertook the task of rereading the works of all the philosophers which I could obtain to learn whether anyone had ever proposed other motions of the universe's spheres than those expounded by the teachers of [mathematical arts] in the schools. And in fact first I found in Cicero that Hicetas supposed the earth to move. Later I also discovered in Plutarch that certain others were of this opinion. I have decided to set his words down here, so that they may be available to everybody:

> Some think that the earth remains at rest. But Philolaus the Pythagorean believes that, like the sun and moon, it revolves around the fire in an oblique circle. Heraclides of Pontus and Ecphantus the Pythagorean make the earth move, not in a progressive motion, but like a wheel in a rotation from west to east about its own center.

Therefore, having obtained the opportunity from these sources, I too began to consider the mobility of the earth. And even though the idea seemed absurd, nevertheless I knew that others before me had been granted the freedom to imagine any circles whatever for the purpose of explaining the heavenly phenomena. Hence I thought that I too would be readily permitted to ascertain whether explanations sounder than those of my predecessors could be found for the revolution of the celestial spheres on the assumption of some motion of the earth.

Having thus assumed the motions which I ascribe to the earth later on in the volume, by long and intense study I finally found that if the motions of the other planets [are brought into a relation with the circular course of the earth, and are reckoned for the revolution of each planet, not only do their phenomena follow therefrom but also the order and size of all the planets and spheres, and heaven itself is so linked together that nothing can be moved from its place without causing confusion in the remaining parts and the universe as a whole.] Accordingly in the arrangement or the volume too I have adopted the following order. In the first book I set forth the entire distribution of the spheres together with the motions which I attribute to the earth, so that this book contains, as it were, the general structure of the universe. Then in the remaining books I correlate the motions of the other planets and of all the spheres with the movement of the earth so that I may thereby determine to what extent the motions and appearances of the other planets and spheres can be saved if they are correlated with the earth's motions. I have no doubt that acute and learned [mathematicians] will agree with me if, as this discipline especially requires, they are willing to examine and consider, not superficially but thoroughly, what I adduce in this volume in proof of these matters. However, in order that the educated and uneducated alike may see that I do not run away from the judgement of anybody at all, I have preferred dedicating my studies to Your Holiness rather than to anyone else. For even in this very remote corner of the earth where I live you are considered the highest authority by virtue of the loftiness of your office and your love for all literature and [mathematics] too. Hence by your prestige and judgment you can easily suppress calumnious attacks although, as the proverb has it, there is no remedy for a backbite.

Perhaps there will be babblers who claim to be judges of [mathematics] although completely ignorant of the subject and, badly distorting some passage of Scripture to their purpose, will dare to find fault with my undertaking and censure it. I disregard them even to the extent of despising their criticism as unfounded. For it is not unknown that Lactantius, otherwise an illustrious writer but hardly [a mathematician], speaks quite childishly about the earth's shape, when he mocks those who declared that the earth has the form of a globe. Hence scholars need not be surprised if any such persons will likewise ridicule me. [Mathematics] is written for [mathematicians]. To them my work too will seem, unless I am mistaken, to make some contribution also to the Church, at the head of which Your Holiness now stands. For not so long ago under Leo X the Lateran Council considered the problem of reforming the ecclesiastical calendar. The issue remained undecided then only because the lengths of the year and month and the motions of the sun and moon were regarded as not yet adequately measured. From that time on, at the suggestion of that most distinguished man, Paul, bishop of Fossombrone, who was then in charge of this matter, I have directed my attention to a more precise study of these topics. But what I have accomplished in this regard, I leave to the judgement of Your Holiness in particular and of all other learned [mathematicians]. And lest I appear to Your Holiness to promise more about the usefulness of this volume than I can fulfill, I now turn to the work itself. . . .

All these statements are difficult and almost inconceivable, being of course opposed to the beliefs of many people. Yet, as we proceed, with God's help I shall make them clearer than sunlight, at any rate to those who are not unacquainted with the science of astronomy. Consequently, with the first principle remaining intact, for nobody will propound a more suitable principle than that the size of the spheres is measured by the length of the time, the order of the spheres is the following, beginning with the highest.

The first and the highest of all is the sphere of the fixed stars, which contains itself and everything, and is therefore immovable. It is unquestionably the place of the universe, to which the motion and position of all the other heavenly bodies are compared. Some people think that it also shifts in some way. A different explanation of why this appears to be so will be adduced in my discussion of the earth's motion. . . .

[The sphere of the fixed stars] is followed by the first of the planets, Saturn, which completes its circuit in 30 years. After Saturn, Jupiter accomplishes its revolution in 12 years. Then Mars revolves in 2 years. The annual revolution takes the series' fourth place, which contains the earth, . . . together with the lunar sphere as an epicycle. In the fifth place Venus returns in 9 months. Lastly, the sixth place is held by Mercury, which revolves in a period of 80 days.

At rest, however, in the middle of everything is the sun. For in this most beautiful temple, who would place this lamp in another or better position than that from which it can light up the whole thing at the same time? For, the sun is not inappropriately called by some people the lantern of the universe, its mind by others, and its ruler by still others. [Hermes] the Thrice Greatest labels it a visible god, and Sophocles' Electra, the all-seeing. Thus indeed, as though seated on a royal throne, the sun governs the family of planets revolving around it. Moreover, the earth is not deprived of the moon's attendance. On the contrary, as Aristotle says in a work on animals, the moon has the closest kinship with the earth. Meanwhile the earth has intercourse with the sun, and is impregnated for its yearly parturition.

In this arrangement, therefore, we discover a marvelous symmetry of the universe, and an established harmonious linkage between the motion of the spheres and their size, such as can be found in no other way. For this permits a not inattentive student to perceive why the forward and backward arcs appear greater in Jupiter than in Saturn and smaller than in Mars, and on the other hand greater in Venus than in Mercury. This reversal in direction appears more frequently in Saturn than in Jupiter, and also more rarely in Mars and Venus than in Mercury. Moreover, when Saturn, Jupiter, and Mars rise at sunset, they are nearer to the earth than when they set in the evening or appear at a later hour. But Mars in particular, when it shines all night, seems to equal Jupiter in size, being distinguished only by its reddish color. Yet in the other configurations it is found barely among the stars of the second magnitude, being recognized by those who track it with assiduous observations. All these phenomena proceed from the same cause, which is in the earth's motion.

Yet none of these phenomena appears in the fixed stars. This proves their immense height, which makes even the sphere of the annual motion, or its reflection, vanish from before our eyes. For, every visible object has some measure of distance beyond which it is no longer seen, as is demonstrated in optics. From Saturn, the highest of the planets, to the sphere of the fixed stars there is an additional gap of the largest size. This is shown by the twinkling lights of the stars. By this token in particular they are distinguished from the planets, for there had to be a very great difference between what moves and what does not move. So vast, without any question, is the divine handiwork of the most excellent Almighty.

Tycho Brahe

Reform of Copernicus and Ptolemy

In the universe of Aristotle and Ptolemy, the region above the moon was believed to be composed of a perfect, unchangeable, transparent, impenetrable substance called aether. *Comets were regarded as sublunary or meteorological phenomena because of their transient nature and the irregular, noncircular shape of their trajectories. In November 1577 an unusual new object appeared in the night skies. Numerous observations were made in different parts of Europe to try to determine by parallactic measurements how far away the object was from the earth. Tycho Brahe (1546–1601), a Danish nobleman who had erected a remarkable astronomical castle and observatory on the tiny island of Hveen near Copenhagen, concluded that the comet must be located above the moon in the aetherial heavens. In this selection, Brahe explains how he tried to find a place for the comet in the midst of the planets. Note, however, that in the course of his account, he describes a new world system, a compromise between Copernicus and Ptolemy, where the earth is at rest in the center of the universe, the sun revolves around the earth, and all the other planets (with the exception of the moon) circumnavigate the sun, just as in the Copernican system.*

It was made obvious and beyond any controversy that our Phaenomenon [the comet of 1577] had nothing in common with the elementary world, but was shown to have a motion in the Aether far up above the Moon, its tail perpetually maintaining an Olympian relationship to certain stars. It remains now, and seems to be especially fitting, that we should assign to it also some particular place in the very wide space of the same Aether, in order that we may establish between which orbs of the Secundum Mobile it will direct its path. Indeed the Aetherial World comprises incredible vastness, so that if we assume that this elementary world [measures] from the centre of the Earth to the nearest limits of the Moon about 52 Earth-radii (each of which contains 860 of our common or German miles) this will be contained 235 times in the rest of the space of the Secundum Mobile, that is to say as far as the extreme distance of Saturn from the Earth. In this enormously vast interval seven planets perform incessantly their wonderful and almost divine periodic motions; so that I can say nothing about that immense distance of the Eighth Sphere, which is beyond doubt greater by far than that of Saturn at his furthest point. On the other hand, according to the Copernican hypothesis, that space between Saturn and the Fixed Stars will be many times greater than the distance of the Sun from the Earth (which however is such that it includes the semidiameter of the elementary world about twenty times). For otherwise the annual revolution of the Earth in the great orb, according to his speculation, will not turn out to be insensible with respect to the Eighth Sphere, as it ought. Because the region of the Celestial World is of so great and such incredible magnitude as aforesaid, and since in what has gone before it was at least generally demonstrated that this comet continued within the limits of the space of the Aether, it seems that the complete explanation of the whole matter is not given unless we are also informed within narrower limits in what part of the widest Aether, and next to which orbs of the planets, [the comet] traces its path, and by what course it accomplished this. So that this may be more correctly and intelligibly understood, I will set out my reflections of more than four years ago about the disposition of the celestial revolutions, or synthesis of the whole system of the world. These were referred to before, but postponed to this point in the Astronomical Work, where they are required.

I considered that the old and Ptolemaic arrangement of the celestial orbs was not elegant enough, and that the assumption of so many epicycles, by which the appearances of the planets towards the Sun and the retrogradations and stations of the same, with some part of the apparent inequality, are accounted for, is superfluous; indeed, that these hypotheses sinned against the very first principles of the Art, while they allow, improperly, uniform circular motions not about [the orbit's] own centre, as it ought to be, but about another point, that is an eccentric centre which for this reason they commonly call an equant. At the same time I considered that newly introduced innovation of the great Copernicus, in these ideas resembling Aristarchus of Samos (as Archimedes shows in his *Sand-Reckoner*), by which he very elegantly obviates those things which occur superfluously and incongruously in the Ptolemaic system, and does not at all offend against mathematical principles. Nevertheless the body of the Earth, large, sluggish and inapt for motion is not to be disturbed by movement (especially three movements) any more than the Aetherial Lights are to be shifted, so that such ideas are opposed to physical principles and also to the authority of Holy Writ which many times confirms the stability of the Earth (as we shall discuss more fully elsewhere). Consequently I shall not speak now of the vast space between the orb of Saturn and the Eighth Sphere left utterly empty of stars by this reasoning, and of the other difficulties involved in this speculation. As (I say) I thought that both these hypotheses admitted no small absurdities, I began to ponder more deeply within myself, whether by any reasoning it was possible to discover an hypothesis, which in every respect would agree with both Mathematics and Physics, and avoid theological censure, and at the same time wholly accord with the celestial appearances. And at length almost against hope there occurred to me that arrangement of the celestial revolutions by which their order becomes most conveniently disposed, so that none of these incongruities can arise; this I will now communicate to students of celestial philosophy in a brief description.

I am of the opinion, beyond all possible doubt, that the Earth, which we inhabit, occupies the centre of the universe, according to the accepted opinions of the ancient astronomers and natural philosophers, as witnessed above by Holy Writ, and is not whirled about with an annual motion, as Copernicus wished. Yet, to speak truth, I do not agree that the centre of motion of all the orbs of the Secundum Mobile is near the Earth, as Ptolemy and the ancients believed. I judge that the celestial revolutions are so arranged that not only the lamps of the world, useful for discriminating time, but also the most remote Eighth Sphere, containing within itself all others, look to the Earth as the centre of their revolutions. I shall assert that the other circles guide the five planets about the Sun itself, as their Leader and King, and that in their courses they always observe him as the centre of their revolutions, so that the centres of the orbs which they describe around him are also revolved yearly by his motion. . . .

. . . So that this our new invention for the disposition of the celestial orbs may be better understood, we shall now exhibit its picture [Figure 5-1].

I have, in truth, constructed a fuller explanation of the new disposition of the celestial orbs, in which are important corollaries of all the present cogitations. I shall add this near the end of the work, and there it will be shown first of all from the motions of comets, and then clearly proved, that the machine of Heaven is not a hard and impervious body stuffed full of various real spheres, as up to now has been believed by most people. It will be proved that it extends everywhere, most fluid and simple, and nowhere presents obstacles as was formerly held, the circuits of the planets being wholly free and without the labour and whirling round of any real spheres at all, being divinely governed under a given law. . . .

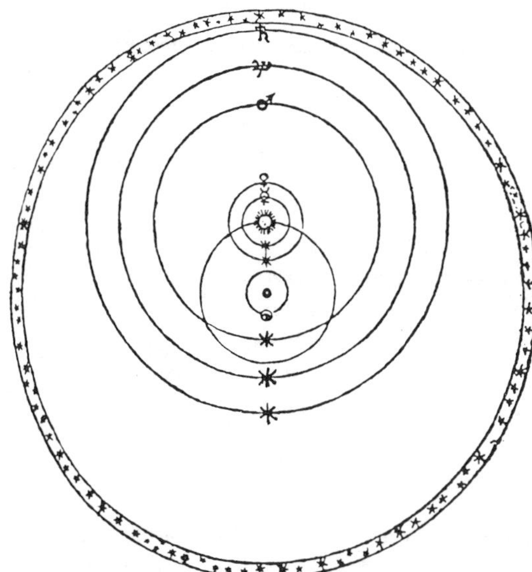

Figure 5-1 *A new hypothesis of a world system newly invented by the author, which excludes both the redundancy and awkwardness of the Ptolemaic system, and the recent physical absurdity of Copernicus in [postulating] the motion of the Earth, and in which all things correspond most conveniently with the celestial appearances.*

. . . To perceive all these things rightly, we must now submit to the eyes a suitable arrangement of the construction of the orbs [Figure 5-2].

By A is understood the globe of the Earth located in the centre of the universe, closest to which revolves the Moon in the orb BEFD, in which all the region of the elements is contained. That the comet can in no way be discovered between these bounds of the Lunar orb was, however, sufficiently demonstrated by us in the Sixth Chapter. Above this let CHIG be the annual orb of the Sun revolving about the Earth, in which the Sun is represented near C, upon which are located the centres of all the orbs of all the rest of the 5 planets, according to our renovation of the celestial hypothesis. And since the star of Mercury revolves closest to the Sun in the orb LKMN and a little above this the star of Venus revolves in the orb OPQR, it happens fitly that the comet revolves in yet a little greater orb described about the Sun.

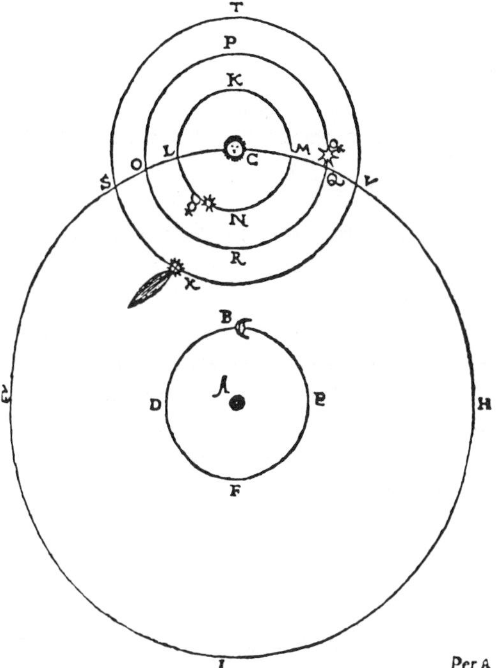

Figure 5-2

[The comet's orb] includes the orbs of Mercury and Venus only; it does not [include] the Lunar orb together with the Earth (as the star of Mars on its revolution does) because it cannot digress from the Sun by more than 60 degrees. And this same orb which we impute to the comet may be understood by the circle STVX, with the comet itself near X, in which sitation it was when seen at the time of our first observation. It has a motion in this orb in the order of the Signs, contrary to the revolutions of Venus and Mercury, so that it goes round from X through S to T. And the centre of the same orb observes it simple motion allied to the Sun perpetually. And this disposition of the revolution of the comet between the celestial orbs being accepted, I assert that it is possible to satisfy its apparent motion, as it is perceived by us dwelling on A, the Earth. . . .

Nor is it the case that anyone may think that our hypotheses are to be overthrown because of the short duration or great inequality of the motion. For it is probable that comets, just as they do not have bodies as perfect and perfectly made for perpetual duration as do the other stars which are as old as the beginning of the world, so also they do not observe so absolute and constant a course of equality in their revolutions—it is as though they mimic to a certain extent the uniform regularity of the planets, but do not follow it altogether. This will be clearly shown by comets of subsequent years, which will no less certainly be located in the Aethereal region of the world. Therefore either the revolution of this our comet about the Sun will not be at all points exquisitely circular, but somewhat oblong, in the manner of the figure commonly called ovoid; or else it proceeds in a perfectly circular course, but with a motion slower at the beginning, and then gradually augmented. However this may be, the comet in fact revolves around the Sun just the same, even though with a certain inequality, which yet is not confused or irregular.

Owen Gingerich

Kepler Breaks the Spell of Circularity

Many historians of science regard Johannes Kepler (1571–1630) as a true revolutionary in the history of early modern astronomy because, unlike Copernicus and Tycho Brahe, it was he who broke decisively with the ancient axiom that all celestial motions are both uniform and circular. In this selection, Owen Gingerich, an astrophysicist and historian of astronomy, explains how Kepler became a follower of Copernicus, how he inherited Tycho Brahe's treasured celestial data, and how he discovered three new laws of planetary motion.

Johannes Kepler was conceived on the 16th of May, 1571, at 4:37 A.M. and born on the 27th of December at 2:30 P.M. Such minutely kept dates remind us that Kepler lived in an age when "astronomer" still meant "astrologer" and when the word *scientist* had not yet been invented. Like many of the world's greatest scientists, including Ptolemy and Copernicus, Kepler had a profound feeling for the harmony of the heavens; although he rejected most of the traditional details of astrology, he believed in a powerful accord between the cosmos and the individual.

There was little in Kepler's youth to indicate that he would become one of the foremost astronomers of all time. A weak and sickly child, but intelligent, he easily won a scholarship to the nearby Tübingen University so that he could study to become a Lutheran clergyman. In recommending him for a scholarship renewal, the University Senate noted that Kepler had "such a superior and magnificent mind that something special may be expected of him."

Yet Kepler himself wrote . . . that although he had done well enough in the prescribed mathematical studies, nothing indicated to him a special talent for astronomy. Hence he was surprised and distressed when, midway through his third and last year as a theology student at Tübingen, he was summoned to Graz, far away in southern Austria, to become an astronomy teacher and provincial mathematician. It was there in one of his astronomy lectures that he hit upon what he believed to be the secret key to the construction of the universe.

This key hung upon a crucial thread: at Tübingen, Kepler had become a Copernican. The astronomy teacher at the University, Michael Mästlin, was remarkably knowledgeable about Copernicus's *On the Revolutions*. In his lectures, Mästlin explained how the new Copernican system accounted for the retrogradations in a most natural way, and how the planets were laid out in an elegant harmonic fashion both with respect to their spacing from the Sun and to their periods.

It was undoubtedly the beautiful harmonic regularities so "pleasing to the mind" that appealed strongly to Kepler's sense of the aesthetic and induced him to become such an enthusiastic Copernican. To Kepler the theologian, such regularities revealed the glory of God. . . .

Because of his preoccupation with the Copernican system, Kepler had begun to ask himself three unusual questions: Why are the planets spaced this way? Why do they move with these regularities? Why are there just six planets? All these questions are very Copernican, the last one particularly so because a traditional geocentrist would have counted both the Sun and the Moon, but not the Earth, thereby listing seven planets.

In illustrating to his class how the great conjunctions of Jupiter and Saturn fall sequentially along the ecliptic, Kepler drew a series of quasi-triangles whose lines began to form an inner circle half as large as the outer ecliptic circle. The proportion between the circles struck Kepler's eye as almost identical with the proportions between the orbits of Saturn and Jupiter. Immediately he began a search for a similar geometrical relation to account for the spacing of Mars and the other planets, but his quest was in vain.

"And then it struck me," he wrote. "Why have plane figures among three-dimensional orbits? Behold, reader, the invention and the whole substance of this little book!" He knew that there were five regular polyhedra, that is, solid figures each with faces all of the same kind of regular polygon. By inscribing and circumscribing these five figures between the six planetary spheres (all nested in the proper order), he found that the positions of the spheres closely approximated the spacings of the planets. Since there are five and only five of these regular or Platonic polyhedra, Kepler thought that he had explained the reason why there were precisely six planets in the solar system.

Kepler published this scheme in 1596 in his *Mysterium cosmographicum,* the "Cosmographic Secret." It was the first new and enthusiastic Copernican treatise in more than fifty years, since *On the Revolutions* itself. Without a Sun-centered universe, the entire rationale of the book would have collapsed.

The young astronomer realized, however, that Copernicus had made the Sun immobile without actually using it as his central reference point; rather, he had used the center of the Earth's orbit. Although the Sun was nearby, it played no physical role. But, Kepler argued, the Sun's centrality is essential to any celestial physics, and the Sun itself must supply the driving force to keep the planets in motion. Not only did he propose this critically significant physical idea, he attempted to describe mathematically how the Sun's driving force diminished with distance. Again, his result was only approximate, but at least the important physical-mathematical step had been taken. This idea, which was to be much further developed in his *New Astronomy,* establishes Kepler as the first scientist to demand physical explanations for celestial phenomena. Although the physical explanation of the *Mysterium cosmographicum* was erroneous, never in history has a book so wrong been so seminal in directing the future course of science.

Kepler sent a copy of his remarkable treatise to the most famous astronomer of the day, Tycho Brahe. Unknown to Kepler, the renowned Danish astronomer was in the process of leaving his homeland. Tycho had boasted that his magnificent Uraniborg Observatory had cost the king of Denmark more than a ton of gold. Now, however, fearing the loss of royal support at home, Tycho had decided to join the court of Rudolph II in Prague.

Kepler describes the sequence of events in his greatest book, the *Astronomia nova* (or *New Astronomy*); . . . Tycho had been impressed by the *Mysterium cosmographicum,* though he was unwilling to accept all its strange arguments; then, Kepler writes, "Tycho Brahe, himself an important part in my destiny, did not cease from then on to urge me to visit him. But since the distance of the two places would have deterred me, I ascribe it to Divine Providence that he came to Bohemia, where I arrived just before the beginning of the year 1600, with the hope of obtaining the correct eccentricities of the planetary orbits. . . . Now at that time Longomontanus had taken up the theory of Mars, which was placed in his hands so that he might study the Martian opposition with the Sun in 9° of Leo [that is, Mars near aphelion]. Had he been occupied with another planet, I would have started with that one. That is why I consider it again an act of Divine Providence that I arrived at the time when he was studying Mars; because for us to arrive at the secret knowledge of astronomy, it is absolutely necessary to use the motion of Mars; otherwise it would remain eternally hidden."

Kepler's *Astronomia nova* was not to be published until nine years later, in 1609. His greatest work, it broke the two-millennium spell of perfect circles and uniform angular motion—it was truly the "New Astronomy." Never had there been a book like it. Both Ptolemy in the *Almagest* and Copernicus in *On the Revolutions* had carefully dismantled the scaffolding by which they had erected their mathematical models. Kepler's book is nearly an order of magnitude more complete and complex than anything that had gone before, but he himself admits that he might have been too prolix.

Kepler had quickly perceived the quality of Brahe's treasure of observations, but he had realized that Tycho lacked an architect for the erection of a new astronomical structure. A devoted Copernican, he nevertheless recognized certain shortcomings in the helio*static* system described in *On the Revolutions,* and he was determined to derive a truly helio*centric* system in which the Sun played a vital physical role in keeping the planets in motion.

In the first three months in Prague, he established two fundamental points: first, the orbital plane of Mars had to be referred to the Sun itself, and not to the center of the Earth's orbit, as Copernicus had assumed; and second, the traditional eccentric circle of the Earth-Sun relation had to be modified . . . Hence, it was of paramount importance to Kepler's physics to prove that the Earth's motion resembled those of the other planets, and this he accomplished by an ingenious triangulation from the Earth's orbit to Mars.

. . . By the spring of 1601 he had been able, with the help of Tycho's accurate observations, to achieve a far better solution for Mars's longitudes than any of his predecessors. By a series of iterations he established a model accurate to 2′; . . .

From the predicted latitudes, however, Kepler realized that his scheme gave erroneous distances; again, unlike previous astronomers, who were satisfied with separate models for longitudes and latitudes, Kepler sought a unified, physically acceptable description. To obtain the correct distances his physical model demanded, he was obliged to reposition his circular orbit . . . but this move destroyed the excellent results he had previously found for the longitudes. The errors now rose to 8′ in the octants of the orbit, and in the *New Astronomy* Kepler goes on to say, "Divine providence granted us such a diligent observer in Tycho Brahe that his observations convicted this Ptolemaic calculation of an error of 8′; it is only right that we should accept God's gift with a grateful mind, because these 8′ have led to a total reform of astronomy."

While Kepler's admiration for Tycho's achievements always remained high, the imperious and high-handed Dane exceedingly frustrated the young German, almost driving him to a nervous breakdown. How well a collaboration between the two very different personalities would have worked in the long run is now impossible to assess, but what actually happened after Kepler had worked with Tycho for only ten months was that his mentor unexpectedly took ill and died. Suddenly Kepler inherited both the use of Tycho's observations and his position as Imperial Mathematician (although at only a third of the salary!). Kepler took time out to complete Tycho's nearly finished book—the *Progymnamata astronomiae instauratae,* "Exercises for the Reform of Astronomy"—and then he returned to his warfare on Mars.

Kepler now revived his earlier speculations on a planetary driving force, something like magnetism, emanating from the Sun. He envisioned a rotating Sun with rotating emanations that continuously pushed the planets in their orbits. His revised model . . . enabled him to formulate what we can call his distance law, that the velocity of a planet is inversely proportional to its distance from the Sun. Finding the angular motion from the distances immediately raised a computational problem of some proportion that could only be solved by tedious numerical summations. Here he had the fortunate inspiration to replace the sums of the lines between the Sun and the planet with the area swept out by the line between the Sun and the planet. This relation is formulated by chapter 40 of his *New Astronomy,* [but nowhere] in the *New Astronomy,* is this relation, the so-called law of areas, clearly stated. By 1621, however, Kepler finally understood its fundamental nature and clearly stated both the area law and a revised distance law. . . .

At this point Kepler had an accurate but physically inadmissable scheme for calculating longitudes and an intuitively satisfactory physical principle (the distance law) that worked well for the Earth's orbit, but which left an unacceptable 8′ error when applied to Mars. . . .

He next turned to a closer examination of the shape of Mars's path. Having established the proper position of the Earth's orbit by triangulation of Mars, he was able to turn the procedure around and investigate a few points in the orbit of Mars itself. Although the method did not yield a quantitative result, it clearly showed that Mars's orbit was noncircular. Kepler recognized that observational errors prevented him from getting precise distances to the orbit. Because of this scatter, he had to use, as he picturesquely described it, a method of "votes and ballots."

Armed with these results, Kepler found . . . a convenient means for generating a simple noncircular path. The resulting curve was similar to an ellipse, but was slightly egg-shaped with the fat end containing the Sun. In working with this ovoid curve, Kepler got himself into a very messy computational problem when he tried to apply his area rule to various segments. As an approximation he used an ellipse, but rather different from the one he was finally to adopt. Then, in the course of his calculations, he stumbled upon a pair of numbers that alerted him to the existence of another ellipse, a curve answering some of his requirements almost perfectly. "It was," he wrote, "as if I had awakened from a sleep."

Nevertheless, Kepler was also searching for a *physical* picture of planetary motion, something quasi-magnetic connected with the Sun that would explain not only the varying speed of Mars but also its varying distances. He fervently hoped that the oscillations of a hypothetical magnetic axis of Mars would satisfy his requirements. "I was almost driven to madness in considering and calculating this matter," he wrote in chapter 58 of his *New Astronomy*. . . . "I could not find why the planet . . . would rather go on an elliptical orbit as shown by the equations. O ridiculous me! As if the oscillation on the diameter could not be the way to the ellipse! So this notion brought me up short, that the ellipse exists because of the oscillation. With reasoning derived from physical principles agreeing with experience there is no figure left for the orbit except a perfect ellipse."

With justifiable pride he could call his book the *New Astronomy;* its subtitle emphasizes its repeated theme: "Based on Causes, or Celestial Physics, Brought out by a Commentary on the Motion of the Planet Mars." Although his magnetic forces have today fallen by the wayside, his requirement for a celestial physics based on causes has profoundly influenced contemporary science, which takes for granted that physical laws operate everywhere in the universe.

The work was completed by the end of 1605, but publication did not follow immediately, for Tycho's heirs demanded censorship rights over materials based on his observations, and they were displeased that Kepler had chosen a Copernican basis rather than Tycho's fixed-Earth, geoheliocentric arrangement, which of course made little sense in the framework of Kepler's physical ideas. Eventually a compromise was reached, primarily with respect to the dedicatory materials at the beginning of the volume, and the book was at last printed in 1609. Kepler added to the work a long introduction defending his physical principles, and he described how the Copernican system could be reconciled with the Bible. This latter part he had already written in 1596 for inclusion in his *Mysterium cosmographicum,* but the Tübingen University Senate, which had been asked to referee his first publication, had objected to such theological material. Now independent, Kepler had no such restrictions; of all Kepler's writings, this section of the introduction was the most frequently reprinted during the seventeenth century, and was the only part to be translated into English. . . .

It is difficult to gauge what impact Kepler's new astronomy would have had if he had stopped at this point. His cleansing and reformulation of the heliocentric system had been worked out in theory for only a single planet, Mars, and he had not provided any practical tables for calculating its motions. As Imperial Mathematician to Rudolph II, Kepler had been explicitly charged with the preparation of new planetary tables based on Tycho's observations, an arduous task that he still faced. "Don't sentence me completely to the treadmill of mathematical calculations," Kepler replied to one correspondent. "Leave me time for philosophical speculations, my sole delight."

Soon after completing his *Mysterium cosmographicum* Kepler had drafted an outline for a work on the harmony of the universe, but his plan had lain dormant while he grappled with the intricacies of Mars. Then, in the fall of 1616, after he had completed the first in a long series of ephemerides [astronomical tables] based on his work, he began to work intermittently on his *Harmonice mundi,* the *Harmonies of the World.* A major work of 225 pages, it was finally completed in the spring of 1618. . . .

Kepler developed his theory of harmony in four areas: geometry, music, astrology, and astronomy. It is the latter treatment, in Book V, which commands the primary attention today . . .

In the *Mysterium cosmographicum* the young Kepler had been satisfied with the rather approximate planetary spacings predicted by his nested polyhedrons and spheres; now, imbued with a new respect for data, he could no longer dismiss its five percent error. In the astronomical Book V of the *Harmonies,* he came to grips with this central problem: By what secondary principle did God adjust the original archetypical model based on the regular solids? Indeed, Kepler now found a supposed harmonic reason not only for the detailed planetary distances but also for their orbital eccentricities. . . .

In the course of this investigation, Kepler hit upon the relation now called his third or harmonic law: The ratio that exists between the periodic times of any two planets is precisely the ratio of the 3/2 power of the mean distances. . . . (This is equivalent to saying that the square of the periodic time is proportional to the cube of the mean distance.) Neither here nor in the few later references to it does Kepler bother to show how accurate the relation really is. Using his own data, we can calculate the table he failed to exhibit:

	Period (years)	*Mean Distance*	*Period Squared*	*Distance Cubed*
Mercury	0.242	0.388	0.0584	0.0580
Venus	0.616	0.724	0.3795	0.3795
Earth	1.000	1.000	1.000	1.000
Mars	1.881	1.524	3.540	3.538
Jupiter	11.86	5.200	140.61	140.73
Saturn	29.33	9.510	860.08	867.69

The harmonic law pleased him greatly for it neatly linked the planetary distances with their velocities or periods, thus fortifying the a priori premises of the *Mysterium* and the *Harmonies*. . . .

At the same time that Kepler was preparing his planetary ephemerides and his *Harmonies of the World,* he also embarked upon his longest and perhaps most influential book, an introductory textbook for Copernican astronomy in general and Keplerian astronomy in particular. Cast in the catechetical form of questions and answers typical of sixteenth-century textbooks, the *Epitome* treated all of heliocentric astronomy in a systematic way, including the three relations now called Kepler's laws. Its seven books were issued in installments; the first three appeared in 1617 and the final three in 1621. . . .

Kepler's *Epitome* can be considered the theoretical handbook to his *Rudolphine Tables,* finally published in 1627; this monumental work furnished working tables based on his reforms of the Copernican system. Kepler's planetary positions were generally about thirty times better than any of his predecessors', and in 1631 (the year following Kepler's death) they provided the grounds for a dramatically successful observation, for the first time, of a transit of the planet Mercury across the face of the Sun. In 1632 a further impetus to the Copernican system came through the brilliant polemic of Galileo's *Dialogue Concerning Two Great World Systems*.

Precisely how influential Kepler's *Epitome* was is difficult to assess. His reputation was considerably enhanced by the publication of his *Rudolphine Tables,* and the Copernican idea was rapidly gaining in acceptability. Thus, by 1635, there was sufficient demand for the *Epitome* to warrant reprinting it, and for many years it remained one of the few accessible sources for the details of the revised Copernican system.

Galileo and the Conflict over Holy Scripture

Questions about the compatibility of the Bible with the Copernican planetary arrangement broke into open conflict in the years after Galileo's telescopic observations were made public in 1610. In 1613 Galileo wrote a letter to his disciple Benedetto Castelli that he published in expanded form as his Letter to the Grand Duchess Christina *(1636). Paolo Foscarini (1580–1616), a Church progressive and sympathizer of Galileo, Copernicus, and Kepler, published a short treatise in 1615 with the object of showing that Copernicus's theory was "not improbable" and that biblical passages pertaining to the physical world could be accommodated to popular discourse. In the first selection below, Cardinal Robert Bellarmine (1542–1621) reminds Foscarini that the meaning of Scripture depends upon agreement of the Church fathers; that the fathers do agree on the matter of the earth's motion; and that, as a result, mathematicians and astronomers ought to speak "hypothetically and not absolutely." Galileo evidently knew Bellarmine's letter because he replies to its charges in a letter to his friend Piero Dini and then later in a series of private notes that clearly show his position on the relationship between the Bible and natural knowledge.*

Cardinal Bellarmine's Letter to Paolo Foscarini, 12 April 1615

"I have gladly read the letter in Italian and the essay in Latin that Your Reverence has sent me, and I thank you for both, confessing that they are filled with ingenuity and learning. But since you ask for my opinion, I shall give it to you briefly, as you have little time for reading and I for writing.

"First. I say that it appears to me that

Your Reverence and Sig. Galileo did prudently to content yourselves with speaking hypothetically and not positively, as I have always believed Copernicus did. For to say that assuming the earth moves and the sun stands still saves all the appearances better than eccentrics and epicycles is to speak well. This has no danger in it, and it suffices for mathematicians. But to wish to affirm that the sun is really fixed in the center of the heavens and merely turns upon itself without traveling from east to west, and that the earth is situated in the third sphere and revolves very swiftly around the sun, is a very dangerous thing, not only by irritating all the theologians and scholastic philosophers, but also by injuring our holy faith and making the sacred Scripture false. For Your Reverence has indeed demonstrated many ways of expounding the Bible, but you have not applied them specifically, and doubtless you would have had a great deal of difficulty if you had tried to explain all the passages that you yourself have cited.

"Second. I say that, as you know, the Council [of Trent] would prohibit expounding the Bible contrary to the common agreement of the holy Fathers. And if Your Reverence would read not only all their works but the commentaries of modern writers on Genesis, Psalms, Ecclesiastes, and Joshua, you would find that all agree in expounding literally that the sun is in the heavens and travels swiftly around the earth, while the earth is far from the heavens and remains motionless in the center of the world. Now consider whether, in all prudence, the Church could support the giving to Scripture of a sense contrary to the holy Fathers and all the Greek and Latin expositors. Nor may it be replied that this is not a matter of faith, since if it is not so with regard to the subject matter, it is with regard to those who have spoken. Thus that man would be just as much a heretic who denied that Abraham had two sons and Jacob twelve, as one who denied the virgin birth of Christ, for both are declared by the Holy Ghost through the mouths of the prophets and apostles.

"Third. I say that if there were a true demonstration that the sun was in the center of the universe and the earth in the third sphere, and that the sun did not go around the earth but the earth went around the sun, then it would be necessary to use careful consideration in explaining the Scriptures that seemed contrary, and we should rather have to say that we do not understand them than to say that something is false which had been proven. But I do not think there is any such demonstration, since none has been shown to me. To demonstrate that the appearances are saved by assuming the sun at the center and the earth in the heavens is not the same thing as to demonstrate that in fact the sun is in the center and the earth in the heavens. I believe that the first demonstration may exist, but I have very grave doubts about the second; and in case of doubt one may not abandon the Holy Scriptures as expounded by the holy Fathers. I add that the words *The sun also riseth, and the sun goeth down, and hasteth to the place where he ariseth* were written by Solomon, who not only spoke by divine inspiration, but was a man wise above all others, and learned in the human sciences and in the knowledge of all created things, which wisdom he had from God; so it is not very likely that he would affirm something that was contrary to demonstrated truth, or truth that might be demonstrated. And if you tell me that Solomon spoke according to the appearances, and that it seems to us that the sun goes round when the earth turns, as it seems to one aboard ship that the beach moves away, I shall answer thus. Anyone who departs from the beach, though to him it appears that the beach moves away, yet knows that this is an error and corrects it, seeing clearly that the ship moves and not the beach; but as to the sun and earth, no sage has needed to correct the error, since he clearly experiences that the earth stands still and that his eye is not deceived when it judges the sun to move, just as he is likewise not deceived when it judges that the moon and the stars move. And that is enough for the present." . . .

Galileo's Letter to Piero Dini in Rome, May 1615.

"Eight days ago I wrote to Your Reverence in reply to yours of the second of May. My answer was very brief, because I then found myself (as now) among doctors and medicines, and much disturbed in body and mind over many things, particularly by seeing no end to these rumors set in motion against me through no fault of mine, and seemingly accepted by those higher up as if I were the originator of these things. Yet for all of me any discussion of the sacred Scripture might have lain dormant forever; no astronomer or scientist who remained within proper bounds has ever got into such things. Yet while I follow the teachings of a book accepted by the church, there come out against me philosophers quite ignorant of such teachings who tell me that they contain propositions contrary to the faith. So far as possible, I should like to show them that they are mistaken, but my mouth is stopped and I am ordered not to go into the Scriptures. This amounts to saying that Copernicus's book, accepted by the church, contains heresies and may be preached against by anyone who pleases, while it is forbidden for anyone to get into the controversy and show that it is not contrary to Scripture.

"To me, the surest and swiftest way to prove that the position of Copernicus is not contrary to Scripture would be to give a host of proofs that it is true and that the contrary cannot be maintained at all; thus, since no two truths can contradict one another, this and the Bible must be perfectly harmonious. But how can I do this, and not be merely wasting my time, when those Peripatetics who must be convinced show themselves incapable of following even the simplest and easiest of arguments, while on the other hand they are seen to set great store in worthless propositions?

"Yet I should not despair of overcoming even this difficulty if I were in a place where I could use my tongue instead of my pen; and if I ever get well again so that I can come to Rome, I shall do so, in the hope of at least showing my affection for the holy Church. My urgent desire on this point is that no decision be made which is not entirely good. Such it would be to declare, under the prodding of an army of malign men who understand nothing of the subject, that Copernicus did not hold the motion of the earth to be a fact of nature, but as an astronomer merely took it to be a convenient hypothesis for explaining the appearances. Thus to admit it to use but prohibit it from being considered true would be to declare that Copernicus's book had not even been read. . . . I should not like to have great men think that I endorse the position of Copernicus only as an astronomical hypothesis which is not really true. Taking me as one of those most addicted to his doctrine, they would believe all its other followers must agree, and that it is more likely erroneous than physically true. That, if I am not mistaken, would be an error."

[When Galileo wrote to Dini there can be little doubt that he had seen Bellarmine's letter to Foscarini. Some rather significant notes jotted down by Galileo have survived. They appear to have been written as comments on Bellarmine's letter to Foscarini and give us deep insight into Galileo's strategy in answering Bellarmine's objections:]

"One reads on the verso of the title page of Copernicus's book a certain preface to the reader which is not by the author, as it speaks of him in the third person and is unsigned.* There it is blandly stated that Copernicus did not believe his system to be true at all, but only claimed to advance it for the calculation of heavenly motions, and finished his reasoning by concluding that it would be foolish to take his theory as real and true. This conclusion is so positively stated that anyone who did not read further, and thought this to have been put there with the author's consent, might well be excused for his mistake. But what value can we place on the opinion of a person who would judge a book by reading no more than a brief preface of the printer and bookseller? I leave this to everyone to judge for himself; and I say that this preface can be nothing but a word from the bookseller to assist the vending of the work, which would have been considered a monstrous

*Kepler had revealed at the beginning of his *Astronomia Nova* that in his own copy of Copernicus's book there was a note, written by Jerome Schreiber of Nuremberg, stating that this preface had been inserted by Andreas Osiander (a Protestant theologian who had supervised the printing) for reasons similar to those set forth here by Galileo.

chimera by people in general if it had not been qualified in some such way—and generally the buyer reads no more than such a preface before purchasing a book. And that this preface was not only not written by the author, but that it was placed there without his knowledge, to say nothing of his consent, is made manifest by the misuse of certain terms in it which the author would never have permitted."

[*As Stillman Drake observes:*] Elsewhere in these notes there is a point-by-point reply to Bellarmine's written opinion. It has the appearance of something intended to be sent to Foscarini for use in the revision and amplification of his book, though its precise date or purpose is not known. In substance it reads as follows:

"1. Copernicus assumes eccentrics and epicycles; not these, but other absurdities, were his reason for rejecting the Ptolemaic system.

"2. As to philosophers, if they are true philosophers (that is, lovers of truth), they should not be irritated; but, finding out that they have been mistaken, they must thank whoever shows them the truth. And if their opinion is able to stand up, they will have cause to be proud and not angry. Nor should theologians be irritated; for finding such an opinion false, they might freely prohibit it, or discovering it to be true they should be glad that others have opened the road to the discovery of the true sense of the Bible, and have kept them from rushing into a grave predicament by condemning a true proposition.

"As to rendering the Bible false, that is not and never will be the intention of Catholic astronomers such as I am; rather, our opinion is that the Scriptures accord perfectly with demonstrated physical truth. But let those theologians who are not astronomers guard against rendering the Scriptures false by trying to interpret against it propositions which may be true and might be proved so.

"3. It may be that we will have difficulties in expounding the Scriptures, and so on; but this is through our ignorance, and not because there really are, or can be, insuperable difficulties in bringing them into accordance with demonstrated truth.

"4. . . . It is much more a matter of faith to believe that Abraham had sons than that the earth moves. . . . For since there have always been men who have had two sons, or four, or six, or none . . . there would be no reason for the Bible to affirm in such matters anything contrary to truth. . . . But this is not so with the mobility of the earth, that being a proposition far beyond the comprehension of the common people. . . .

"5. As to placing the sun in the sky and the earth outside it, as the Scriptures seem to affirm, etc., this truly seems to me to be simply . . . speaking according to common sense; for really everything surrounded by the sky is in the sky. . . .

"6. Not to believe that a proof of the earth's motion exists until one has been shown is very prudent, nor do we demand that anyone believe such a thing without proof. Indeed, we seek, for the good of the holy Church, that everything the followers of this doctrine can set forth be examined with the greatest rigor, and that nothing be admitted unless it far outweighs the rival arguments. If these men are only ninety per cent right, then they are defeated; but when nearly everything the philosophers and astronomers say on the other side is proved to be quite false, and all of it inconsequential, then this side should not be deprecated or called paradoxical simply because it cannot be completely proved. . . .

"7. It is true that to prove that the appearances may be saved with the motion of the earth . . . is not the same as to prove this theory true in nature; but it is equally true, or even more so, that the commonly accepted system cannot give reasons for those appearances. That system is undoubtedly false, just as . . . this one may be true. And no greater truth may or should be sought in a theory than that it corresponds with all the particular appearances.

"8. No one asks that in case of doubt the teachings of the Fathers be abandoned, but only that the attempt be made to gain certainty in the matter questioned. . . .

"9. We believe that Solomon and Moses and all the other holy writers knew the constitution of the universe perfectly well, as they also knew that God did not have hands or feet or wrath or prevarication or regret. We cast no doubt on this, but we say that . . . the Holy Ghost spoke thus for the reasons set forth.

"10. The mistake about the apparent motion of the beach and stability of the ship is known to us after we have frequently stood on the beach and observed the motion of the boat, as well as in the boat to observe the beach. And if we could stand thus now on the earth and again on the sun or some other star, we might gain positive and sensory knowledge as to which moved. Yet looking only from these two bodies, it would always appear that the one we were on stood still, just as to a man who saw only the boat and the water, the water would always seem to run and the boat to stand still. . . . It would be better to compare two ships, of which the one we are on will absolutely seem to stand still whenever we can make no other comparison than between the two ships. . . .

"Besides, neither Copernicus nor his followers make use of this appearance of the beach and the ship to prove that the earth moves and the sun stands still. They use it only as an example that serves to show . . . the lack of contradiction between the simple sense-appearance of a stable earth and a moving sun if the reverse were really true. For if nothing better than this were Copernicus's proof, I believe no one would endorse him."

Descartes's Universe and the New Astronomy

In 1644 Descartes set forth publicly the details of his universe, a world in which all natural phenomena—light, planetary motion, magnetism, and so on—were explained by mechanical interactions of material particles. The heavens were composed of three sorts of matter: (1) subtle, fast-moving particles that emit light and of which the sun and fixed stars are constituted; (2) small, spherical particles that transmit light and of which the heavens are composed; and (3) bulkier, gross matter, of which the earth, planets and comets are constituted and that reflect light. In the selection provided, Descartes tries to reconcile his physics with a planetary arrangement that preserves the best features of the theories of Copernicus and Tycho Brahe. It is interesting that he wants his theory of the earth to be taken "only as an hypothesis {of supposition which may be false}."

That various hypotheses may be used to explain the phenomena of the Planets.

Just as a man at sea in calm weather and looking at several other fairly distant vessels, which seem to him to be changing position, is frequently unable to say whether the change is caused by the movement of the vessel on which he is or by that of the other vessels; when, from our situation, we observe the course of the Planets and their various positions, even careful observation does not always bring sufficient understanding to enable us to determine, {from what we see}, to which bodies we ought properly to attribute {the cause of} these changes. And since these changes are very unequal and complicated, it is not easy to explain them, unless we choose one of the various ways in which they can be understood, in accordance with which we then suppose these changes to occur. To this end, Astronomers have devised three different hypotheses or suppositions; which they have merely attempted to make capable of explaining all the phenomena, without considering whether they conformed to the truth.

That Ptolemy's hypothesis is not in conformity with appearances.

Ptolemy devised the first of these; but, as it is already commonly rejected by all Philosophers, because it is contrary to several {recent} observations (especially to the changes in light, similar to those which occur on the Moon, which we observe on Venus), I shall not speak further of it here.

That those of Copernicus and Tycho do not differ if considered only as hypotheses.

The second is that of Copernicus and the third that of Tycho Brahe; considered purely as hypotheses, these two explain the phenomena equally well, and there is not much difference between them. Nevertheless, that of Copernicus is somewhat simpler and clearer; so that Tycho's only reason for altering it was that he was attempting not merely a hypothetical explanation but an account of how [he thought] this matter really was.

That Tycho in words attributes less motion to the Earth than does Copernicus, but that in fact he attributes more.

Seeing that Copernicus had not hesitated to attribute motion to the Earth; Tycho, to whom this opinion seemed not only absurd in Physics but contrary to the common sense of men, tried to correct it. However, because he did not give sufficient consideration to the true nature of motion, he asserted only verbally that the Earth was at rest and in fact granted it more motion than had his predecessor.

That I deny the motion of the Earth more carefully than Copernicus and more truthfully than Tycho.

That is why, although I do not differ at all from these two except on this one point, I shall deny the movement of the Earth more carefully than Copernicus and more truly than Tycho. I shall set forth here the hypothesis which seems to me the simplest and most useful of all; both for understanding the phenomena and for enquiring into their natural causes. And yet I give warning that I do not intend it to be accepted as entirely in conformity with the truth, but only as an hypothesis {or supposition which may be false}. . . .

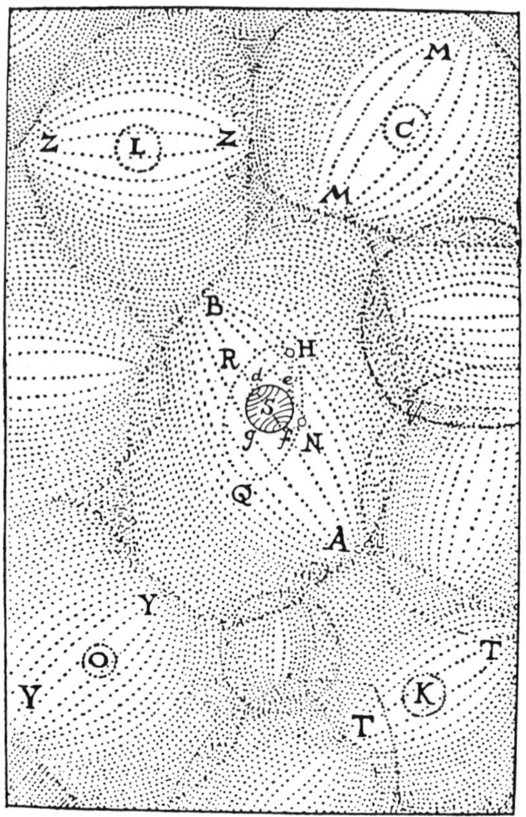

Figure 5-3 *Descartes's universe, with its whirlpools of heavenly matter in which the planets are suspended.*

That the Earth is at rest in its heaven which nevertheless carries it along.

Fourth, since we see that the Earth is not supported by columns or suspended in the air by means of cables but is surrounded on all sides by a very fluid heaven, let us assume that it is at rest and has no innate tendency to motion, since we see no such propensity in it. However, we must not at the same time assume that this prevents it from being carried along by {the current of} that heaven or from following the motion of the heaven without however moving itself: in the same way as a vessel, which is neither driven by the wind or by oars, nor restrained by anchors, remains at rest in the middle of the ocean; although it may perhaps be imperceptibly carried along by {the ebb and flow of} this great mass of water.

That the same is to be believed of all the Planets.

And just as the other Planets resemble the Earth in being opaque and reflecting the rays of the Sun, we have reason to believe that they also resemble it in remaining at rest, each in its own part of the heaven, and that the variation we observe in their position results solely from the motion of the matter of the heaven which contains them. . . .

That all the Planets are carried around the Sun by the heaven.

Now that we have, by this reasoning, removed any possible doubts about the motion of the Earth, let us assume that the matter of the heaven, in which the Planets are situated, unceasingly revolves, like a vortex having the Sun as its center, and that those of its parts which are close to the Sun move more quickly than those further away; and that all the Planets (among which we {shall from now on} include the Earth) always remain suspended among the same parts of this heavenly matter. For by that alone, and without any other devices, all their phenomena are very easily understood. Thus, if some straws {or other light bodies} are floating in the eddy of a river, where the water doubles back on itself and forms a vortex as it swirls; we can see that it carries them along and makes them move in circles with it. Further, we can often see that some of these straws rotate about their own centers, and that those which are closer to the center of the vortex which contains them complete their circle more rapidly than those which are further away from it. Finally, we see that, although these whirlpools always attempt a circular motion, they practically never describe perfect circles, but sometimes become too great in width or in length, {so that all the parts of the circumference which they describe are not equidistant from the center}. Thus we can easily imagine that all the same things happen to the Planets; and this is all we need to explain all their remaining phenomena.

CHAPTER 6

Mechanical and Mathematical Visions of the Universe

Introduction by Richard Olson

In 1949 the British historian Herbert Butterfield wrote a marvelous little book entitled *The Origins of Modern Science 1300–1800,* in which he popularized the term "Scientific Revolution" to characterize the rapid development of European science between about 1500 and 1800. "It changed the character of men's habitual operations," he wrote, ". . . and looms so large as the real origin, both of the modern world and the modern mentality," that "it outshines everything since the rise of Christianity and reduces the Renaissance and Reformation to the rank of mere episodes . . . within the system of medieval Christendom."

What makes the Scientific Revolution so central to all subsequent Western cultural history? It is not simply that under the impetus of magical and artistic orientations Western science began to be viewed as having practical, technological implications. Nor is it merely that printing and the growth of support for science within European universities, courts, academies, and societies encouraged the search for and the rapid dissemination and expansion of scientific knowledge within a growing community of scholars devoted to scientific pursuits. Nor is it even the fact that rapid, dramatic advances were made in fields such as astronomy, mathematics, mechanics, and anatomy.

What is critical is that a major transformation occurred in the dominant intellectual framework within which new scientific discoveries were incorporated and interpreted; and that the new framework, which emerged in connection with mathematics and natural science, rapidly invaded virtually all domains of intellectual life—including theology and political theory.

This chapter first characterizes and illustrates the new framework of thought or "world view," distinguishing it from that of the Renaissance. It then examines the extension of the new perspectives beyond the domains of natural science.

In trying to explain what happened to the fundamental framework of Western intellectual life during the Scientific Revolution, Butterfield used a very illuminating metaphor. He argued that scientific progress involved a transformation of perspective that was very much like "picking up the opposite end of the stick." Others have claimed that it involved the kind of Gestalt switch, so famous in psychology, which allows one to see either two faces in profile or the outline of a vase delineated by the two faces.

What makes Butterfield's metaphor so apt is that when scientists changed their perspective in the seventeenth century they did not merely begin to see a *different* world, they began to see one that was *diametrically opposed* to the old in a number of crucial ways. The world of Plato, of the Scholastic scholars, and of the Renaissance Magi was filled with a hierarchy of spirits who were responsible for all types of natural phenomena. For the follower of the "new" philosophy of the seventeenth century, by contrast, the world was an inanimate machine, devoid of intelligence or intelligences, and capable of being moved but never of moving itself. Johannes Kepler showed this transformation in a very dramatic way, for he began his scholarly life with a typical Renaissance belief in a living cosmos filled with spirits but ended it as an advocate of the new "mechanical" philosophy. Writing to a friend in 1605, just after his conversion, he abandoned the discussion of the "souls which move the planets" that had filled his *Mysterium Cosmographium* of 1596 and wrote instead:

> I am now much engaged in investigating *physical* causes. My goal is to show that the celestial machine is *not* the likeness of the divine being, but is the likeness of [a] clock. . . . In this machine nearly all the variety of movements flow from one very simple magnetic force just as in a clock all the motions flow from a simple weight.

Kepler's comparison of the cosmos to a human artifact—a clock—illustrates one of the central reversals of perspective that culminated in the seventeenth century. For Aristotle all art was an imitation of nature, and for the premodern scientist human artifacts were imperfect copies of God's natural productions. Whenever men wanted to learn how to construct something, they had first to find some natural or living model. Thus, for example, Hugh of St. Victor (c. 1130) described the artificial products of man in the following way:

> The human work, because it is not nature but only imitates nature, is fitly called mechanical or adulterate. . . . The founder who casts a statue has gazed upon man as his model. The builder who has constructed a house has taken into consideration a mountain, for, as the Prophet declares, "Thou sendest forth springs in the vales, between the midst of the hills the waters shall pass; as the ridges of mountains retain the water, even so does the house require to be framed into a high peak that it may safely discharge the weight of pouring rains. He who invented the use of clothes considered how each of the growing things has its proper covering by which to protect itself from offense. Bark encircles the tree, feathers cover the bird, scales encase the fish, fleece clothes the sheep, hair garbs cattle and . . . a shell protects the tortoise.

By the middle of the seventeenth century an almost total inversion of the relationship between nature and artifice had occurred among European scientific intellectuals. Before the seventeenth century someone would naturally have written of a clock being made the "likeness" of the heavens, but he would hardly have thought of the reverse: that God created the cosmos as the likeness of a clock. By 1664, however, Henry Power expressed a widespread view when he claimed that any acceptable understanding of natural phenomena must be expressed in terms of human artifacts. "These are the days," he wrote, "that must lay a new foundation of a more magnificent philosophy, never to be overthrown . . . deducing the causes of things from such originals in nature as we observe are producible by art, and by infallible demonstrations of mechanics."

Two of the selections below attempt to explore the central characteristics of the new mechanical philosophy and to explain some of the reasons for its rise to dominance in seventeenth-century Europe. Richard Olson emphasizes the importance of the revival of ancient atomist philosophies and discusses a series of religious issues which made the mechanical philosophy—with its insistence on the inertness of mere matter and the need for an intelligent artisan external to any artifact—particularly appealing to those scientists who had begun to be concerned that if the world were animate it would have no need for a God to create, govern, and energize it. The mechanical philosophy, which had initially been developed at least partly as a support for traditional Christian doctrine, eventually came to be seen by many as dangerous to Christianity and as the foundation for atheistic tendencies; but that changed perspective did not begin to develop until the eighteenth century.

Paolo Rossi argues that the mechanical philosophy emerged in close connection with artisan activity and that it was a conscious response to disputes about the foundations of human knowledge. These arguments were fueled by the Protestant Reformation, the recovery of the texts of ancient Greek philosophers known as the skeptics, and by European contacts with other cultures, including those of the New World. In the face of competing religious beliefs, philosophical systems, and patterns of morality, it seemed critical to discover some way to distinguish truth from error. But the more Renaissance thinkers had grappled with how to do this, the more confused they became; consequently, many became convinced that humans can truly know only their own creations. This new perspective on the boundaries of human knowledge reversed the assumptions of the Renaissance alchemists, astrologers, and practitioners of mathematical magic, for whom the ability to know the hidden powers in nature was the necessary precondition for human productive activity. Now it came to be suspected that if humans were to have any knowledge of natural things at all, then it could only be to the extent that natural objects were like human mechanical productions.

The same crisis of knowledge which stimulated the "mechanical" philosophy also caused seventeenth-century thinkers to pick up the other end of the stick in their attitudes toward mathematics. For the Renaissance believer in Neoplatonic or Pythagorean number mysticism, like the early Kepler, mathematics was critically important because it allowed humans to disclose a set of hidden relationships which God had incorporated into the universe. And for those who applied mathematics to practical matters in navigation, architecture, commerce, and engineering, or to theoretical questions in astronomy, it came to be important primarily because of its obvious utility in solving a wide range of technical problems. But for a critical group of seventeenth-century scientific intellectuals, including René Descartes and Thomas Hobbes, mathematics became important primarily because it was a domain in which the kinds of doubt and uncertainty that interfered with attempts to draw conclusions in theology, politics, morality, and even natural science, did not seem to exist.

Both Hobbes (1588–1679) and Descartes (1596–1650), struck by the logical rigor of mathematical arguments, sought to extend the methods of mathematical argumentation into other domains to replace the traditional appeals to authority and metaphorical rhetoric which filled Renaissance treatises. Hobbes was particularly insistent that it was because of their "scrupulous ratiocinations" that mathematicians were able to produce a system of knowledge that was both certain and useful; and he attacked all those who used traditional kinds of arguments on the grounds that "the use of metaphors, tropes, and other figures [of speech] in the search for truth can lead only to absurdity."

During the seventeenth century Euclid's *Elements* was used as a model for discourse in almost every sphere of intellectual life. Even in such fields as music (Marin Mersennes's *Harmonie Universelle,* 1636), ethics (Benedict de Spinoza's *Ethics,* 1674), and politics (Thomas Hobbes's *Leviathan,* 1651), arguments were presented in the form of Definitions, Postulates, Axioms, and Theorems with purportedly formal deductive demonstrations. By the end of the century Bernard Fontenelle, the great popularizer of Descartes's views, could write:

> A work on politics, on morals, a piece of criticism, even a manual on the art of public speaking would, other things being equal, be all the better for having been written by a geometrician. The order, the precision and the accuracy which have distinguished the

> worthier kind of books for some time now, may well have been due to the geometrical method which has been continuously gaining ground, and which somehow or other has an effect on people who are quite innocent of geometry.

It thus is no accident that the Scientific Revolution was also a time when the flowery rhetoric of the Renaissance was abandoned in favor of a clear, plain style and when prose supplanted poetry as the dominant literary form.

Logical rigor, though important, was not enough to account for the characteristic certainty of mathematics. Both Descartes and Hobbes insisted that one also needed to consider the origin and nature of the basic definitions and postulates that formed the starting point for mathematical systems; and they both insisted that if, in the same way, one could discover first principles for other domains of knowledge, they should have the same kind of certainty. On precisely how one got the first principles, the two men diverged, but the ideas of each were to have critical long-term impact on Western culture.

Hobbes saw the issue simply. We can have complete and certain knowledge in mathematics for the same reason that we can have complete and certain knowledge of any mechanical artifact—that is, because mathematical entities are created and defined by men. ''Geometry is demonstrable,'' he argued, ''for the lines and figures from which we reason *are drawn and described by ourselves.*'' For Descartes the problem of discovering unimpeachable first principles was more complex but, he held, equally illuminating. In his *Rules for the Direction of the Mind,* excerpted below, Descartes explained the method which mathematicians use to establish their principles and suggested how that method can be generalized. Rule V, which insists that all complex propositions must be broken down into their simplest components, became particularly important because through it the mathematical and mechanical philosophical tendencies became fused into the idea that every complex phenomenon or object is capable of being *analyzed* into a set of simple, self-explanatory ones. The whole method of mathematical analysis seemed to be embodied in machines; for in them, complex combinations of individually simple gears, springs, and weights could give rise to an incredible variety of subtle and sophisticated behaviors. The short passage from Étienne de Condillac's *Logic* demonstrates how the analytic process from mathematics became fused with the mechanical philosophy.

During the seventeenth century some combination of mathematical methods and the mechanical philosophy came to shape virtually every new subject. In his *On the Motion of the Heart and the Blood,* for example, William Harvey (1578–1657) showed that the heart acts like a pump and calculated the volume of blood which it moves each hour, thus extending the methods to the study of living things. Descartes used mechanistic theories to account for how experience is ''received'' and transferred from the sense organs to the brain, thus extending them to the study of psychology. And William Petty (1623–1687) was able to demonstrate mathematically that most of the annual income of the English citizenry came from the products of human labor, rather than from the products of the land, thus initiating a new science which he called ''political arithmetic'' and which we now call economics. Short passages from both Descartes's mechanistic treatment of sensation and Petty's quantitative treatment of the relationship between land and labor are included to illustrate this trend.

Of all the seventeenth-century extensions of the mechanical-mathematical perspective, two stand out as uniquely important to Western culture. First, its application to psychology and political theory by Thomas Hobbes has shaped, for better or for worse, almost all subsequent liberal interpretations of human nature, morality, and politics.

In the selection from Hobbes's *Leviathan,* we see first how he sets up the deductive procedures of mathematics as the model for all scientific thought, including thought about ethics and politics. Then we see how he deduces (from the natural competition among men for goods, safety, and reputation) the need for men to surrender all rights—except that of self-defense—to an absolute sovereign power. Sheldon Wolin, in his selection from *Hobbes and the Epic Tradition in Politics,* points out one of the grave dangers associated with the Hobbesian reduction of politics to a set of rules divorced from "the moralizing and civilizing influence of the philosophical way of life." Politics then becomes little more than the exercise of power with almost no concern for the ends to which it *should* be directed.

Perhaps even more important than Hobbesian "social" science were Isaac Newton's spectacular successes in optics, mechanics, and astronomy. By fusing the mathematical-mechanical philosophy with a renewed emphasis on the role of experiment, Isaac Newton was able to develop dramatic new discoveries. He demonstrated in the *Mathematical Principles of Natural Philosophy* (1687) that all celestial *and* terrestrial objects were subject to a single universal law of gravitation. Since Newton's universal gravitation could be used to explain virtually all astronomical phenomena and many terrestrial ones, its success led to a widespread expectation that the mathematical-mechanical-experimental philosophy might be capable of almost indefinite extension. In his "The Significance of the Newtonian Synthesis," the eminent French historian of science Alexandre Koyre explains the great appeal of Newton's methods and suggests that their successful application and imitation has given rise to a split, unintended by Newton, between an "objective" world of science and a "subjective" world of life experience. Newton's success so reinforced the changes in world view which had already taken place, and so shaped the thought of the centuries to come, that Butterfield's grand assessment of the importance of the Scientific Revolution in the history of Western thought seems entirely justified.

Richard Olson

The Mechanical Philosophy and Anglican Theology

Natural theology—the attempt to discover and demonstrate God's existence and attributes through study of the "book of nature"—has been a continuing part of the Christian tradition since the time of Origen (c. A.D. 200), but never was it more important than in England during the seventeenth century. Faced with a bewildering variety of internal and external sectarian critics, liberal Anglicans hoped to unify the Church by appealing to natural theology on the grounds that what we learn of God from nature is both accessible and acceptable to all. In the following selection Richard Olson explains why such scientists as Robert Boyle and Walter Charleton were especially active in trying to demonstrate that the mechanical philosophy, more than any other, warranted belief in an all-wise and powerful God who not only created the world but continues active in its governance.

During the first three decades of the Seventeenth Century a family of new philosophical or scientific theories, often lumped together under the terms "atomical," "mechanical" or "corpuscular" philosophy, became available to natural philosophers, gradually to dominate natural philosophy by the middle of the century. In part, the new theories drew their explanatory power and appeal from the use of increasingly familiar and sophisticated mechanical devices—including the ubiquitous clock—as analogs to natural entities. Anatomists, for example, began comparing the values of the veins to locks in canals and the heart to a water pump, and astronomers constantly likened the motions of the heavens to those of a mechanical clock.

The second critical source of the mechanical and corpuscular philosophies is more important for our present purposes. It lay in the recovery and republication of the writings of the ancient atomists, Epicurus and Lucretius, whose works had been explicitly antireligious and purely materialist—denying the relevance of immaterial spiritual entities to the events of the natural and human world. There is no question that some of those who adopted the atomic philosophy were sympathetic to its original purely materialist emphasis and at least insensitive to its overtly irreligious thrust. In England, for instance, Thomas Harriot, Lady Margaret Cavendish, and, most importantly, Thomas Hobbes, all seemed to adopt pure materialism; and in doing so, they drew intense criticism upon the new mechanical philosophies from Christian apologists. But for other Christian natural philosophers the mechanical philosophy seemed to offer unique support to the Christian Religion, and in a slightly modified form, spread initially through the works of Pierre Gassendi, it provided the grounds for a powerful movement within Anglican natural theology at the hands of Walter Charleton and Robert Boyle. . . .

In his early work, *On the Usefulness of Natural Philosophy* (1662) Boyle cited Plutarch who advised us to ". . . not venerate the elements, the heaven, the sun, the moon, etc., these are but mirrors, wherein we may behold his excellent art who framed and adorned the world." He rejected the ideas of those who "reject from the production and preservation of things, all but nature," and then he explained how the mechanical philosophy encourages us in a more proper appreciation of nature and God:

> When . . . I see in a curious clock, how orderly every wheel and other part performs its own motions, and with what seeming unanimity they conspire to show the hour, and accomplish the other

> designs of the artificer; I *do not* imagine that any of the wheels, etc., or the engine itself is endowed with reason, but commend that of the workman, who framed it so artificially. So when I contemplate the action of those several creatures, that make up the world, I *do not* conclude the inanimate species, at least, that it is made up of, or the vast engine itself, to act with reason or design, but admire and praise the most wise author, who by his admirable contrivance, can so regularly produce effects, to which so great a number of successive and conspiring causes are required. . . .

[Also,] the mechanical philosophy as it was being developed by Gassendi, seemed to Boyle to offer a great advantage over the organicist philosophy of the Renaissance naturalists in connection with its modest claims regarding the source and extent of human knowledge. The Anglican theologian, Richard Hooker, had sought to undercut the intensity of sectarian bickering by showing that most religious claims rested upon merely probable foundations rather than certain ones. Most Puritans and radical sectarians, on the other hand, claimed the certainty of "divine illumination" through some kind of inner light; and they often supported their claims by reference to such writings as the Hermetic *Pimander.* Boyle specifically wrote, "I dare not affirm, with some of the Helmontians and Paracelsans, that God discloses to men the great mystery of chemistry by good angels or by nocturnal visions." The secrets of natural philosophy must be approached, as the new mechanical philosophy insisted, empirically and hypothetically. One could not expect "a sudden and total revelation of nature's secrets." No more could one expect a total revelation of spiritual secrets. Human reason and industry provide the only proper path to such limited knowledge as may be had in both the natural and the spiritual realms. Indeed Boyle appealed directly to the Hookerian tradition to insist upon the appropriateness of less-than-certain knowledge as a foundation for religious commitment:

> The choosing or refusing to embrace the Christian religion, which is not proposed to us only as a System of Speculative Doctrines, but also as a Body of Laws, according to which it teaches us that God commands us to worship him and regulates our lives; the embracing I say, or not embracing of this religion is an act of human *choice.*

Since certain knowledge *forces* assent and leaves no place for choice, our grounds for choosing to embrace Christian doctrine must be merely probable. Only then is choice possible.

This emphasis on the necessity of limits to human insight into God's works and will was also a central consideration in the conversion of Walter Charleton to the mechanical philosophy sometime between 1650 and 1652. Charleton was convinced that the intellectual pride associated with the notion that some private inner light could provide individuals with certainty led directly to the overthrow of ecclesiastical and civil authority and to the spread of "the most execrable *Heresies,* blasphemous *Enthusiams,* nay even professed Atheism." The mechanical philosophy, on the other hand, involved an insistence that our knowledge is fundamentally conjectural and probable rather than certain. Concerning even its most fundamental assumptions it retains an appropriate humility. Thus, in his most extensive presentation of the mechanical philosophy, *Physiologia Epicuro-Gassendo-Charltoniana: or a Fabric of Science Natuaral, Upon the Hypothesis of Atoms,* published in 1654, he writes:

> It is most *possible* and *verisimilous* that every physical continuum should consist of atoms; yet not absolutely necessary. For insomuch as the true idea of nature is proper only to that *eternal intellect* which first conceived it; it cannot but be one of the highest degrees of madness for dull and unequal men to pretend to an exact, or adequate comprehension thereof.

Over and over again, especially in the aftermath of the civil wars which seemed to confirm the destructiveness of sectarian confrontations, Latitudinarian Anglican scholars appealed to the anti-dogmatic cast of the mechanical philosophy, as set forth by Charleton and Boyle, to help eradicate sectarian bickering as Hooker had proposed:

"The Mechanik philosophy yields no security to *irreligion,*" Joseph Glanvil wrote in his *Scepsis Scientifica: or Confest Ignorance the Way to Science: In an Essay of the Vanity of Dogmatizing.* On the contrary, it may "dispose mens spirits to more *calmness* and *mod-*

esty, charity and *prudence* in their differences of *religion.*" And this in turn could bring agreement on essentials. Similarly, Bishop Thomas Sprat emphasized that the new science was creating "a race of young men . . . who were invincibly armed against the enchantments of enthusiasms," and who were thus, by implication, capable of a more stable and well-founded religious commitment.

If the mechanical philosophy supported a natural theology which was more modest in its claims than that of the older Platonic and scholastic traditions, it was, however, one which had a great deal to say about God's ongoing providential activity in the world. Hooker had insisted that God was both the *creator* and the *worker* of the events of the universe through his providential acts. But the organicist character of Renaissance naturalism seemed to undermine confidence in the necessity of God's ongoing working in nature. Charleton and Boyle both returned to Hooker's insistence on God's multiple and continuing relation to the universe. "God *made, conserves,* and *regulates* nature," in Charleton's terms.

In order to understand how each of these relations of God to the universe is demonstrated by the mechanical philosophy, we must recall the basic premises of that philosophy. As stated by Charleton they are:

1. That every effect must have its cause;
2. That no cause can act but by motion;
3. That nothing can act upon a distant subject or upon such whereunto it is not actually present . . . ; and consequently, that no body can move another except by contact. . . ." To this we must add that "bodies" are themselves inert or passive and do not contain within themselves any way to initiate notions, i.e., no mere body can move itself.

First let us consider now how this set of assumptions helps us to establish the existence and attributes of God as creator or "maker" or "author" of the universe. First, natural causes can do nothing but "mould an old matter into a new figure." Within the mechanical philosophy, then, shorn of its Epicurean assumption of the eternal existence and motion of all atoms, there is no possibility for an origin or beginning of the universe without appeal to God. Some first cause beyond the bodily elements of the physical world was necessary to turn "nothing into all things." So the role of God as first cause of the creation is established. Just as the physical existence of the atoms or corpuscles of the mechanical philosophy seems to demand some transcendent first cause, so too does the initial motion with which those atoms were created. Since there clearly is motion in the world and since all atoms are intrinsically passive, "the Hypothesis of atoms . . . fitly declares the radical cause of motion, activity, or energie . . . [to be] God."

According to critics of ancient atomism, the atomic philosophy presumed that the universe came into existence by chance. But Charleton's atomic-mechanical philosophy appeals to the new mechanistic imagery—especially that associated with the ubiquitous watches and clocks—to suggest the improbability of this claim.

> If fortune had the power and skill to make the *World,* why can she not make the more rude and facile movement of a *watch*? If atoms could [make] immense battlements of the World: why not also . . . the narrower structure of a *Castle*? If . . . the mighty bulwarks of an Island, why not a fort? If . . . wide campania's of Herbs and veriegated *flowers,* why not a piece of Landskip Tapestry?

In each comparison made above, an immensely complex natural entity is related to some presumably similar but simpler creation of some human designer and artisan. No one in his right mind, implies Charleton, would suppose that the fabricated object could have been produced by chance—without the skill and intelligence of its human creator. How much less likely, then, would it be that the immeasurably more complex natural entities came into existence without an equally immeasurably intelligent and skillful creator—God. This is a typical expression of the immensely important argument from design; and we should note that in this form it quite clearly implies that *probable* rather than demonstrative reasonings have a place in natural theology.

Next, let us turn to the ways in which God conserves and regulates the universe after its initial creation. Charleton insists, not that the machine of the world, once started, goes on running by itself forever, but rather that "this vast machin[e] depends on God in every minute freshly to create it, or to conserve it in being by a continual communication." Elsewhere he insists that nature depends on the "vigilancy and moderation" of a divine providence that continuously operates "through all the independant subdivisions of time" and in "the most minute, . . . trivial and contemptable events."

To bring this notion home in terms of the mechanical philosophy, Charleton focuses on the inertness of matter and the necessity of some activating principle. Just as a watch cannot run without a mainspring, so the world cannot run without God as an "energetical principle" or as the "*Spring* in the Engine of the world." Thus, the "Hypothesis of atoms . . . fitly declares the radical cause of all Motion, activity, or energie in second causes [to be] God." Though Boyle was less extensive in his stipulations regarding the precise way in which God's continuing activity is manifested through his ordinary providence, he was no less insistent than Charleton that the *preservation* of the material universe was no less problematic without God than was its creation, and he agreed that God was responsible for the "efficiency of every physical agent".

Paolo Rossi

Rationality and the Scientific Revolution

The Italian historian of science Paolo Rossi has been particularly active in studying the relations between technology and science in the early modern era. In the following selection he first lists a series of fourteen major changes which constitute the Scientific Revolution before he isolates one, "the theory that man can only know what he does or what he himself constructs," for special attention. This notion, which was central to the rise of the mechanical philosophy, developed as European machine technology became increasingly sophisticated and economically important; and it came to transform not merely our attitude toward nature but our understanding of ethics, politics, and history as well.

Between the second half of the sixteenth and the end of the seventeenth century a new cosmology and a new astronomy came into being. It was the time of the first observations with the microscope and the telescope and of experiments on the vacuum. A new science of motion was born and the principle of inertia formulated, the circulation of the blood was discovered and a decisive step forward was taken in the study of anatomy and physiology. The theory of spontaneous generation was disproved and theories were formulated to explain how the earth was formed. A new distinction was drawn between the "subjective" world of everyday experience based on the senses and the "objective" reality of corpuscles moving according to definable laws.

Each of the innovations I have listed (with the exception of the last, which is usually included in histories of philosophy as well) would form an important chapter in any history of astronomy, physics, biology or geology. But we can only talk of "the scientific revolution" when we go beyond specialized histories of individual sciences and the arbitrary grouping of scientists into categories made to fit the subject matter of our university curriculum. . . .

I believe that the period between [Copernicus's] *De revolutionibus* and [Isaac Newton's] *Principia* is quite rightly regarded as a turning point in the history of the world, not only because of important discoveries, new theories and novel experiments. It was a time when certain ideas and themes that are inextricably bound up with "science" came to the fore. These allow us to see the sudden break, the discontinuity that separates the new science from the old and helps us to understand some of the essential and decisive factors of what we usually call *modern thought.* I should like to list the following. First, the refutation of the priestly idea of knowledge inherent in hermeticism, in the alchemical literature, and in much of natural philosophy in the Renaissance. Second, a new appreciation of technical skills and mechanical arts, which as Leibnitz himself pointed out, led to consider the work carried out by 'empiricists' in their workshops and arsenals as a *kind of knowledge* that enriches man's true cultural heritage. Third, the new importance of scientific instruments as a means of making precise measurements and reproducing certain phenomena under controllable conditions. Fourth, the discovery of celestial bodies, plants, animals and men unknown before. Fifth, the birth, after a period of speculation on the existence of "savages" and the discovery of different cultures, of cultural relativism. Sixth, the idea of the plurality of other habitable planets and its consequence for the position and significance of man in a universe whose outer walls had crumbled. Seventh, the attempt, based on the new science of nature, to shape a new kind of ethics and politics. Eighth, the notion of the world as a machine not necessarily designed to suit man's standards, of a world therefore in which hierarchy had been abolished because all phenomena, like the component parts of a machine, have the same relation to the whole. Ninth, the conception of God as an engineer or watchmaker who follows a logical pattern in creating the world, but who does not intervene in its functioning. Tenth, the introduction of the dimension of time into the study of natural phenomena and the demise of the theory of an unchanging universe. Eleventh, the rise of the comparative anatomy that would lead to the "death of Adam". Twelfth, the notion of progress as the expansion of mankind and the idea of knowledge as the outcome of the efforts of several generations slowly accumulating results capable of integration and perfection. Thirteenth, the idea of collaboration and of publicizing the results of scientific research that lies at the heart of the first great scientific societies; and last of all, the theory that man can only know what he does or what he himself constructs.

The last topic on this list, so often neglected by those who study the *mechanization of the world picture,* deserves a moment's attention. The affirmation that there is no substantial difference between the products of art and those of nature—supported by many seventeenth century thinkers—runs contrary to the Aristotelian definition of art as completing the work of nature or as imitating its products. Aristotelian philosophy (as well as Hippocratic medicine) sees nature as an *ideal* that art must imitate, indeed as a *norm* whose precepts art must follow. Should art assert that it can equal the perfection of nature then it is treated—by the medieval doctrine of *imitatio naturae*—as an example of ungodliness: art is trying to counterfeit nature. The mechanical arts are necessarily *"adulterinae"* because they must borrow their movement from nature. According to Francis Bacon, this doctrine is connected with the Aristotelian theory of species, according to which a product of nature (e.g. a tree) has a *primary form,* whereas a product of art (e.g. a table) has only a *secondary form.* This is why Bacon's projected *History of the Arts* which was to complete the *Natural History* is so important:

> And I am the more induced to set down the History of the Arts as a species of Natural History, because an opinion has long been prevalent, that art is something different from nature, and things artificial different from things natural; whence this evil has arisen, that most writers of Natural History think they have done enough when they have given an account of animals or plants or minerals, omitting all mention of the experiments of mechanical arts. But there is likewise another and more subtle error which has crept into the human mind; namely that of considering art as merely an assistant to nature, having the power indeed to finish what nature has begun, to

> correct her when lapsing into error or to set her free when in bondage but by no means to change, transmute, or fundamentally alter nature. And this has bred a premature despair in human enterprise. Whereas men ought on the contrary to be surely persuaded of this; that the artificial does not differ from the natural in form or essence, but only in the efficient.

Within the mechanistic picture of the world, a machine, whether real or merely imagined, functions as an explicative model. It becomes an adequate representation of reality based on quantitatively measurable data, in which each element fulfils a function that is dependent on the configuration and the motion of the whole. To know reality means to understand how the world-machine functions. And a machine can always, at least in theory, be taken apart and put together again. "We enquire into things in nature," Gassendi writes,

> in the same way as we enquire into those things of which we are ourselves the authors. . . . Wherever possible in the study of nature we make use of anatomy, chemistry and other aids so as to understand, by breaking down the bodies as far as possible and dividing them as it were into their component parts, what these elements are and what manner of criteria helped in their composition, and to see whether, by following different criteria, they would have or still could become other than what they are.

The end of Gassendi's statement is particularly significant. The world of phenomena that can be reconstructed with the aid of scientific analysis, and the world of artificial products, which have been constructed or reconstructed intellectually or manually, are *the only realities of which we are able to have true knowledge:* the new science is . . . based on a phenomenological knowledge of the world.

We can only have full knowledge of machines (the artificial products of man) and of what can be interpreted mechanically. Several basic principles of Aristotle's notion of the relationship between art and nature, which had dominated the whole of European culture, were thus deliberately rejected. Descartes, like Bacon, draws no distinction between natural and artificial objects. Lightning, which according to the ancients could not be copied, has, in fact, been imitated in modern times. Art does not ape nature, nor does it "kneel before nature" as a popular medieval tradition asserted. Descartes is explicit on this point: "There is no difference between the machines constructed by artisans and the divers bodies that nature composes". The only difference is that the mechanisms of man-made machines are visible, whereas "the tubes and springs of natural objects are often too small to be perceived by the human senses".

The Platonic idea of God as a geometer was modified by the idea of God as a mechanic, artisan of this perfect clock that is the world. Knowledge of ultimate causes and essences, which is denied to man, is the prerogative of God as creator and *constructor* of the world-machine. The criterion of *knowledge-as-making* or the *identity of knowledge and construction (or reconstruction)* holds therefore not only in the case of man, but of God also. The human intellect is finite and limited, it can apprehend only those truths that have been constructed by man. We are really able to know only that which we make ourselves, or rather that which is *artificial.*

Broadly speaking, in so far as nature is not conceived of as an artifact, it is unknown and unknowable. . . .

After the second half of the sixteenth century, philosophers and scientists began to look upon 'experience' in a different light. As manual labour and the mechanical arts acquired a new status and a new importance within the encyclopaedia of knowledge, a new understanding of the relations between *knowing* and *doing* was born. Once the identification of *knowing* and *doing* had forced researchers to give up the idea of ever understanding the "essential" structure of nature, it was bound to influence ethics, politics and history, with consequences that can hardly be underrated.

René Descartes

Rules for the Direction of the Mind

Often called the father of modern philosophy, René Descartes was obsessed with the problem of how to establish knowledge which was certain beyond any conceivable doubt. In his Rules for the Direction of the Mind *(1628), he took mathematics as a model of certain knowledge, seeking to discover both a set of criteria by which to judge those characteristics of knowledge which make it certain and a way to generalize mathematical thought so as to make it possible to expand our domain of certainty. That part of the* Rules *reprinted below argues that the analytic procedure by which mathematicians decompose a complex proposition into a set of intuitively given simple statements must be an essential part of the attempt to achieve certainty on any topic.*

Rule I

The end of study should be to direct the mind towards the enunciation of sound and correct judgments on all matters that come before it. . . .

Rule II

Only those objects should engage our attention, to the sure and indubitable knowledge of which our mental powers seem to be adequate.

. . . There is scarce any question occurring in the sciences about which talented men have not disagreed. But whenever two men come to opposite decisions about the same matter one of them at least must certainly be in the wrong, and apparently there is not even one of them who knows; for if the reasoning of the second was sound and clear he would be able so to lay it before the other as finally to succeed in convincing *his* understanding also. Hence apparently we cannot attain to a perfect knowledge in any such case of probable opinion, for it would be rashness to hope for more than others have attained to. Consequently if we reckon correctly, of the sciences already discovered, Arithmetic and Geometry alone are left, to which the observance of this rule reduces us. . . .

. . . now let us proceed to explain more carefully our reasons for saying . . . that of all the sciences known as yet, Arithmetic and Geometry alone are free from any taint of falsity or uncertainty. We must note then that there are two ways by which we arrive at the knowledge of facts, viz. by experience and by deduction. We must further observe that while our inferences from experience are frequently fallacious, deduction, or the pure illation of one thing from another, though it may be passed over, if it is not seen through, cannot be erroneous when performed by an understanding that is in the least degree rational. . . .

But one conclusion now emerges out of these considerations, viz. not, indeed, that Arithmetic and Geometry are the sole sciences to be studied, but only that in our search for the direct road towards truth we should busy ourselves with no object about which we cannot attain a certitude equal to that of the demonstrations of Arithmetic and Geometry.

Rule III

In the subjects we propose to investigate, our inquiries should be directed, not to what others have thought, nor to what we ourselves conjecture, but to what we can clearly and perspicuously behold and with certainty deduce; for knowledge is not won in any other way.

. . . We shall here take note of all those mental operations by which we are able, wholly without fear of illusion, to arrive at the knowledge of things. Now I admit only two, viz. intuition and induction.

By *intuition* I understand, not the fluctuating testimony of the senses, nor the misleading judgment that proceeds from the blundering constructions of imagination, but the conception which an unclouded and attentive mind gives us so readily and distinctly that we are wholly freed from doubt about that which we understand. Or, what comes to the same thing, *intuition* is the undoubting conception of an unclouded and attentive mind, and springs from the light of reason alone; it is more certain than deduction itself, in that it is simpler, though deduction, as we have noted above, cannot by us be erroneously conducted. Thus each individual can mentally have intuition of the fact that he exists, and that he thinks; that the triangle is bounded by three lines only, the sphere by a single superficies, and so on. Facts of such a kind are far more numerous than many people think, disdaining as they do to direct their attention upon such simple matters.

This evidence and certitude, however, which belongs to intuition, is required not only in the enunciation of propositions, but also in discursive reasoning of whatever sort. For example consider this consequence: 2 and 2 amount to the same as 3 and 1. Now we need to see intuitively not only that 2 and 2 make 4, and that likewise 3 and 1 make 4, but further that the third of the above statements is a necessary conclusion from these two.

Hence now we are in a position to raise the question as to why we have, besides intuition, given this supplementary method of knowing, viz. knowing by *deduction,* by which we understand all necessary inference from other facts that are known with certainty. This, however, we could not avoid, because many things are known with certainty, though not by themselves evident, but only deduced from true and known principles by the continuous and uninterrupted action of a mind that has a clear vision of each step in the process. It is in a similar way that we know that the last link in a long chain is connected with the first, even though we do not take in by means of one and the same act of vision all the intermediate links on which that connection depends, but only remember that we have taken them successively under review and that each single one is united to its neighbour, from the first even to the last. . . .

These two methods are the most certain routes to knowledge, and the mind should admit no others. All the rest should be rejected as suspect of error and dangerous. . . .

Rule V

Method consists entirely in the order and disposition of the objects towards which our mental vision must be directed if we would find out any truth. We shall comply with it exactly if we reduce involved and obscure propositions step by step to those that are simpler, and then starting with the intuitive apprehension of all those that are absolutely simple, attempt to ascend to the knowledge of all others by precisely similar steps.

Étienne de Condillac

Analysis and Accurate Minds

Étienne de Condillac (1715–1780) was one of the great teachers of the eighteenth century and a popularizer of the analytic method as practiced by Descartes and Newton. La Logic *[Logic], from which the following excerpt comes, was written in 1780 in response to a request from the Polish government for a textbook on "the art of reasoning." In this passage Condillac explains the analytic method and illustrates it by considering how we come to know about a machine.*

Each one of us can notice that he knows sensible objects only through the sensations which he receives from them: these are the sensations which represent them to us.

If we are sure that when these objects are present we see them only in the sensations which they concurrently produce in us, we are no less sure that when they are absent we see them only in the recollection of the sensations which they produced. Thus the principle of all the knowledge which we can have concerning sensible objects is, and can only be, our sensations.

Sensations, considered as representing sensible objects, are called *ideas;* a figurative expression which literally signifies the same thing as *images*.

We distinguish as many kinds of ideas as we distinguish different sensations; and these ideas are either current sensations or else they are only a recollection of the sensations which we have had.

When we acquire ideas by the analytic method . . . they are arranged in an orderly way within the mind; there they retain the order which we have given them, and we can easily reproduce them for ourselves with the same clarity as that with which we acquired them. But if, instead of acquiring ideas by this method, we accumulate them at random, they will be in a state of great confusion and they will remain that way. This confusion will no longer permit the mind to recall them in a distinct manner; and if we want to speak of the knowledge which we think we have acquired, no one will understand a word of what we say because we will not understand a word of it ourselves. To speak intelligibly, we must conceive and present our ideas in the analytic order, which decomposes and recomposes each thought. This order is the only one which can give ideas all the clarity and all the precision which they are capable of having; and, as we have no other means for educating ourselves, so too we have no other means for communicating our knowledge. I have already proven this, but I return to it, and I shall return to it again; for this truth is not sufficiently known; it is even contested, although it is simple, evident and fundamental.

In fact, if I wish to know about a machine, I shall decompose it in order to study each part of it separately. When I have an exact idea of each part, and when I can replace them all in the same order as they previously were in, then I shall understand this machine perfectly because I have decomposed it and recomposed it.

What is it, then, to understand this machine? It is to have a thought which is composed of as many ideas as there are parts in the machine itself—ideas which represent each part exactly and which are arranged in the same order.

When I have studied the machine according to this method, which is the only one, then my thought offers me nothing but distinct ideas; and it analyses itself, whether I wish to make myself aware of it or to give an account of it to others.

Each of us can be convinced of this truth by his own experience; even the youngest seamstresses are convinced of it: for if you give them a dress of an unusual style as a model and ask them to make another one like it, they will naturally think of taking this model apart and putting it together again, in order to learn how to make the dress which you have requested. Thus they know the method of analysis as well as the philosophers do, and they know the utility of it much better than do those who persist in maintaining that there is another method of educating oneself.

René Descartes

Man a Machine

Although Descartes had begun to think of animal and human bodies as like *machines as early as 1619, it was not until his final work, the* Treatise of Man, *published posthumously, that he insisted that the body is not merely like a machine, it* is nothing but *a machine. In the following passage he explains how the human machine reacts when its foot is placed in a fire.*

Men [are] composed . . . of a soul and a body; and I must first separately describe for you the body. . . .

I assume [the] body to be but a statue, an earthen machine formed intentionally by God. . . . Thus not only does He give it externally the shapes and colors of all the parts of our bodies; He also places inside it all the pieces required to make it walk, eat, breathe, and imitate whichever of our own functions can be imagined to proceed from mere matter and to depend entirely on the arrangement of our organs.

We see clocks, artificial fountains, mills, and similar machines which, though made entirely by man, lack not the power to move, of themselves, in various ways. And I think you will agree that the present machine could have even more sorts of movements than I have imagined and more ingenuity than I have assigned, for our supposition is that it was created by God.

Now I shall not pause to describe to you the bones, nerves, muscles, veins, arteries, stomach, liver, spleen, heart, brain, nor all the other different pieces of which the machine must be composed. . . . If you do not already know them sufficiently, you can have them shown to you by some learned anatomist, those at least that are large enough to be seen. As for those which because of their smallness are invisible, I shall be able to make them known to you most simply and clearly by speaking of the movements which depend upon them; so that it remains only for me to explain these movements to you here in proper order and by that means to tell you which of the machine's [latent] functions these [patent] movements represent. . . .

To understand . . . how external objects that strike the sense organs can incite [the machine] to move its members in a thousand different ways: think that

[*a*] the filaments (I have already . . . told you that these come from the innermost part of the brain and compose the marrow of the nerves) are so arranged in every organ of sense that they can very easily be moved by the objects of that sense and that

[*b*] when they are moved, with however little force, they simultaneously pull the parts of the brain from which they come, and by this means open the entrances to certain pores in the internal surface of this brain; [and that]

[*c*] the animal spirits in its cavities begin immediately to make their way through these pores into the nerves, and so into muscles that give rise to movements in this machine. . . .

Thus [in Figure 6-1], if fire *A* is near foot *B*, the particles of this fire (which move very quickly, as you know) have force enough to displace the area of skin that they touch; and thus pulling the little thread *cc*, which you see to be attached there, they simultaneously open the entrance to the pore [or conduit] *de* where this thread terminates [in the brain]: just as, pulling on one end of a cord, one simultaneously rings a bell which hangs at the opposite end. . . .

Now the entrance of the pore or small conduit *de*, being thus opened, the animal spirits from cavity *F* enter and are carried through it—part into the muscles that serve to withdraw this foot from the fire, part into those that serve to turn the eyes and head to look at it, and part into those that serve to advance the hands and bend the whole body to protect it.

But they can also be carried through the same conduit *de* into many other muscles.

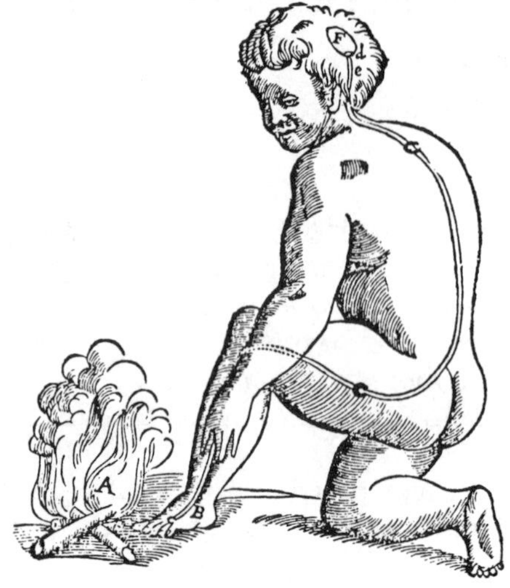

Figure 6.1

William Petty

On the Value of People

William Petty (1623–1687), a close friend of Hobbes, gave up a comfortable living as a London physician during the late 1640s to become physician to the King's (soon to be Parliament's) Army in Ireland. While in this position he made a fortune by applying the analytic procedure of Descartes to the practical process of surveying. He broke the surveyor's task down into dozens of subroutines, trained workers to concentrate on a single simple task, and completed the work of surveying most of Ireland in just over thirteen months, about one-tenth of the time expected by surveyors using traditional methods. With his profits Petty purchased huge Irish estates, and through the rest of his life he used the leisure which his income purchased in public service and in developing the science which we now call economics.

In the following selection Petty demonstrates that most of the value of the annual production of England comes not from land or stock, but from people; and he goes on to estimate the "value" of an Englishman.

There are of Men, Women, and Children, in *England* and *Wales,* about six Millions, whose Expence . . . for Food, Housing, Cloaths, and all other necessaries, amount to 40 Millions, *per Annum.*

There are in *England* and *Wales,* of Acres of Land . . . 24 Millions, that is, which yields 8 Millions [pounds] *per Annum* Rent, and which are worth 144 Millions to be sold. . . .

Now if the Land worth 144 Millions, yield 8 Millions *per annum,* the other Estate [houses, furniture, personal property] converted into the like Species must yield $5\frac{8}{9}$ more; but because Money and other personal Estates yield more *per annum* than Land; (that is) doubles it self under 17 years . . . then instead of $5\frac{8}{9}$, suppose it to yield 7, making the whole Annual Proceed 15.

Now if the Annual proceed of the Stock, or Wealth of the Nation, yields but 15 millions, and the expence be 40. Then the labour of the People must furnish the other 25. . . .

Whereas the Stock of the Kingdom, yielding but 15 Millions of proceed, is worth 250 Millions; then the People who yield 25, are worth 416⅔ Millions. . . .

If 6 Millions of People be worth 417 millions of pounds *Sterling,* then each head is worth 69 *l.* [pounds] or each of the 3 millions of Workers is worth 138 *l.* . . .

From whence it follows, that 100,000 persons dying of the Plague, above the ordinary number, is near 7 Millions loss to the Kingdom; and consequently how well might 70,000 *l.* have been bestowed in preventing this Centuple loss?

Thomas Hobbes

Leviathan

Thomas Hobbes, born in 1588, was one of the most brilliant and most hated intellectuals of his age. Trained as a humanist, he served for a time as secretary to Francis Bacon and published a translation of Thucydides' History of the Peloponnesian War. *Then around 1631 he underwent a "conversion" to the mathematical and mechanical philosophies, seeking in* De Cive *(1643) and* Leviathan *(1651) to extend the mathematical-mechanistic approach to the study of social and political phenomena. In the passages below he explains the nature of science and deduces the need for men to lay down their natural rights to a sovereign power.*

It appears that reason is not, as sense and memory, born with us; nor gotten by experience only, as prudence is; but attained by industry: first in apt imposing of names; and secondly by getting a good and orderly method in proceeding from the elements, which are names, to assertions made by connexion of one of them to another; and so to syllogisms, which are the connexions of one assertion to another, till we come to a knowledge of all the consequences of names appertaining to the subject in hand; and that is it, men call *science.* And whereas sense and memory are but knowledge of fact, which is a thing past and irrevocable, science is the knowledge of consequences, and dependence of one fact upon another; by which, out of that we can presently do, we know how to do something else when we will, or the like, another time: because when we see how anything comes about, upon what causes, and by what manner; when the like causes come into our power, we see how to make it produce the like effects. . . .

But yet they that have no science are in better and nobler condition with their natural prudence than men that, by misreasoning, or by trusting them that reason wrong, fall upon false and absurd general rules. For ignorance of causes, and of rules, does not set men so far out of their way as relying on false rules, and taking for causes of what they aspire to, those that are not so, but rather causes of the contrary.

To conclude, the light of humane minds is perspicuous words, but by exact definitions first snuffed, and purged from ambiguity; reason is the *pace;* increase of science, the *way;* and the benefit of mankind, the *end.* And, on the contrary, metaphors, and senseless and ambiguous words are like *ignes fatui* [will o' the wisps]; and reasoning upon them is wandering amongst innumerable absurdities; and their end, contention and sedition, or contempt.

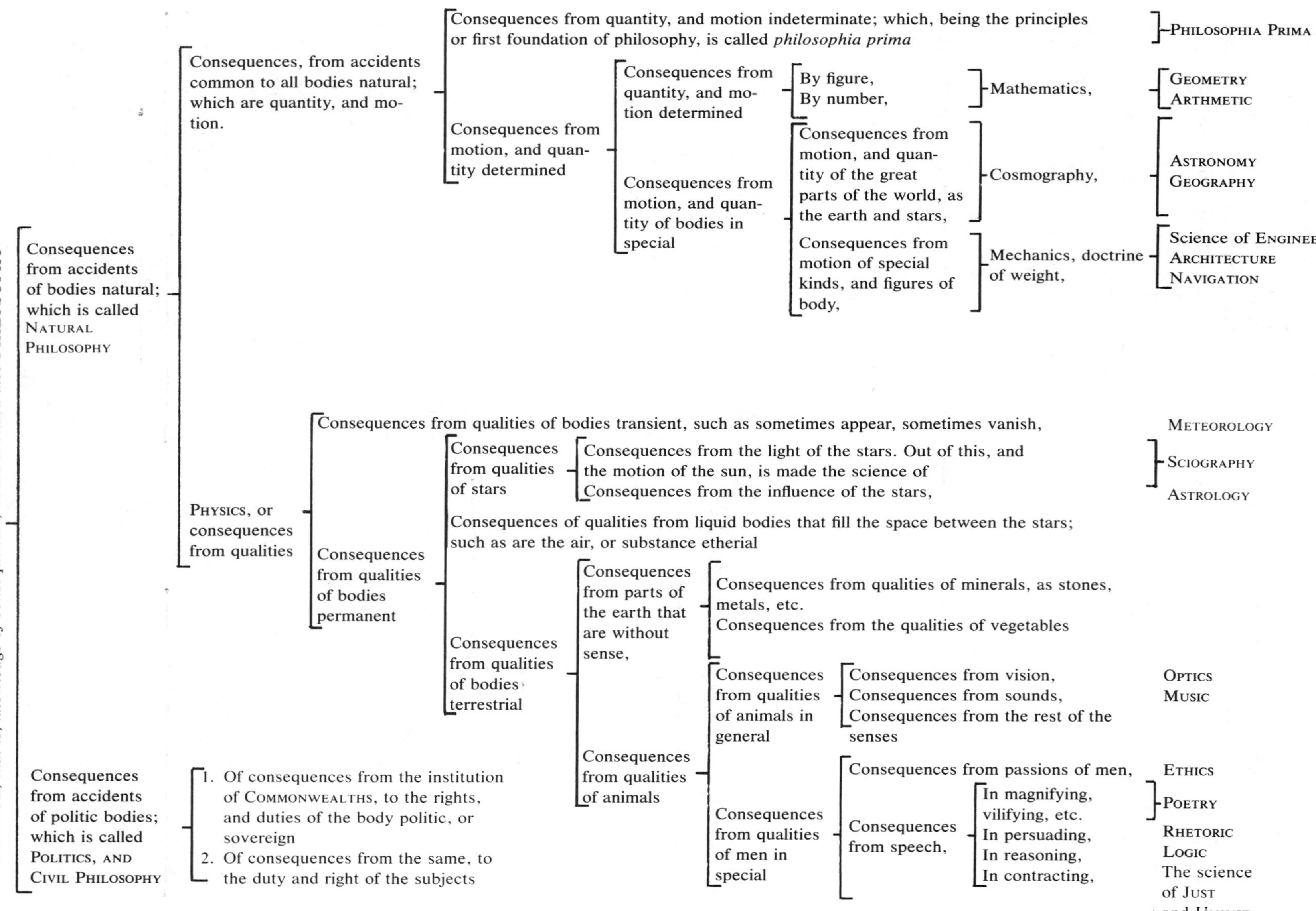
SCIENCE, that is, knowledge of consequences; which is called also PHILOSOPHY
Consequences from accidents of bodies natural; which is called NATURAL PHILOSOPHY
Consequences, from accidents common to all bodies natural; which are quantity, and motion.
Consequences from quantity, and motion indeterminate; which, being the principles or first foundation of philosophy, is called philosophia prima
PHILOSOPHIA PRIMA
Consequences from motion, and quantity determined
Consequences from quantity, and motion determined
By figure,
By number,
Mathematics,
GEOMETRY
ARTHMETIC
Consequences from motion, and quantity of bodies in special
Consequences from motion, and quantity of the great parts of the world, as the earth and stars,
Cosmography,
ASTRONOMY
GEOGRAPHY
Consequences from motion of special kinds, and figures of body,
Mechanics, doctrine of weight,
Science of ENGINEERS
ARCHITECTURE
NAVIGATION
PHYSICS, or consequences from qualities
Consequences from qualities of bodies transient, such as sometimes appear, sometimes vanish,
METEOROLOGY
Consequences from qualities of bodies permanent
Consequences from qualities of stars
Consequences from the light of the stars. Out of this, and the motion of the sun, is made the science of
SCIOGRAPHY
Consequences from the influence of the stars,
ASTROLOGY
Consequences of qualities from liquid bodies that fill the space between the stars; such as are the air, or substance etherial
Consequences from qualities of bodies terrestrial
Consequences from parts of the earth that are without sense,
Consequences from qualities of minerals, as stones, metals, etc.
Consequences from the qualities of vegetables
Consequences from qualities of animals
Consequences from qualities of animals in general
Consequences from vision,
OPTICS
Consequences from sounds,
MUSIC
Consequences from the rest of the senses
Consequences from qualities of men in special
Consequences from passions of men,
ETHICS
Consequences from speech,
In magnifying, vilifying, etc.
POETRY
In persuading,
RHETORIC
In reasoning,
LOGIC
In contracting,
The science of JUST and UNJUST
Consequences from accidents of politic bodies; which is called POLITICS, AND CIVIL PHILOSOPHY
1. Of consequences from the institution of COMMONWEALTHS, to the rights, and duties of the body politic, or sovereign
2. Of consequences from the same, to the duty and right of the subjects

There are of knowledge two kinds, whereof one is knowledge of fact; the other, knowledge of the consequence of one affirmation to another. The former is nothing else but sense and memory, and is *absolute knowledge;* as when we see a fact doing, or remember it done; and this is the knowledge required in a witness. The latter is called *science,* and is conditional; as when we know that: *if the figure shown be a circle, then any straight line through the center shall divide it into two equal parts.* And this is the knowledge required in a philosopher; that is to say, of him that pretends to reasoning.

The register of knowledge of fact is called *history,* whereof there be two sorts: one called *natural history;* which is the history of such facts, or effects of Nature, as have no dependence on man's will; such as are the histories of metals, plants, animals, regions, and the like. The other is *civil history,* which is the history of the voluntary actions of men in Commonwealths.

The registers of science are such books as contain the demonstrations of consequences of one affirmation to another; and are commonly called *books of philosophy;* whereof the sorts are many, according to the diversity of the matter; and may be divided in such manner as I have divided them in the table opposite.

[The Application of Science to Politics]

So that in the nature of man, we find three principal causes of quarrel. First, competition; secondly, diffidence; thirdly, glory.

The first maketh men invade for gain; the second, for safety; and the third, for reputation. The first use violence, to make themselves masters of other men's persons, wives, children, and cattle; the second, to defend them; the third, for trifles, as a word, a smile, a different opinion, and any other sign of undervalue, either direct in their persons or by reflection in their kindred, their friends, their nation, their profession, or their name.

Hereby it is manifest that during the time men live without a common power to keep them all in awe, they are in that condition which is called *war;* and such a war as is of every man against every man. For war consisteth not in battle only, or the act of fighting, but in a tract of time, wherein the will to contend by battle is sufficiently known: and therefore the notion of *time* is to be considered in the nature of war, as it is in the nature of weather. For as the nature of foul weather lieth not in a shower or two of rain, but in an inclination thereto of many days together: so the nature of war consisteth not in actual fighting, but in the known disposition thereto during all the time there is no assurance to the contrary. All other time is *peace.*

Whatsoever therefore is consequent to a time of war, where every man is enemy to every man, the same is consequent to the time wherein men live without other security than what their own strength and their own invention shall furnish them withal. In such condition there is no place for industry, because the fruit thereof is uncertain: and consequently no culture of the earth; no navigation, nor use of the commodities that may be imported by sea; no commodious building; no instruments of moving and removing such things as require much force; no knowledge of the face of the earth; no account of time; no arts; no letters; no society; and which is worst of all, continual fear, and danger of violent death; and the life of man, solitary, poor, nasty, brutish, and short. . . .

It may peradventure be thought there was never such a time nor condition of war as this; and I believe it was never generally so, over all the world: but there are many places where they live so now. For the savage people in many places of America, except the government of small families, the concord whereof dependeth on natural lust, have no government at all, and live at this day in that brutish manner. . . .

A *law of nature, lex naturalis,* is a precept, or general rule, found out by reason, by which a man is forbidden to do that which is destructive of his life, or taketh away the means of preserving the same, and to omit that by which he thinketh it may be best preserved. . . .

And because the condition of man . . . is a condition of war of every one against every one, in which case every one is governed by his own reason, and there is nothing he can make use of that may not be a help unto him in preserving his life against his enemies; it followeth that in such a condition every man has a right to every thing, even to one another's body. And therefore, as long as this natural right of every man to every thing endureth, there can be no security to any man, how strong or wise soever he be, of living out the time which nature ordinarily alloweth men to live. And consequently it is a precept, or general rule of reason: *that every man ought to endeavour peace, as far as he has hope of obtaining it; and when he cannot obtain it, that he may seek and use all helps and advantages of war.* The first branch of which rule containeth the first and fundamental law of nature, which is: *to seek peace and follow it.* The second, the sum of the right of nature, which is: *by all means we can to defend ourselves.*

From this fundamental law of nature, by which men are commanded to endeavour peace, is derived this second law: *that a man be willing, when others are so too, as far forth as for peace and defence of himself he shall think it necessary, to lay down this right to all things; and be contented with so much liberty against other men as he would allow other men against himself.* For as long as every man holdeth this right, of doing anything he liketh; so long are all men in the condition of war. But if other men will not lay down their right, as well as he, then there is no reason for anyone to divest himself of his: for that were to expose himself to prey, which no man is bound to, rather than to dispose himself to peace. This is that law of the gospel: *Whatsoever you require that others should do to you, that do ye to them.*

Sheldon S. Wolin

Hobbes and Political Theory

Sheldon Wolin is one of the most articulate present-day spokesmen for the need to retain the classical tradition of moral and political philosophy in the face of the scientific tradition of political discourse represented by Hobbes. In this selection he argues that Hobbesian theory leads to a politics which involves the exercise of power to no well-established ends.

". . . The utility of moral and civil philosophy," Hobbes states, "is to be estimated, not so much by the commodities we have by knowing these sciences, as by the calamities we receive from not knowing them." This seems a modest claim when compared with the list of benefits produced by "natural philosophy and geometry." The latter have to their credit "the greatest commodities of mankind," such as "measuring matter and motion; moving ponderous bodies; architecture; navigation; making instruments for all uses; calculating the celestial motions, the aspects of the stars, and the great parts of time; geography, etc." Yet without peace and security none of these arts can flourish. The justification of civil philosophy, then, is its ability to state the conditions which make possible the pursuit of science and the arts. "Commodious living," which is the promise contained in science, presupposes that men know and practice "the rules of civil life." Just as practical arts and techniques serve as the middle term which translates science and geometry into "the greatest commodities," so "civil rules" are the means for translating civil philosophy into civil order.

Once the point is reached where it is proper to ask how and by what means the promise of political theory is to be realized, the hero, for the first time, falters. The nature of the task is such that he is no longer self-sufficient, for he must rely on others to give practical substance to his "infallible rules." In the history of political theory, Hobbes was not the first uncertain hero. From Plato onwards, theorists had recognized the need for some kind of alliance between theory and power. The difficulty was not to persuade theorists that by some means they must gain control of power; this was obvious, and most theorists devised stratagems for the purpose. The main stumbling-block was to persuade rulers that their business was to effect theories and then, having persuaded them, to insure that the practice of the ruler would remain in accord with the teachings of the theory. Plato's solution had been the most thorough-going and demanding. He had proposed the complete re-education of rulers, including the requirement that they become proficient in philosophy, the most difficult type of knowledge. Hobbes, however, believed that Plato had demanded too much of rulers and that the virtue of his own method was to have simplified a difficult subject so as to make it comprehensible and useful to rulers. In place of the "depth of moral philosophy," he put "theorems of moral doctrine." In place of the abstruse and difficult political philosophies of the past and the endless disquisitions on civility, he was prepared to offer in handy and convenient form "the true rules of politics":

". . . But when I consider again, that the science of natural justice, is the only science necessary for sovereigns and their principal ministers; and that they need not be charged with the sciences mathematical, as by Plato they are, farther than by good laws to encourage men to the study of them; and that neither Plato, nor any other philosopher hitherto, hath put into order, and sufficiently or probably proved all the theorems of moral doctrine, that men may learn thereby, both how to govern, and how to obey; I recover some hope, that, one time or other, this writing of mine may fall into the hands of a sovereign who will consider it himself (for it is short, and I think clear,) without the help of any interested, or envious interpreter; and by the exercise of entire sovereignty, in protecting the public teaching of it, convert this truth of speculation, into the utility of practice."

To have succeeded in making political knowledge useful and accessible was no mean achievement, but it raised another kind of difficulty, one that Hobbes seemed totally unaware of and that, in his unawareness, affirmed his modernity. Hobbes overlooked the difficulty which Plato had never forgotten: if political knowledge were available in digest form, rulers might be armed without being educated. If rulers were simply handed the power of knowledge without prolonged exposure to the moralizing and civilizing influence of the philosophical way of life, they would naturally use that knowledge for the ends of *Machtpolitik* [Power politics]. Unless men have a contervision of a better way of life, one that transcends politics, they will destroy society and themselves. "All goes wrong," Plato had declared, "when starved for lack of anything good in their own lives, men turn to public affairs hoping to snatch from thence the happiness they hunger for. They set about fighting for power, and this internecine conflict ruins them and their country. The life of true philosophy is the only one that looks down upon offices of state; and access to power must be confined to men who are not in love with it . . ."

Plato's full formula provided not only that philosophers should become kings and kings philosophers, but also that access to power should be denied to the lovers of power, just as access to knowledge should be opened to those who loved it. In contrast, Hobbes had sought to detach "rules" from the broader, civilizing context of philosophical education, because only in that form would political knowledge make a difference to human life.

Hobbes was thus the first to practice the "technical fallacy" in its truly modern form, the fallacy of concentrating exclusively upon knowledge that will work immediately and leaving moral culture to take care of itself. In this he was being perfectly consistent: if knowledge is only another form of power, then it behooves us to seek its most compact, efficient form, trusting either that men will somehow acquire the tacit moral understandings needed if rules are to be adapted to life, or that men will develop a mode of life in which those understandings are no longer important.

Alexandre Koyre

The Significance of the Newtonian Synthesis

The great success of Isaac Newton in using mathematical reasoning and observation to "discover" the law of universal gravitation and in employing experiments to determine the various colors in a ray of sunlight convinced many that his method was capable of solving virtually all problems. Alexander Pope expressed a widely held feeling with his famous couplet:

> *Nature and Nature's laws lay hid in night:*
> *God said Let Newton be! and all was light.*

But the French physicist and philosopher, Jean d'Alembert (1717–1783), shrewdly remarked that while the new method threw light on many topics it cast others into shadows. In his essay "The Significance of the Newtonian Synthesis," Alexandre Koyre discusses the broad appeal that Newton's philosophy held for the eighteenth century. But, like d'Alembert, he points out that Newton's successes were purchased at the cost of creating a barrier between our "objective" and "subjective" worlds.

It has often been said that the unique greatness of Newton's mind and work consists in the combination of a supreme experimental with a supreme mathematical genius. It has often been said, too, that the distinctive feature of the Newtonian science consists precisely in the linking together of mathematics and experiment, in the mathematical treatment of the phenomena, that is, of the experimental or (as in astronomy, where we cannot perform experiments) observational data. Yet, though doubtless correct, this description does not seem to me to be a complete one: thus there is certainly much more in the mind of Newton than mathematics and experiment; there is, for instance—besides religion and mysticism—a deep intuition for the limits of the purely mechanical interpretation of nature. And as for Newtonian science, built . . . on the firm basis of corpuscular philosophy, it follows, or, better, develops and brings to its utmost perfection, the very particular logical pattern (by no means identical with mathematical treatment in general) of atomic analysis of global events and actions, that is, the pattern of reducing the given data to the sum total of the atomic, elementary components (into which they are in the first place dissolved).

The overwhelming success of Newtonian physics made it practically inevitable that its particular features became thought of as essential for the building of science—of any kind of science—as such, and that all the new sciences that emerged in the eighteenth century—sciences of man and of society—tried to conform to the Newtonian pattern of empirico-deductive knowledge, and to abide by the rules laid down by Newton in his famous *Regulae philosophandi,* so often quoted and so often misunderstood. The results of this infatuation with Newtonian logic, that is, the results of the uncritical endeavor mechanically to apply Newtonian (or rather *pseudo* Newtonian) methods to fields quite different from that of their original application, have been by no means very happy. . . . We have to dwell for a moment upon the more general and more diffuse consequences of the universal adoption of the Newtonian synthesis, of which the most important seems to have been the tremendous reinforcement of the old dogmatic belief in the so-called "simplicity" of nature, and the reintroducing through science into this very nature of very important and very far-reaching elements of not only *factual* but even *structural* irrationality.

In other words, not only did Newton's physics use *de facto* such obscure ideas as power and attraction (ideas suggesting scholasticism and magic, protested the Continentals), not only did he give up the very idea of a rational deduction of the actual composition and formation of the choir of heaven and furniture of earth, but even its fundamental dynamic law (the inverse-square law), though plausible and reasonable, was by no means necessary, and, as Newton had carefully shown, could be quite different. Thus, the law of attraction itself was nothing more than a mere fact. . . .

The intricate and subtle machinery of the world seemed obviously to require a purposeful action, as Newton did not fail to assert. Or, to put it in Voltaire's words: the clockwork implies a clockmaker. . . .

Thus the Newtonian science, though as *mathematical philosophy of nature* it expressedly renounced the search for causes (both physical and metaphysical), appears in history as based on a dynamic conception of physical causality and as linked together with theistic or deistic metaphysics. This metaphysical system does not, of course, present itself as a constitutive or integrating part of the Newtonian science; it does not penetrate into its formal structure. Yet it is by no means an accident that not only for Newton himself, but also for all the Newtonians—with the exception only of Laplace—this science implied a reasonable belief in God.

Once more the book of nature seemed to reveal God, an engineering God this time, who not only had made the world clock, but who continuously had to supervise and tend it in order to mend its mechanism when needed (a rather bad clockmaker, this Newtonian God, objected Leibniz), thus manifesting his active presence and interest in his creation. Alas, the very development of the Newtonian science which gradually disclosed the consummate skill of the Divine Artifex and the infinite perfections of his work left less and less place for divine intervention. The world clock more and more appeared as needing neither rewinding nor repair. Once put in motion it ran for ever, . . . whereas for the first generation of Newtonians, as well as for Newton himself, God had been, quite on the contrary, an eminently active and present being, who not only supplied the dynamic power of the world machine but positively "ran" the universe according to his own, freely established, laws.

It was just this conception of God's presence and action in the world which forms the intellectual basis of the eighteenth century's world feeling and explains its particular emotional structure: its optimism, its divinization of nature, and so forth. Nature and nature's laws were known and felt to be the embodiment of God's will and reason. Could they, therefore, be anything but good? To follow nature and to accept as highest norm the law of nature, was just the same as to conform oneself to the will, and the law, of God.

Now if order and harmony so obviously prevailed in the world of nature, why was it that, as obviously, they were lacking in the world of man? The answer seemed clear: disorder and disharmony were man-made, produced by man's stupid and ignorant attempt to tamper with the laws of nature or even to suppress them and to replace them by man-made rules. The remedy seemed clear too: let us go back to nature, to our own nature, and live and act according to its laws.

But what is human nature? How are we to determine it? Not, of course, by borrowing a definition from Greek or Scholastic philosophers. Not even from modern ones such as Descartes or Hobbes. We have to proceed according to pattern, and to apply the rules which Newton has given us. That is, we have to find out, by observation, experience, and even experiment, the fundamental and permanent faculties, the properties of man's being and character that can be neither increased nor diminished; we have to find out the patterns of action or laws of behavior which relate to each other and link human atoms together. From these laws we have to deduce everything else.

A magnificent program! Alas, its application did not yield the expected result. To define "man" proved to be a much more difficult task than to define "matter," and human nature continued to be determined in a great number of different, and even conflicting, ways. Yet so strong was the belief in "nature," so overwhelming the prestige of the Newtonian (or pseudo-Newtonian) pattern of order arising automatically from interaction of isolated and self-contained atoms, that nobody dared to doubt that order and harmony would in some way be produced by human atoms acting according to their nature, whatever this might be—instinct for play and pleasure (Diderot) or pursuit of selfish gain (A. Smith). Thus return to nature could mean free passion as well as free competition. Needless to say, it was the last interpretation that prevailed.

The enthusiastic imitation (or pseudo-imitation) of the Newtonian (or pseudo-Newtonian) pattern of atomic analysis and reconstruction that up to our times proved to be so successful in physics, in chemisty, and even in biology, led elsewhere to rather bad results. Thus the unholy alliance of Newton and Locke produced an atomic psychology, which explained (or explained away) mind as a mosaic of "sensations" and "ideas" linked together by laws of association (attraction); we have had, too, atomic sociology, which reduced society to a cluster of human atoms, complete and self-contained each in itself and only mutually attracting and repelling each other.

Newton, of course, is by no means responsible for these, and other, *monstra* engendered by the overextension—or aping—of his method.

. . . No man can ever be made responsible for the misuse of his work or the misinterpretation of his thought, even if such a misuse or misinterpretation appears to be—or to have been—historically inevitable.

Yet there is something for which Newton—or better to say not Newton alone, but modern science in general—can still be made responsible: it is the splitting of our world in two. I have been saying that modern science broke down the barriers that separated the heavens and the earth, and that it united and unified the universe. And that is true. But, as I have said, too, it did this by substituting for our world of quality and sense perception, the world in which we live, and love, and die, another world—the world of quantity, of reified geometry, a world in which, though there is place for everything, there is no place for man. Thus the world of science—the real world—became estranged and utterly divorced from the world of life, which science has been unable to explain—not even to explain away by calling it "subjective."

True, these worlds are every day—and even more and more—connected by the *praxis* [application of science]. Yet for *theory* they are divided by an abyss.

Two worlds: this means two truths. Or no truth at all.

This is the tragedy of modern mind which "solved the riddle of the universe," but only to replace it by another riddle: the riddle of itself.

CHAPTER 7

Enlightenment and Industrialization

Introduction by Richard Olson

We saw in Chapter 6 that new attitudes toward mathematics and a "mechanical" philosophy began to reshape ideas in a wide range of intellectual disciplines during the early and mid-seventeenth century; but until the end of the century scientific influences tended to be confined to small learned elites centered in the courts, commercial centers, and universities of Europe.

Similarly, we saw in Chapter 3 that as early as the Renaissance there was a growing feeling that scientific knowledge might provide the foundation for productive activities. But by the late seventeenth century there were still only a few limited domains in which science had begun to play a significant role in economic life. Even in such fields as navigation and military engineering, where the scientific dimension was relatively great, it continued to offer more promise than performance. Where technological developments had any relationship to science they still tended to stimulate scientific activity rather than vice versa. And when Samuel Johnson (1709–1784) wrote of the early Royal Society that "great expectations were raised of the sudden progress of the useful arts . . . [but] the truth is that little had been done compared with what fame had suffered to promise . . . ," he was essentially correct.

Against this background, the impact of science on European life from about 1686 to 1800 was spectacular. During no period before or since has science so openly and obviously impinged on almost every aspect of elite *and* popular culture in both its intellectual *and* economic dimensions.

Prior to the eighteenth century, the centrality of science to intellectual life had been very great; but its role in shaping the economy through technological innovation was small, and the literate audience for scientific ideas was a tiny fraction of the population. After the eighteenth century, the centrality of scientific developments to technological innovation, and thus to economic life, is unquestionable; but in spite of a general spread of literacy in Europe, science soon became so specialized and esoteric that with rare exceptions scientific arguments were and are no longer directly accessible to any but a professional elite. During the eighteenth century, however, most scientific ideas and theories were still *relatively* simple. At the same time, the educational levels of those whom

Molière (1622–1673) had termed "bourgeois" and Daniel Defoe (1660–1731) called the "middling sort" were becoming sufficiently high that almost every moderately educated middle-class person was capable of being well informed about scientific developments; and, as we shall soon see, a remarkable number chose to exercise that capability by reading scientific literature and by attending public scientific lectures.

On one hand, at least in part because scientific ideas and attitudes were becoming more central to the common pool of knowledge available to people whose primary interests were in producing goods and services and in lining their own pockets, there was an ever-increasing tendency to try to exploit scientific knowledge to economic advantage. Even more important, there was a tendency to submit practical issues to a new kind of empirical, rational, and quantitative analysis promoted by the popularizers of the sciences. As a consequence of the widespread use of scientific knowledge and method, the Baconian promise of material progress driven by experimental science gradually began to be realized in dramatic fashion. On the other hand, the increasing popularity of natural science and confidence in its methods created an ever-widening expectation that scientific methods, with their rejection of arguments based on mere tradition or "superstition" and their demands for empiricism and reasoned arguments, might illuminate *all* domains of human interest and provide the foundation for an almost limitless progress that would be not only material, but also intellectual, social, and moral. Jean d'Alembert (1717–1783), coeditor of the great French *Encyclopedie,* expressed a widely held attitude in his *Elements of Philosophy* of 1749:

> The discovery and application of a new method of philosophizing, the kind of enthusiasm which accompanies discoveries, a certain exaltation of ideas which the spectacle of the universe produces in us; all these causes have brought about a lively fermentation of minds. Spreading . . . in all directions, this fermentation has swept everything before it which stood in its way with a sort of violence like a river which has burst its dams. . . . Thus, from the principles of the secular sciences to the foundations of religious revelation, from music to morals, from the scholastic disputes of theologians to matters of commerce, from natural law to the arbitrary laws of nations . . . everything has been discussed, analyzed, or at least mentioned.

Thus it seems that the spread of scientific knowledge and confidence in scientific method lay at the very root of two of the great historical movements of the eighteenth century—the Enlightenment and the Industrial Revolution. And as we shall also see, a strong argument can be made that they even played a very significant role in the third great eighteenth-century movement—the democratizing movement which culminated in the American and French revolutions.

Science had already begun to filter into relatively popular culture during the late seventeenth century through such works as Bernard Fontenelle's simplified and lively depiction of Cartesian cosmology, *The Conversations on the Plurality of Worlds* (1686), and John Ray's homey exploration of natural history in *The Wisdom of God as Evidenced in the Creation* (1690), and through courses in applied mathematics and medicine taught at a few institutions such as Gresham College in London. But science virtually exploded onto the social scene with the publication and popularization of the ideas of Isaac Newton (1642–1727) and John Locke (1632–1704), whose works were the great symbolic foundations of Enlightenment thought. Newton provided the model in natural philosophy, while Locke seemed to offer a completely new understanding of human nature, drawing from the same empirical approach.

Unknown to almost anyone, Locke had first introduced the Newtonian philosophy on the continent through a long French language essay review of the *Principia* published in Jean LeClerc's widely read literary review journal, the *Bibliothèque Universelle;* but it was the clever and brilliant Voltaire (1694–1778) who served as the chief continental popularizer of both of the great English philosophers. In the selections from Voltaire's *Lettres philosophiques* printed below, one can get a sense of Voltaire's deep admiration for his English predecessors and yet detect his irreverent attitude toward traditional authority, in both philosophy and theology, which he associated with their works.

Locke and Newton had certainly emphasized the primacy of empiricism and reason for the discovery of truth: but each had retained a deep commitment to revealed religion and traditional morality, which was shared by most of their English followers. Voltaire and his witty French *philosophe* colleagues, on the other hand, tended to turn Reason into a weapon against all tradition and "superstition."

By the middle of the eighteenth century the popularity of science knew almost no limits. Works like Voltaire's *Elements de la physique de Newton* (1738) and Francesco Algarotti's *Newtonianismo per le Dame (Newtonianism for Ladies)* (1737) were published in many editions in all major languages, and Buffon's magnificent multivolume *Natural History* became so popular that, as historian Daniel Mornet has shown, no other book of any kind was to be found in as many late-eighteenth-century French private libraries. At the same time, scientific lecturers such as the Abbé Nollet in Paris and Peter Shaw in London drew huge audiences; and a whole raft of itinerant lecturers like Benjamin Martin plied their trade throughout provincial France, England, and the American colonies. Popular periodicals like *Gentleman's Magazine* featured mathematical puzzles and reviews of scientific publications. In 1750 even ladies' fashions in Paris were influenced by science with the appearance of lightning rod hats—an item of apparel suggested by the device invented by Benjamin Franklin (1706–1790), who was lionized in Paris, not for his political acumen, but for his work with electricity.

James Secord suggests how deeply science penetrated the institutions of eighteenth-century culture in the selection from "Newton in the Nursery: Tom Telescope and the Philosophy of Tops and Balls." In this article Secord mentions the role of Locke and that of the scientific lecturers in spreading the word about science; and he emphasizes the close linkages that Newtonian science had with traditional theology and morality in eighteenth-century England.

In spite of attempts like that discussed by Secord to appropriate science in support of religious and social stability, it does seem that the spread of scientific ideas and attitudes during the eighteenth century became increasingly linked with liberal, democratic, and even socialist tendencies in social thought. Scientific theorists since Descartes and Hobbes had emphasized the universality of reason, and Locke tried to convince his readers both that humans were created nearly equal in their sensory and intellectual abilities and that every human at birth was untainted by original sin and uninformed by any innate knowledge. Virtually all that men and women became resulted from their experience or education, so that there was no natural foundation for the distinctions of rank and wealth which existed in society and no justification for laws which assigned different rights to different classes.

It is certain that Locke did not intend his psychological theorizing to provide the foundation for fundamental criticisms of the social order, but it is equally clear that it rapidly did become associated with utilitarian social theories in England and with democratic movements in America and France. David Thomson, in his article, "Scientific Thought and Revolutionary Movements," discusses this association, with special reference to what he calls the "dream of unity" and the "natural harmony of interests."

The selection from Charles Gillispie's "The Natural History of Trades" begins a shift of focus from the place of popular science in the Enlightenment to the role of science in the Industrial Revolution. Central figures of the Industrial Revolution—men such as James Watt, John Smeaton, Josiah Wedgwood, and James Brindley in Britain, and Nicolas Leblanc, J. A. C. Chaptal, and the Montgolfier brothers in France—unquestionably felt that science was central to eighteenth-century industrial innovations and advances. But twentieth-century scholars have been engaged in an extended and open-ended debate about how accurately contemporary perceptions represent the "real" state of affairs; for in almost no cases have scholars been able to establish in detail how any new eighteenth-century scientific theory led directly to a commercially significant new process or product. Those who persist in thinking that eighteenth-century industrialists probably did have a reasonable understanding of what they were doing argue that the "realist" skeptics have been misled by nineteenth- and twentieth-century patterns of science-technology interactions into wrongly supposing that the specific-theory-to-specific-product pattern is the only possible way for science to shape technology. Gillispie agrees in his recent works that, "whatever the interplay between [eighteenth-century] science and agrarian or industrial production may have been, it did not consist in the application of up-to-date theory to techniques for making or growing things." He does, however, argue convincingly that, for France especially, two key elements associated with the scientific thrust of Enlightenment thought—that is, the replacement of tradition by experiment and reason as a foundation for effective action, and the emphasis on spreading knowledge throughout all levels of society—played a major role in technological and industrial development.

The Industrial Revolution was at first largely a British phenomenon, and in England one can make a very strong case for the importance of scientific attitudes and activities in producing those technological innovations which in turn increased agricultural and industrial productivity and led to immense increases in profits and population. One of the most successful entrepreneurs of the Industrial Revolution was Josiah Wedgwood (1730–1795). Although he was perhaps best known as a master of marketing strategies, Wedgwood's success also depended on significant technical innovations in the glazing and firing of pottery and in the highly efficient organization of workers at his Etruria factories. Short selections from two articles by Neil McKendrick illuminate the role of science in the Industrial Revolution. The first shows how Wedgwood set about improving his glazes and firing practices in a systematic and scientifically well-informed manner; the second discusses Wedgwood's attempts to break down traditional work patterns and rationalize the process of production so as to increase its efficiency.

This last point is of special concern; for as McKendrick suggests, it demonstrates a tendency of eighteenth-century science which was not as unambiguously admirable as most that we have so far discussed—the tendency to treat humans as if they are mere cogs in some great productive machine. Indeed, as Wedgwood boasted, one of his great goals was to "make such *machines* of the *Men* as cannot err." Wedgwood and many of his efficiency-minded colleagues intended only the benefit of their workers, but as Adam Smith (1723–1790) and his close colleague Adam Ferguson (1723–1816) soon began to fear, the division of labor and "rationalization" of production which they sought might be purchased at the serious cost of creating a class of uninformed and alienated workers. Ferguson complained that the mechanical arts "succeed best under a total suppression of sentiment and reason; and ignorance is the mother of industry as well as of superstition. . . . Manufacturers, accordingly, prosper most where the mind is least consulted, and where the workshop may, without any great effort of imagination be considered as an engine, the parts of which are men." Smith was even more uneasy about these tendencies which, he argued, would leave the intellectual faculties of the inferior ranks of people "mutilated and deformed" and would leave "all of the nobler parts of the human character . . . obliterated and extinguished in the great body of the people." The concerns of Smith and Ferguson were largely ignored by the industrial entrepreneurs of the nineteenth century; but they formed the foundation of Karl Marx's powerful critique of the "alienation" of labor under commercial capitalism.

Voltaire

On Mr. Locke and On the System of Gravitation

Voltaire was one of eighteenth-century France's most brilliant men of letters. A dramatist, epic poet, essayist, and author of one of the most witty and popular novels of all time, Candide, *Voltaire visited England for three years between 1726 and 1729. He returned home filled with enthusiasm for British parliamentary government and for British intellectual life, which he saw as grounded in Newtonian scientific method and in Locke's sensationalist psychology. His fascination with Locke's psychology is most fully illustrated in the short story,* Micromegas, *and his commitment to the Newtonian philosophy is most fully expressed in the* Elements of Newtonian Physics, *which Voltaire wrote with his mistress and mathematics tutor, Madame du Chatelet. But Voltaire first introduced Locke and Newton to his readers in the* Letters on England, *published in 1733.*

Never, perhaps, has a wiser, more methodical mind, a more precise logician existed than Mr. Locke, yet he was not a great mathematician. He had never been able to put up with the tedium of calculations or the sterility of mathematical truths which do not at first offer anything appreciable to the mind, and nobody has proved better than he that one can possess the geometrical mind without the help of geometry. Before him great philosophers had decided exactly what the human soul is, but since they knew nothing whatever about it it is only to be expected that they all had different views. . . .

Locke has expounded human understanding to mankind as an excellent anatomist explains the mechanism of the human body. At all points he calls in the light of physics, he sometimes ventures to speak affirmatively, but he also dares to doubt. Instead of defining at one fell swoop what we don't know, he examines by degrees what we want to know. He takes a child at the moment of birth and follows step by step the progress of its understanding, he sees what the child has in common with the animals and in what it is superior to them, he consults especially his own experience, the consciousness of his own thought.

> 'I let those who know more about it than I do,' he says, 'argue whether our souls exist before or after the organization of our bodies, but I confess that I have been endowed with one of those coarse souls that are not thinking all the time, and am even unfortunate enough not to think it any more necessary for the soul to be always thinking than it is for the body to be always in motion.'

For myself I am proud of the honour of being as stupid as Locke in this matter. Nobody will persuade me that I am always thinking, and I feel no more disposed than he to imagine that a few weeks after my conception I was a most sapient soul, already knowing a thousand things that I forgot at birth, having quite pointlessly possessed in the uterus knowledge that escaped me as soon as I was in a position to need it and which I have never been able to regain properly since.

Locke, having destroyed innate ideas, having abandoned the vanity of believing that we are always thinking, establishes that all our ideas come to us via the senses, examines our simple ideas and those that are complex, follows the human mind in all its operations, shows how imperfect are the languages men speak and how we misuse terms at every moment.

Finally he considers the extent, or rather the non-existence, of human knowledge. It is in this chapter that he modestly ventures these words: *We shall probably never be able to know whether a purely material being thinks or not.*

This wise remark seemed to more than one theologian a scandalous declaration that the soul is material and mortal.

Some Englishmen, devout after their own fashion, sounded the alarm. The superstitious in society are what cowards are in an army: they feel and pass on panic terrors. There was an outcry that Locke sought to overthrow religion. But religion had nothing to do with this affair; it was a purely philosophical question quite independent of faith and revelation, simply a matter of examining without rancour whether there is any contradiction in saying: *matter can think,* and whether God can communicate thought to matter. But theologians too often begin by saying that God is outraged when you don't share their point of view. . . .

It must never be feared that any philosophical sentiment can harm the religion of a country. Our mysteries may well clash with our demonstrations, but they are no less revered by Christian philosophers, who know that the aims of reason and faith are different by nature. Philosophers will never form a religious sect. Why? Because they do not write for the people and are devoid of emotional fire. . . . Indeed, all the books of the modern philosophers put together will never make as much noise in the world as a simple dispute made some time ago among the Franciscans about the cut of their sleeves and cowls. . . .

The discoveries of Sir Isaac Newton, which have earned him such a universal reputation, concern the system of the world, light, infinity in mathematics and finally chronology, with which he toyed for relaxation.

I am going to tell you . . . the little I have been able to gather about all these sublime ideas.

Concerning the system of our world, arguments had been going on for a long time on the basic cause which makes all the planets turn and keeps them in their orbits, and on that which in our own world makes all bodies fall on to the surface of the earth. . . .

Having retired to the country near Cambridge in 1666, [Newton] was walking in his garden, saw some fruit falling from a tree and let himself drift into a profound meditation on this weight, the cause of which all the scientists have vainly sought for so long and about which ordinary people never even suspect there is any mystery. He said to himself: 'From whatever height in our hemisphere these bodies might fall, their fall would certainly be in the progression discovered by Galileo, and the spaces traversed by them would be equal to the square of the time taken. This force which makes heavy bodies descend is the same, with no appreciable diminution, at whatever depth one may be in the earth and on the highest mountain. Why shouldn't this force stretch right up to the moon? And if it is true that it reaches as far as that, is it not highly probable that this force keeps the moon in its orbit and determines its movement? But if the moon obeys this principle, whatever it may be, is it not also very reasonable to think that the other planets are similarly influenced?

'If this force exists it must (as is proved, moreover) increase in inverse ratio to the squares of the distances. So it only remains to examine the distance covered by a heavy body falling to the ground from a medium height, and that covered in the same time by a body falling from the orbit of the moon. To know this it only remains to have the measurements of the earth and the distance from the moon to the earth.'

That is how Newton reasoned. But in England at that time there existed only very erroneous measurements of our globe. People relied on the faulty reckoning of pilots, who counted sixty English miles to a degree, whereas they should have counted nearly seventy. As this false calculation did not agree with the conclusions Newton wanted to draw, he abandoned them. A mediocre scientist, motivated solely by vanity, would have made the measurements of the earth fit in with his system as best he could. Newton preferred to abandon his project for the time being. But since M. Picard had measured the earth accurately by tracing this meridian, which is such an honour for France, Newton took up his first ideas again and found what he wanted with the calculations of M. Picard. This is a thing that still seems admirable to me; to have discovered such sublime truths with a quadrant and a bit of arithmetic.

The circumference of the earth is 123,249,600 Paris feet. From that alone the whole system of gravitation can follow.

The circumference of the earth is known, that of the orbit of the moon is known and the diameter of this orbit. One revolution of the moon in this orbit takes twenty-seven days, seven hours, forty-three minutes. So it is clear that the moon in its mean path travels at 197,960 Paris feet per minute and by a known theorem it is clear that the central force that would make a body fall from the height of the moon would make it fall only fifteen Paris feet in the first minute.

Now if the law by which bodies weigh, gravitate, attract each other in inverse ratio to the squares of the distances is true, if it is the same force that acts according to this law throughout nature, it is evident that, the earth being sixty half-diameters distant from the moon, a heavy body falling to the earth must cover fifteen feet in the first second and 54,000 feet in the first minute.

Now, it does happen that a heavy body falls fifteen feet in the first second and in the first minute travels 54,000 feet, which number is the square of sixty multiplied by fifteen, so the weight of bodies is in inverse ratio to the squares of distances, so the same force causes weight on the earth and holds the moon in its orbit.

As it is proved that the moon is drawn towards the earth which is the centre of its own particular movement, it is proved that the earth and the moon are drawn towards the sun, which is the centre of their annual movement.

The other planets must be governed by this general law, and if this law exists these planets must follow the laws discovered by Kepler. All these laws and relationships are indeed adhered to with the utmost accuracy by the planets. Therefore the force of gravity draws all the planets towards the sun, like our own globe. Finally, the reactions of all bodies being proportional to their action, it is certain that the earth is in its turn drawn towards the moon and that the sun is drawn towards both, that each of the satellites of Saturn is drawn towards the other four and the other four towards it, and all five towards Saturn and Saturn towards all of them, that the same applies to Jupiter and that all these globes are drawn towards the sun, which in its turn is drawn towards them.

This force of gravity acts in proportion to the matter contained in the bodies—this is a truth Newton has established by experiments. This new discovery has shown that the sun, centre of all the planets, draws them all in direct proportion to their mass combined with their distance. From there, rising by degrees to knowledge that seemed beyond the human mind, he ventures to calculate the amount of matter in the sun and how much there is in each planet. Thus he shows, by the simple laws of mechanics, that each celestial globe must of necessity be at the place where it is. His one principle of the laws of gravitation explains all the apparent inequalities in the paths of the heavenly bodies. The variations of the moon become a necessary result of these laws. Moreover it is clear why the nodes of the moon take nineteen years to perform a revolution and those of the earth about twenty-six thousand years. The high and low tides of the sea are another very simple effect of this attraction. The proximity of the moon when it is full or when it is new and its greater distance at the quarters, together with the action of the sun, provide a convincing reason for the rise and fall of the ocean. . . .

That is not all. If this force of gravity or attraction acts in all heavenly spheres, it acts no doubt on all the parts of those spheres, for if bodies attract each other in proportion to their masses, it can only be in proportion to the quantity of their parts. And if this power is present in the whole, it certainly is in the half, the quarter, the eighth and so ad infinitum. Moreover if this power were not equally in each part, there would always be certain parts of the globe that would gravitate more than the others, which is not the case. So this power actually exists in all matter, and in the smallest particles of matter.

Thus attraction is the mainspring which keeps the whole of nature in motion.

James A. Secord

Newton in the Nursery: Tom Telescope and the Philosophy of Tops and Balls 1761–1838

One sign of the deep penetration of science into eighteenth-century culture was the popularity of juvenile encyclopedias of natural history and treatises on natural philosophy. Among the most popular of such works was The Newtonian System of Philosophy, adapted to the capacities of young gentlemen and ladies, *first published in 1761 and probably written by the famous children's bookseller John Newbery. In this article James Secord discusses the intellectual background to this work, the nature of the audience it reached, and the linkages it tried to make between science, religion, and traditional social values.*

In the seventeenth century, the Scientific Revolution was for adults only. Copernicus, Kepler, Harvey, Galileo and Descartes never appeared in juvenile versions, and generally remained unread by all but a few experts. But in England by the middle of the next century, the New Philosophy had been popularized as the Newtonian, and could be taught even to the youngest children. *The Newtonian system of philosophy, adapted to the capacities of young gentlemen and ladies,* a book which first appeared in April 1761, could be advertised as a "Philosophy of Tops and Balls". This tiny gold-covered volume was ostensibly the published result of scientific lectures at the "Lilliputian Society" delivered by an enterprising (if imaginary) lad named Tom Telescope. In the space of one hundred and twenty-six pages, his discourses survey the entire range of eighteenth century natural philosophy, from the solar system to the five senses of man. Tom's lectures provide not only one of the most convenient surveys of Enlightenment science, but also illustrate the ways in which natural philosophy had become securely established as a pillar in the established order of Georgian England. . . . The present case shows that by the middle of the eighteenth century science had entered the family circle, and could be employed in teaching moral codes and good manners. The Newtonian cosmology, predigested by a host of commentators, had become fit food for babies, an innocuous pabulum that could be served up to the most impressionable child. The enterprising commercial publishers' "new world of children in eighteenth century England" could include Isaac Newton along with Goody Two-shoes and Wog-log the Giant. . . .

Like many other innovations in juvenile publishing, Tom Telescope's lectures came from the press (and probably the pen) of John Newbery, a London bookseller active during the middle years of the eighteenth century. . . .

Whether or not Newbery wrote the lectures, there can be no doubt of his importance in the development of a commercial culture centred around children. He aimed his books explicitly at the middling orders, men like James Watt, Matthew Boulton, and Thomas Barnes anxious to see their sons and daughters escape the dangers of "the tavern, the gaming table or the brothel". To meet the needs of this expanding market among the mercantile and professional classes, a whole range of goods and services became available during the period. There were jigsaw puzzles, private academies, museums, lectures, dollhouses, panoramas—all formulated with a double function of entertainment and enlightenment. Many of the new schools were beginning to include instruction in mathematics, the use of globes, and experimental philosophy. "Such are many of the diversions," Tom Telescope says, "at the school where I am placed". Other indicators of the increasing availability of science for children are provided by trends within the book trade, especially the growing number of juvenile natural histories and encyclopedias. Like Thomas Boreman's *Description of three hundred animals* of 1730 (a great boyhood favourite of the Wernerian mineralogist Robert Jameson), the books in natural history derived from Renaissance bestiaries and Aesop's fables. Later, publishers quarried Buffon's *Histoire naturelle* and Goldsmith's *History of the Earth and animated nature* for suitably moralized descriptions of birds and beasts. Both these works enjoyed an immense popularity among children in their own right. . . .

Natural history possessed obvious attractions for a juvenile audience. With their strong visual appeal and emphasis on concrete description, even adult works could be rapidly passed on to children. Publishers recognized this very rapidly, and miniature herbals and bestiaries were among the very earliest children's books. But experimental natural philosophy, chemistry, and astronomy presented more difficult problems. Very probably certain popular texts, particularly those intended for women, could also be read to children; Newbery himself published a few such books in the 1750s, notably works by the itinerant lecturer Benjamin Martin. Another important pioneer was the Abbé Pluche's *Spectacle de la nature* (1732–51), a French physico-theological work repeatedly translated into English and intended "for the rational amusement of young noblemen".

There was, however, one precedent of particular significance for Newbery's *Newtonian system*. John Locke, while engaged in the late 1690s as a tutor to the twelve year old son of Sir Francis and Lady Masham, had composed a brief manuscript treatise on the "Elements of natural philosophy". In his 1693 *Some thoughts concerning education,* Locke had strongly emphasized the need for young gentlemen to obtain a cultured acquaintance with science, and his ten-thousand-word outline provided his own tutee with precisely that, from "Matter and motion" to four pages abstracting the *Essay concerning human understanding*. . . . From the 1750s onwards it began to be published as a separate pamphlet, often from provincial towns (Glasgow, Whitehaven, Berwick-upon-Tweed) and always with Locke's "Some thoughts concerning reading and study for a gentleman". So nearly not included by its early eighteenth century editor, the "Elements" eventually proved the most popular item in the collection.

The most lasting beneficiary of this popularity was none other than Tom Telescope, for Locke's "Elements" served directly as the basis for Newbery's "Philosophy of Tops and Balls". Obviously there was a demand for a general treatise on natural philosophy, suitable for children, that had barely existed in Locke's time. Newbery did not singlehandedly create this demand, but he certainly went a long way towards satisfying it. In his hands Locke's treatise was expanded threefold, and although the organization remained unchanged, the style and contents are transformed. Tom's six lectures touch on all aspects of contemporary natural philosophy and natural history, using Locke's outline to give a continuous picture of the natural world rather than the isolated descriptions characteristic of the juvenile encyclopedias. A glance at the chapter headings, so closely based on Locke's original treatise, shows that the *Newtonian system* aimed to give much more than a simple redaction of the *Principia* and the *Opticks:*

I. "Of Matter & Motion"
II. "Of the Universe, and particularly of the Solar System"
III. "Of the Air, Atmosphere, and Meteors"
IV. "Of Mountains, Springs, Rivers, and the Sea"
V. "Of Minerals, Vegetables, and Animals"
VI. "Of the five Senses of Man, and of his Understanding"

Indeed, one of the most notable features of the work is the way in which all aspects of natural knowledge have been subsumed under the 'Newtonian' designation. At the level of elementary popularization, Newtonianism had become virtually synonymous with natural science. Given that Newton could cast so long a shadow, it is hardly surprising that historians have recognized the difficulties of determining "where the statue stood" for practising natural philosophers. The point, of course, is not that all Enlightenment science somehow derived from one man, but rather that Newton's reputation possessed a remarkable drawing power for the eighteenth century audience. As an astute publisher, Newbery took advantage of the obvious: the name of Newton sold books.

The debt of "The Philosophy of Tops and Balls" to Locke is obviously more direct and substantial. Not only did his "Elements" give Tom's lectures their basic framework, but *Some thoughts concerning education* provided the enterprising Newbery with a widely accepted pedagogical theory to underpin his sales campaign. Newbery's books for children ring with manifestos that might be taken straight out of Locke:

> It has been said, and said wisely, that the only way to remedy these Evils, is to begin with the rising Generation, and to take the Mind in its infant State, when it is uncorrupted and susceptible of any Impression; To represent their Duties and future Interest in a Manner that shall seem rather intended to amuse than instruct, to excite their Attention with Images and Pictures that are familiar and pleasing; To warm their Affections with such little Histories as are capable of giving them Delight, and of impressing on their tender Minds proper Sentiments of Religion, Justice, Honour, and Virtue.

Without doubt the greatest single difference between Locke's "Elements" and the *Newtonian system* lies in the effort Newbery makes to render science as a mode of play. As a lecturer Tom Telescope is enviably successful in making natural philosophy entertaining. In *Some thoughts* Locke had insisted on the need to make learning pleasant, for the circumstances of education were associated by the child with knowledge itself, and could foster both an initial interest in a subject and its later recall. Newbery carries this emphasis much further. Throughout the book, the very toys of the nursery are revealed as philosophical apparatus: a candle, a cricket ball, and a fives ball illustrate the workings of an eclipse, while the double rotation of the Earth—on its axis and around the Sun—is explained by analogy with the motion of a carriage wheel circling on a nearby drive. And as Tom says, "every boy who can whip his top knows what motion is"; the fascination engendered at the Lilliputian Society by his "Lecture on Matter & Motion" is evident. . . . Further adding to the appeal are constant references to current events. The Lisbon earthquake, the deaths of the British in the Black Hole of Calcutta, and (in later editions) the cross-Channel balloon flight of Jean-Phillip Blanchard and John Jeffries, are all used to enliven the lecture on the properties of air. The more difficult points, as in most early children's books, are clarified by numerous copperplate engravings and woodcuts. There are pictures of an air gun and an orrery, an engraving of a carriage on fire from the friction of its wheels, and an illustration of the observatory where the Lilliputian Society holds several of its meetings. As with the use of everyday objects as scientific apparatus, these illustrations make the wonders of nature immediately available to the child. Toys are imbued with the marvellous, while the marvellous becomes a comprehensible part of daily life in the nursery.

Thus by the mid-eighteenth century an interest in natural philosophy had entered the home itself and had become an activity for the entire family. This is shown just as surely by the *Newtonian system* as it is by the familiar paintings of the orrery and the air pump by Joseph Wright of Derby. The audience for Tom's lectures includes children from a wide variety of ages, although the appeal seems to have centred around ten to twelve year olds. "A very pretty book for Children it is", commented the *Monthly review* on its first appearance. "Pretty *old Children* too, may read it with improvement." It seems likely that the book was chiefly intended, not for private study, but rather to be read aloud by a parent or guardian to an eager group of budding natural philosophers. In the illustration of Tom's first lecture, the children are surrounded by a bevy of ladies and gentlemen, who participate in the discussions throughout the text. The Lilliputian Society, never designed to be a conspiracy of youth, brought adults and children together in a common pursuit of scientific understanding.

Tom's discourses are obviously intended to recreate one of the most typical means for introducing science to children, attendance with their parents at the spectacular demonstrations by itinerant lecturers in natural philosophy. . . .

The circumstances of Tom's lectures are presented with wit and imagination, and the result is far from being a dry catalogue of facts. Tom can barely get two sentences out before he is interrupted by one of his youthful auditors. There is the abrasive sceptic Tom Wilson; the persistent Lady Caroline; young Master Blossom; the rich heir Sir Harry. The liveliness of the repartee and the sharpness of their questioning brings the philosophy into relief, making the book a series of dramatic episodes rather than a set of static speeches. This loose, almost theatrical style, as well as the remarkable range of subjects covered, facilitates the merging of moral and religious sentiment with technical descriptions of natural phenomena. These wider lessons of science emerge both through the content of the lectures and through the circumstances of their delivery. The edifying exchanges between lecturer and audience frame the Newtonian cosmology in a domestic world ordered by good manners and social deference. Tom Telescope, expositor of the Newtonian system, becomes a paragon of virtue advocating an order of human relationships as regular as the movements of the planets.

The wider implications of Tom's philosophy are present from the very beginning of the book. The lectures are intended in the first place to save Tom's young companions from the insidious pastime of card playing. "Master *Telescope,* a young Gentleman of distinguished abilities, sat silent, and heard all with complacency and temper till this diversion was proposed; but then he started from his seat, and begged they would think of some more innocent amusement." The message becomes even more apparent as Tom Telescope, the well-mannered, God-fearing Newtonian, is explicitly contrasted with Tom Wilson, an impolite leering Aristotelian in the mode of Galileo's Simplicio. Tom Telescope is hardly able to begin his first lecture with a description of inertia before Tom Wilson interrupts him with a laugh. "What, says he, shall any body tell me that my hoop or my top will run for ever, when I know by daily experience that they drop of themselves, without being touched by any body?" Wilson is fooled by the evidence of his senses, and constantly questions any attempt to penetrate the mysteries of mundane objects. The possibility of people at the antipodes and the explanation of gravity give further occasion for rude interruptions.

The turning point in the relation between the contrasting Toms comes at the beginning of the lecture on the solar system, as the Lilliputian Society gathers at the observatory on a moonlit night. Extolling the beauties of the "glorious canopy" of the heavens, Tom Telescope shows that the possibility of other inhabited worlds was far from having dangerous ideological overtones. Rather than wondering how extraterrestrials could have participated in God's historical plan, he demonstrates that the probable existence of people on other planets simply gives one more illustration of the glorious plenitude of God. Tom Telescope raises his eyes to the heavens, lost in boundless wonder, his piety silencing the company. Naturally this gives "infinite satisfaction" to one of the adults present, and Tom Wilson begins "to consider himself a fool in comparison of our philosopher." Wilson's further questionings are markedly subdued. Through his mastery of the realm of scientific knowledge, Tom Telescope obtains an understanding of God and his handiwork, a high degree of moral virtue, approved by adult authority, and conversion of his archenemy. . . .

One by one, other potentially dangerous aspects of science are shown to be fully safe for juvenile consumption. The air pump, for example, initially threatens to raise an acute moral dilemma, as Tom suggests observing the effect of placing a living animal in an evacuated bell-jar. The advance of science is pitted against Lady Caroline, an ardent anti-vivisectionist who opposes performing the cruel experiment. Fortunately a compromise is made available in the form of a recently caught rat. If the rat survives the trial, he will be freed. The compromise is accepted, the rat lives, and the purposes of science and sentiment are jointly served. Another potentially distasteful aspect of scientific inquiry has been successfully defused.

The young lecturer has further teachings

to impart. As one would expect, he gives little emphasis to natural philosophy as a direct source of technological or medical benefit, although connections between chemistry and new drugs are suggested, and techniques of mouth-to-mouth resuscitation are explained. Rather, science carries the cultural and social values of the high Enlightenment. Tom's discussion of botany becomes a discourse on the virtues of empire; the acorn gives rise to the oak, which in turn is the material foundation for Britain's maritime supremacy. "Hence *Britain* boasts her wide extensive reign", Tom quotes from some verses, "And by th' expanded acorn rules the Main".

But from Tom's perspective science is valued above all as an education in good manners and virtuous citizenship, and the most important lessons in the book deal with deference and social position. Because of the theatrical quality of the lectures, these lessons are acted out rather than merely preached, so that manners are taught both by precept and by practice. The setting itself opens up opportunities for teaching patterns of deference. Although the actual market for the book almost certainly consisted of the professional and mercantile middling orders, the text closely ties knowledge of nature with the landed social elite. . . .

The blend of moral instruction and rational entertainment achieved in the *Newtonian system* evidently proved tremendously successful. Tom's book was to have a remarkable life, for various editions were on booksellers' shelves for the next eight years, with nine editions in the eighteenth century alone. There were also Irish printings, Dutch translations in 1768 and 1783, Swedish ones in 1782 and 1786, American editions in 1803 and 1808, and even an Italian version by "Tommaso Telescopio" in 1832. Certainly the book was one of Newbery's best-selling publications. J. H. Plumb has estimated that a minimum of twenty-five to thirty thousand copies must have been issued during the eighteenth century in England alone, and the actual figure may be far higher. The survival of a tiny handful of this vast output indicates the extent to which the books were used, coloured in, and passed down from generation to generation—until the tiny volumes fell into separate sheets and torn pages, to be thrown into the fire. In 1761 the exposition of natural philosophy as a mode of religious and moral instruction, fit for children, met with ready acceptance among the new mercantile and professional orders. Newbery, ever the astute business man, had judged his market correctly once again.

David Thomson

Scientific Thought and Revolutionary Movements

At the height of the French Revolution and in its immediate aftermath, there was an understandable desire among intellectuals to fix credit or blame for the tumultuous events of the 1790s. In general, proponents (such as Tom Paine (1737–1809)) and opponents (such as Edmund Burke (1729–1797) and Joseph DeMaistre (1754–1821) alike saw the intellectual movement of the Enlightenment, with its emphasis on the power of reason and the poverty of "superstition," as a central factor in shaping political sensibilities and events. In this article historian David Thomson tries to indicate those aspects of Enlightenment thought which were derived from its scientific dimension and which encouraged the growth of the liberal and democratic ideals that fueled revolutionary movements.

This broad revolutionary movement of the last two centuries flows from the great tide of liberal-democratic ideas, generated in the eighteenth century and first breaking upon the *ancien régime* in the American and French revolutions. In combination with the economic and social transformations wrought by the industrial revolution, this tide of liberal-democratic ideas produced a further current of social-democratic ideas; and both, mingling with the insurgent dynamic forces of nationalism which were released by the French Revolutionary and Napoleonic Wars and their aftermath in Europe can be watched surging tumultuously through one country after another during the following 150 years. When the tide of revolutionary ideas lapped the borders of the East it produced the Russian revolutions of 1917: fired the Liberal Revolution which overthrew Tsardom, and within six months the Bolshevik Revolution.

The gravitational pull which began this tide of revolution came first from the rationalist philosophy of the eighteenth century. In the ancient and highly traditionalist society of Western Europe, and particularly in France as the greatest and most influential country of Western Europe, religious and ethical beliefs, no less than political and social institutions, were subjected to new tests of value and purpose. Voltaire and the Encyclopaedists, Montesquieu, Rousseau and all the *philosophes,* brought to bear upon the *ancien régime* a devastating rationalist critique which made inevitable first a reorientation of opinion and then a reorganization of society. Much has been written about how this rationalist philosophy came to transform ethics and politics. But remarkably little has been written about the precise connexions between this rationalist philosophy itself and the changes in scientific thought which began with the so-called 'scientific revolution' of the seventeenth century. It is clear enough that this scientific revolution, hinging upon the work of Galileo, Descartes and Newton, encouraged a fresh intellectual outlook, a new bias of mind which valued the collection of verifiable data, and promoted a spirit of inquiry, a critical habit of mind and the search for underlying 'laws of nature'. It is unlikely to be mere coincidence that France, the main fountain-head of rationalist thought during the eighteenth century and of revolutionary ideas and movements at the end of that century, was also the home of Cartesianism and the leader of the world in the progress of science. Some direct connexion and affiliation is to be expected. But can it be shown more precisely what this connexion was, and how it affected the character of the revolutionary movement which was born of rationalist philosophy? . . .

It was not that the *philosophes* were themselves revolutionaries. It would be wrong to think of the rationalist movement in the eighteenth century as consciously directed towards producing revolution. The rationalist intellectuals who contributed to the *Encyclopédie* of Diderot contemplated no political action which would destroy the regime. Some expected further reforms from the regime itself; others, like Rousseau, who envisaged an idyllic new order put it into an indefinitely remote future. They were propagandists against superstition and the power of priests, on behalf of the new ideas suggested by the discoveries of science and propounded by rationalist thinkers. In this respect their influence was similar in kind to that of the scientists themselves; and indeed science had not yet come to be regarded as a specialized way of seeking truth, distinct from moral philosophy or theology. Even the French word *science* retained its wider meaning of 'Knowledge' in general. Knowledge was not yet departmentalized, and 'philosophy' comprised any sort of examination of the physical world, of man as a biological or moral or economic entity, and of man's organization into societies and states. The great *Encyclopédie* was itself a symbol of their continued belief in the unit of knowledge, and a conscious effort to prevent its disintegration. It appeared both natural and inevitable to apply similar principles to the investigation of human thought and activity in all their forms; it would scarcely have occurred to them to do otherwise. . . . Yet, given their rationalist hypotheses, this very insistence on the unity of human knowledge made the inherent clash between rationalism and the presuppositions of the social and political order all the more inevitable, and all the more revolutionary in its consequences. . . .

One of the most important of the mediations by which scientific thought affected the democratic movement was also one of the most concealed. Scientific knowledge, it was noticed, gained acceptance by a kind of consensus of informed opinion. A scientific theory was accepted as true not only because the man who advanced it produced mathematical demonstration or experimental evidence to support it, but also because his fellow-scientists, having heard his demonstration and his experimental evidence, were convinced and lent it their support. Scientific truths ultimately elicited the unanimous agreement of all informed and enlightened opinion. It was natural, for the reasons suggested above, that rationalist thinkers should expect ethics and politics to be capable of eliciting similar unanimity. Reason was universal, and reason was one. The more the social order approximated to a rational or 'natural' order, the more it would reflect such unity and unanimity. Once men's understanding was undistorted by superstition and privilege, and all were recognized as equal in rights and freedom, then reason would reveal their natural harmony of interests. Little room would remain for honest differences of opinion, which would all be capable of being resolved by rational demonstration and persuasion. There would be a consensus of opinion in democracy closely akin to the consensus of opinion which lay behind scientific truth; and education would open the door to the necessary enlightenment and to ultimate unanimity. This startling and unduly neglected assumption of revolutionary democratic thought has its modern counterpart in the present-day insistence that, in British trial by jury, the members must be unanimous in their verdict. It is taken for granted that when an objective truth is being established—a question of guilt or innocence—then twelve honest persons, having listened to all the available evidence, will necessarily reach complete agreement.

There is abundant proof that such was the underlying expectation of the democratic movement in Britain, America and France. In his *Fragment of Government* in 1776 Jeremy Bentham declared that: 'Men, let them but once clearly understand one another, will not be long ere they agree.' In the same year the fathers of American Independence not only proclaimed that 'all men are created equal' and that man's inalienable rights include 'life, liberty, and the pursuit of happiness', but added the even more startling assertion that such 'truths' were 'self-evident'. William Godwin, in 1793, remarked of truth that: 'The more it be discovered, especially that part of it which relates to man in society, the more simple and self-evident will it appear.' In the same year Condorcet formulated the same idea: 'Every correctly reasoning being will be led to the same conclusions, in ethics as in geometry.' Rousseau, in a revealing footnote to his *Contrat social,* noted the proviso that: 'To be general, a will need not *always* be unanimous'; and he added that for the general will of the community to find full expression 'each citizen should think only his own thoughts'.

This belief seems remarkable to a modern democrat accustomed to regard a free society as involving diversity of opinion, multiplicity of parties and associations, and a constant effort to find a working compromise between contrary viewpoints. It vitally affected the course of the revolutionary movement itself. It produced a certain fear of party politics, and a widespread distrust of political parties as mere factions hostile to freedom. . . . Universal suffrage was expected to make organized opposition unnecessary; and in politics, as in science, all honest and well-informed men would spontaneously arrive at a wide consensus of opinion. . . .

The democratic revolution was destined not only to change the purposes of government, to insist that government should express the will of the community as a whole and serve the material interests of the community as a whole. It also transformed the methods of government and administration—a revolution no less radical and no less important in its implications. No man with experience of eighteenth-century administration could help believing in the merits of rationalizing it. But reforms in the judicial and penal systems, in the fiscal and financial systems, in the institutions of administration itself, seemed increasingly certain to be frustrated without a prior political revolution. The utilitarian reforms which Jeremy Bentham and his disciples were to fight for and so largely achieve in Britain demanded, in most European states, some form of republican or at least representative government; failing that, they could be achieved only by the centralized authority of a Napoleon, liquidating the remnants of the *ancien régime* along with the excesses of the revolution. The methods of measurement, statistical analysis, political arithmetic or what Condorcet called *la mathématique sociale,* were introduced by the rationalist and democratic revolutionary movement, along with the ideals of republicanism. They were equally a consequence of the impact of science on the revolutionary movement; though Napoleon utilized these new resources of power to serve his own ambitions and despotism. They were nevertheless indispensable to any liberal-democratic regime existing to serve efficiently the welfare and interests of the whole community. . . .

The Natural Harmony of Interests

The conclusion to which all the considerations so far described irresistibly led the democratic revolutionaries was that in a 'natural society', just as there would be spontaneous unanimity about politics, so there would also be a natural harmony of interests. For a century to come no idea was to prove a more constant or a more revolutionary element in the thinking of liberal and social democrats alike than this many-sided concept.

The problem of how the interests of the community are related to the interests of all the individuals within it was raised by the novel task of designing a government to serve efficiently the needs of the whole community. Could individual interests be completely reconciled with the interests of society as a whole? If so, would individual interests be automatically harmonized in a free society, or must it be the task of legislation to achieve harmony artificially? A medley of different answers could be devised. But the whole impact of scientific thought was directed towards strengthening and encouraging assurance that a harmony could, in one way or another, be achieved. . . .

Just as Newton had discerned, behind the complex working of the universe, certain general laws or principles which governed the motion of all material bodies, and just as Lavoisier had revealed a new elemental pattern within the structure of nature, was it not likewise probable that great general principles of synthesis could be discovered behind the working of human society? Nature seemed to operate on many such laws, waiting only to be discovered by human intelligence and reason and used for the benefit of mankind. And in the notion of a general harmony of interests it was thought that such a 'law of nature' had been found.

In the context of the prevailing psychological theories inherited from Locke and Hume, individual interests were understood in terms of happiness. Individuals were assumed to seek their self-interest in the sense of desiring pleasure and avoiding pain. Such interests could be reconciled with the interests of others, and with those of society as a whole, in three possible ways. If one postulated a natural sympathy for others or a 'moral sense', one could argue that harmony came from the moral action of each individual in adjusting his conduct so as to promote the interests of the whole: and as Francis Hutcheson, the master of Adam Smith, put it, 'that action is best which accomplishes the greatest happiness for the greatest number'. Alternatively, one might deny the existence of such a personal 'moral sense', but argue that harmony is produced artificially by the action of the legislator, who by a rational and ingenious system of rewards and punishments so plays upon the impulses of self-interest in each individual that he will be induced to behave in a way which serves the interests of society as a whole. This was the view most characteristic of Jeremy Bentham and the school of philosophic radicals or utilitarians in Britain. Thirdly, in much more optimistic vein, one could deny that either the individual or the state had to manipulate individual behaviour so as to produce harmony; and maintain that harmony would emerge automatically from the natural reasonableness of free men living in a rational or 'natural' society. This was the view mainly urged by Adam Smith in relation to economic life, and by all later liberal *laissez-faire* theorists about social life in general. It postulated a natural identity of interests, and was expressed paradoxically by Bentham in the argument that if each individual were always to promote the interests of his neighbour at the expense of his own, humanity could not survive for a single moment. The operation of the principle of self-interest was, in short, the great cohesive principle, comparable to the law of gravitation or attraction of masses, which held society together, and society rested on a self-operating mechanism. . . .

It was in such a faith that early socialism was born. When Robert Owen in England planned his 'New Harmony' and looked to education to remould mankind into a completely new species of creature, he was drawing one set of inferences from the doctrine. When Charles Fourier in France depicted the future fantastic phase of Harmony, when lions would be replaced by anti-lions with all the qualities lacking in present lions, and the globe would be inhabited by totally different species of animal life, he was carrying the same idea to its extremest absurdities. His drastic criticisms of existing civilization, in which each person in economic life is at war with the mass, and commerce breeds a multitude of vices, are well known. Conflicts of individual interests derived from the competitive commercial system, in which producers profited most from supplying inferior goods to consumers. On this contention Fourier built up his plea for co-operative principles in social life and buttressed it by his theory of 'attractions'. The discovery of this theory, he claimed, entitled him to rank above Newton. God has so arranged our passions that they serve socially useful ends. The care of infants, otherwise repulsive, becomes attractive to a mother because she is endowed with 'a passionate attraction for this dirty work'. His projected phalanges were self-contained co-operative associations of workers, the units on which the era of Harmony would be based. When Proudhon, a possibly even greater influence on socialist thought than Fourier, propounded his theory of *mutualisme,* he placed himself in the same line of descent. His *système mutualiste* is a society which holds together its members by assuring 'service for service', and is in essence a voluntary, anarchistic version of co-operation.

To indicate the long-term revolutionary repercussions of the concept, it is necessary only to mention the names of Mazzini and of Woodrow Wilson. Mazzini, the champion of Italian unification through liberalism, believed in a sort of division of labour among nations. Each nation had special aptitudes, peculiar to itself, which fitted it to make a unique contribution to the welfare and progress of humanity. To contribute most to mankind and to international harmony, each nation had therefore to be true only to itself; and nationalism was automatically compatible with internationalism and indeed essential to it. Similarly, in 1919 President Woodrow Wilson, and many other liberal-democratic makers of the peace settlement, saw national self-determination as the key to world peace. Liberal-democratic revolutionary movements merged into nationalist revolutionary movements; and together they transformed the nature of European states and governments during the nineteenth century.

Charles C. Gillispie

The Natural History of Industry

In his recent award-winning book, Science and Polity in France at the End of the Old Regime *(Princeton, 1980), Charles C. Gillispie has provided a detailed analysis of the place of science in the economic life of eighteenth-century France. His major conclusions—that science played a* different *role in industrialization in France than it did in Britain, but that its role was nonetheless important—were anticipated in his brief article on "The Natural History of Trades" nearly twenty-five years earlier. In France, scientific work was both more highly centralized and more highly professionalized than in Britain, so there were few industrialists with scientific backgrounds and even fewer who actively sought to apply scientific methods to technological problems. In France, instead, professional state-supported scientists attempted to develop a scientific understanding of current industrial practice and then to spread knowledge of the best practices through government publications. Under these conditions it was in industries owned and encouraged by the state and in military engineering that scientific developments had their greatest economic impact in eighteenth-century France.*

In histories of eighteenth-century and imperial France, the assertion is often encountered that science was revolutionizing manufacturing, and Napoleon's encouragement of this process is frequently described as a major reason for the success of his industrial policy until the crisis of 1810. Historians have obviously drawn this judgement from contemporary writings by scientists themselves. That science ought to receive public credit for the unprecedented progress of the arts since the 1770s or 1780s is perhaps the most protean reflection in the technological literature of the time. . . . No one, it is true, specifies exactly how science was changing the face of industry. But knowledge is power; yesterday's banality is today's source material; and on the basis of all this authoritative testimony, the modern historian has very naturally supposed that theoretical science exerted a fructifying and even a causative influence in industrialization. The proposition is inherently persuasive, and there must be something in it.

The problem, however, is to know what there is in it. For if the question be approached in detail . . . it proves extraordinarily difficult to trace the course of any significant theoretical concept from abstract formulation to actual use in industrial operations. . . .

The true mode in which science was actively applied to industry has been obscured by the tendency of modern historical writers to suppose that the framing of basic theories is the main business of science, and that if science was related to industry, it must have been through the medium of advancing theory. They have failed to pay attention to the language of their texts, where the relationship of science to industry is not only clear but so clearly a corollary of the eighteenth-century conception of scientific explanation that it required no explicit formulation. For the eighteenth-century scientists do not write of the application of theory. What they say is that science illuminates the arts, that it enlightens the artisans, and that this process honours the century, and in holding this language they are simply considering industry in the light of [the] pervasive notion of the function of scientific explanation. . . . In this light, science itself is positive knowledge, of course. Its function in the world is essentially an educational one, however, and its mode of procedure is analytical. First, it seeks to discern the essential elements of a complex subject. These once found, it ranges and classifies them according to the logical connexions which subsist underneath all the welter of phenomena. Next, it establishes a systematic nomenclature designed to fix the thing in the name, fasten the idea to its object, and cement the memory to nature. In this fashion, the human understanding will be led descriptively towards a rational command over every department of nature by following its inherent order. Scientific explanation, then, consists in resolving a subject into its elements in the objective world in order that it may be reassembled in the mind according to the principles of the associationist psychology. The inspiration was algebra. But the model was botany.

It is consistent, therefore, that when scientists turned to industry, it was to describe the trades, to study the processes, and to classify the principles. In this taxonomic fashion science was indeed applied to industry, and very widely. What were those enormous ventures, the Academy's *Description des arts et métiers,* the *Encyclopédie* itself, the *Encyclopédie méthodique,* if not attempts to lift the arts and trades out of the slough of ignorant tradition and by rational description and classification to find them their rightful place within the great unity of human knowledge? The eighteenth-century application of science to industry, then, was little more and nothing less than the attempt to develop a natural history of industry. In this sense, the scientific development of an industry is measured, not by the degree to which new theory is used to change it, but by the extent to which science can explain it theoretically. 'We are frequently able', writes Berthollet (the quotation is from the first translation of his *Éléments de l'art de la teinture*), 'to explain the circumstances of an operation which we owe entirely to a blind practice, improved by the trials of many ages; we separate from it everything superfluous; we simplify what is complicated; and we employ analogy in transferring to one process what has been found useful in another. But there are still a great number of facts which we cannot explain, and which elude all theory: we must then content ourselves with detailing the processes of the art; not attempting idle explanations, but waiting till experience throws greater light upon the subject.' Similarly, the metal industries were not at first much changed by the development of the science of metallurgy; they simply began to be understood. But that processes will be altered for the better if their principles are understood, that artisans will improve their manipulations if they know the reasons for them, are simply illustrations of the eighteenth-century faith in progress through classification and industrial examples of the eighteenth-century belief in scientific explanation as a kind of cosmic education.

Accordingly, the revolutionary manufacture of saltpetre consisted essentially in subjecting the French people to mass instruction in a simple technical process. What Monge, Fourcroy Guyton, Berthollet, and the others did when they were brought into this enterprise, was to give popular courses. The military crisis of 1793–4 created the greatest practical incentive ever experienced for the application of science to production in general. The Committee of Public Safety urged it forward with all its fearful authority. And inevitably the project assumed an educational form; science was mobilized in defence of the Republic, and at great speed scientists produced a series of textbooks, instructing practitioners, not so much in new methods, as in the best methods. . . .

And there is ample confirming evidence that a new generation of scientifically instructed entrepreneurs and managers came into control of industrial operations during the Revolutionary period, and that the Revolution saw the culmination and the end of the very real hostility between scientists and artisans. As a result, it was no longer necessary to complain constantly of the obstruction of rational procedures by the ignorance and traditionalism of the ordinary artisan, always singled out as the greatest barrier to progress. Popular superstition was the *bête noire* of the rational writer, and whether he looked to religion or technology, he found it flourishing in ignorance and secrecy. To publicize processes, therefore, to get them out in the light of day, must be the business of science.

Neil McKendrick

The Role of Science in the Industrial Revolution

In England scientific activity remained both decentralized and largely amateur until well into the nineteenth century. Scientific literacy and attitudes were, as we have seen, widely spread through the middle levels of society. And the Baconian expectation that systematic experimentation should lead to material improvements, coupled with the general emphasis on quantification which pervaded Newtonian science, led to widespread and self-conscious attempts by artisans, engineers, and "improvers" in general to apply an empirical approach to a wide range of productive activities. Neil McKendrick characterizes this approach as "persistent," "exhaustive," "precise," "rational," "logical," "qualitative," and "economically motivated." It was such an approach, informed by the latest scientific knowledge, that led to Josiah Wedgwood's great success in producing commercially successful pottery. And it was such an approach that produced many of the dramatic technological innovations that underlay the Industrial Revolution.

The whole controversy over the contribution of science to economic growth in the late eighteenth century can be studied in microcosm in the career of Josiah Wedgwood. So many of the ingredients of the case for scientifically-induced economic growth are present in this case study—the industrialist's friendship with scientists, the mutual discussion of scientific and industrial problems, the unquestioned belief in experiment and the apparent familiarity with theory, the contribution of itinerant scientific lecturers, the central contribution of scientific societies, the purchase of scientific publications, the correspondence with scientists abroad, the publication of scientific papers, the production of scientific instruments—these and many other of the characteristics identified . . . as indicative of the fruitful rôle of science in industry are all prominently displayed. . . .

Wedgwood believed that industrial innovation should ride close on the heels of scientific discoveries, and any new idea which came his way was quickly scrutinized for its potential usefulness. . . .

Few potentially useful ideas were ignored. For Wedgwood was as quick to exploit the possibilities of scientific and technological developments as he was to exploit the potentialities of fluctuations in demand. A man who could market black basalt to exploit the female fashion for bleached white hands, a man who could produce piecrust ware in response to the Flour Tax of the Napoleonic Wars, was not a man to miss a commercial opportunity provided by a new technique or a new scientific theory.

Obviously such knowledge was not always commercially helpful in any direct sense, but a specific example of Wedgwood benefiting from the scientific work of others can be found in his Commonplace Book where he records his method of purifying cobalt. Cobalt was a vital colouring agent in Wedgwood's famous blue jasper, and a reliable method of procuring it in pure form was an important step in Wedgwood's ability to control the colour of his wares. . . .

Wedgwood often acquired new techniques from the chemical journals which provided him with technical short cuts in his experiments, but it is not always explicitly stated that the new process proved economically valuable at the production state. When he read

M. Achard's 'On the changes which different earths undergo, when mixed with metallic calces, and exposed to a fire of fusion' which was printed in 1781 in the *Memoirs of the Royal Academy of Sciences at Berlin,* he noted 'I separated the earths by precipitation with alcaline lye, from Solutions of the Salts which I had previously purified by repeated crystallisations'. But whether the new method proved worthy of adoption on a more permanent basis is not recorded. . . .

What is certain is that the possibilities of any new scientific technique or instrument were eagerly taken up and submitted to experiment: when he learnt of William Parker's burning lenses (which Priestley used in his famous experiment in which he isolated 'dephlogisticated air') he was quick to examine their potentialities for his own experiments and soon recorded a variety of 'Clays tried by Mr. Parker's Burning Glass'.

What is equally certain is that the possibilities of any experiment were optimistically explored for any useful application either to Wedgwood's business or to the world. 'May not Dr. Priestley's Experiments in producing pure air be applied to the improving or changing the air of sick rooms or hospitals? Possibly the fumes of vitriolic acid might answer this purpose, and precipitate the phlogistic matter in the state of sulphur'.

It is clear too that Wedgwood benefited industrially from the accurate control of temperature made possible by his pyrometer. Indeed, he clearly stated that his pursuit of an accurate method of measuring temperatures above the range of the mercury-in-glass thermometer . . . was inspired by industrial need. 'In a long course of experiments for the improvement of the manufacture I am enaged in, some of my greatest difficulties and perplexities have arisen from not being able to ascertain the heat to which the experiment pieces had been exposed.' . . .

Wedgwood's pyrometer continued to serve several generations of potters, and in a modified form it is still used in one large pottery firm today. Wedgwood's confidence in its utility proved to be as justified as his willingness to accept that it was not perfect. Perhaps more significant is that it derived from Wedgwood's wide reading of the earlier scientific work on the subject, that it originated in Wedgwood's industrial needs, and that it was sufficiently effective not merely to serve Wedgwood's production purposes, but also to meet the needs of chemists of the calibre of Priestley and Lavoisier. . . .

Obviously many of the major characteristics of the Industrial Revolution—the canals and the textile inventions—owed nothing to science; and science's contribution to the steam engine is still very much in doubt. But that does not invalidate a science-aided Industrial Revolution. Few serious scholars would now expect any *single* factor—certainly not science—to be a sufficient explanation of Britain's rapid economic growth. All models of growth . . . now include a multiplicity of causes, and the catalytic action of science could play a vital contributory rôle in that complex of causes—even if the science involved made itself felt less in terms of the straightforward transfer of scientific knowledge and more in terms of a growing interest and popularity of science, and a growing belief in the contribution that science could make to industry. . . . This contribution can be detected first—and most easily—in the language of assertion, motivation and endeavour: in the pervasive interest in science, in the attempt to harness it to industry, in the belief that 'all things yield to experiment'. It can be measured in the spread of the experimental method, and the growing acceptance of the idea that knowledge and progress would stem from research, logically organized and ruthlessly pursued and empirically tested. Wedgwood's career offers striking proof of the existence of eighteenth-century business men who passionately believed that 'progress would come by experiment, by discovering more facts, by measuring, by analysing; spurred on by the faith that progress *would* come, that the unknown would become knowable'. . . .

In fact Wedgwood's . . . writings reveal (like the other evidence in the Wedgwood archive) a . . . pattern of empiricism co-existing with a knowledge (if no profound grasp) of other scientists' theoretical findings, and with the empirical approach always dominant.

That the empiricism, and the pursuit of scientific knowledge to aid his experiments, were economically purposeful can be seen from his published papers as clearly as it can be discerned in his experiments. He firmly stated, in an address to the Royal Society, that 'the progress which the manufacturer makes in the improvement of his manufacture will generally be in proportion to his knowledge of the nature and properties of his raw materials, singly and compounded together'. Again in his 'Observations upon Coloured Stones' he began with the view that 'It would be of great importance in several arts and manufactures to find means of introducing beautiful colours into hard and durable materials' and he stressed that his 'attempt to investigate the means by which . . . Nature . . . fixes both plain colours, and infinite variegations, in bodies of the very hardest clay that she produces . . . [was] more than a matter of *mere* curiosity'. . . .

'If one can establish that an industrial leader like Wedgwood benefited in his ceramic inventions from both the stimulus of scientific theory and the short cuts of improved scientific techniques, then one can confidently claim that science influenced hundreds of his industrial competitors. For Wedgwood's industrial secrets—despite all his efforts to preserve them—were quickly stolen and successfully imitated. By such a process of linkage and diffusion through espionage and imitation, the contribution which chemistry made to Wedgwood would be transmitted throughout the potteries of Staffordshire—and with a growing timelag—eventually to the potteries of the rest of England and of Europe. . . .

One of the reasons why the British Industrial Revolution occurred without significant research costs, was that industrial espionage could so quickly carry a new technique or invention to the eagerly awaiting horde of followers and imitators. Men whose knowledge of science was minimal, whose interest was scarcely more, and whose capacity for sustained experiment (if it existed at all) was never exercised or explored, flourished because the fertile mind of Josiah Wedgwood kept them supplied with perhaps the most concentrated list of major ceramic breakthroughs ever achieved by one man. It would border on nihilism to suggest that Wedgwood could have achieved so many new ways of exploiting his native clays—green glaze, cream-colour, black basalt, jasper, to mention only the most famous—without the aid of either scientific methods or scientific knowledge. . . .

Wedgwood stood alone amongst Staffordshire potters in the quality of his experimental work, but the results were shared. So if one accepts that Wedgwood alone was influenced or aided by science, one can quickly multiply that scientific contribution to the economy on behalf of the whole industry. The queue of Wedgwood imitators ensured that what science had spawned would soon be utilized by the Potteries as a whole. They needed to know nothing of science, nothing of scientific method and technique. With the identification and proportion of the ingredients, and a bribed workman for the method, they could reproduce an approximate copy of Wedgwood's results.

Wedgwood's career, in fact, not only exemplifies the routes by which scientific knowledge was diffused by means of scientific societies, itinerant lecturers, and scientific correspondence; it also exemplifies the means by which the results of science were spread and released and multiplied throughout the industry as a whole.

Neil McKendrick

Josiah Wedgwood and Factory Discipline

Among the key process innovations that underlay the productivity increases of the Industrial Revolution was the rationalization of production in the factory setting. Wedgwood was an innovator in this field too; and Neil McKendrick uses Wedgwood to illuminate the emergence of a new emphasis on factory discipline.

Etruria served as a model to its own industry. It ushered in the factory system and the age of industrialization to the Potteries. But in basing his works on the division of labour Wedgwood merely followed and speeded up a process already in motion. In the middle of the eighteenth century the whole of the Potteries, together with industry in general, was moving towards greater specialization. Vital technical changes were altering the whole economic organization of the Staffordshire pot-banks. The family craftsman stage had already given way to the master potter with his journeymen and apprentices recruited from outside the family, and this in turn was becoming inadequate to deal with the growing complexity of potting production. New wares demanded new techniques: moulded ware required specialized block cutters, flat and hollow ware pressers, and casters. Specialization bred further improvement and variety: old shapes were reformed and old methods refined; new bodies were discovered, new glazes evolved, and new clays imported. Potters of limited means and restricted ambition began to find the range of products too large to handle, and the effort of reorganization too great to accept. For not only were new wares required to compete for the growing market, but new methods of production were needed to exploit this increasing demand. The simple pot-bank was no longer adequate.

Wedgwood refused to be deterred by the difficulties of reorganization. . . . He was determined to improve *his* methods of production. He was convinced from the outset that the only efficient means were the division of labour and the separation of different processes. Between 1767 and 1769 the plans for the layout of Etruria changed frequently, but Wedgwood never contemplated abandoning 'the scheme of keeping each workshop separate, which I have much set my heart on'. The 'Useful Works' and the 'Ornamental Works' were to be kept completely apart, each with their own kilns and their own sets of hands. At this stage he planned 'five portions', each to house a major stage in the production of earthenware. It was the plan that all potters would eventually follow. . . . 'The various shops were so arranged that in the course of its metamorphosis the clay travelled naturally in a circle from the ship house by the canal to the packing house by the canal: there was no carrying to and fro.'

He adopted the same system in planning the enamel works at Greek Street. His designs aimed at a conveyor belt progress through the works: the kiln room succeeded the painting room, the account room the kiln room, and the ware room the account room, so that there was a smooth progression from the ware being painted, to being fired, to being entered into the books, to being stored. Yet each process remained quite separate. He organized his men on the same basis, for he believed that 'the same hands cannot make *fine & coarse—expensive & cheap articles* so as to turn to any good account to the Master'. The 'fine figure Painters are another ord(e)r of beings' compared with the common 'flower painters' and must be treated accordingly—paid higher wages, set to work in a different workshop, and encouraged to specialize. His workmen were not allowed to wander at will from one task to another as the workmen did in the pre-Wedgwood potteries. They were trained to one particular task and they had to stick to it. Wedgwood felt that this was the only way to improve the quality of the ware—'We are preparing some hands to work at red & black . . . (ware) . . . *constantly & then we shall make them good,* there is no such thing as making now & then a few of any article to have them tolerable.' . . .

The analysis of workers at Etruria drawn up by Alexander Chisholm, Wedgwood's secretary and amanuensis, in the early 1790's, shows to what extent the division of labour had been developed. . . . Out of the 278 men, women and children that Wedgwood employed in June 1790, only five had no specified post. These five were listed simply as 'Odd men', the lowest in the hierarchy and the first to go in bad times. The rest were specialists. However humble their task they did it constantly, and therefore they did it well. To pretend as some do that the division of labour destroyed skill is to deny the superiority of Wedgwood's products over his rivals, and to sentimentalize the crude Staffordshire salt glaze of his predecessors. Division of labour did not destroy skill: it limited its field of expression to a particular task, but within those limits it increased it. . . .

This great change in the organization of labour had not occurred without difficulty. Having designed his system, Wedgwood had to train men to fit it, and to regiment them to exploit its potentialities. His twin task was, in his own words, first, 'to make *Artists* . . . (of) . . . mere *men'*, and second, to 'make such *machines* of the *Men* as cannot err'. . . .

Wedgwood had not only to train a new generation of skilled potters, he had also to mould these workers to the needs of his factory system. It was not an easy task, for he had centuries of local tradition to oppose him. The potters had enjoyed their independence too long to take kindly to the rules which Wedgwood attempted to enforce—the punctuality, the constant attendance, the fixed hours, the scrupulous standards of care and cleanliness, the avoidance of waste, the ban on drinking. They did not surrender easily. The stoppages for a wake or a fair or a three-day drinking spree were an accepted part of the potter's life—and they proved the most difficult to uproot. When they did work, they worked by rule of thumb; their methods of production were careless and uneconomical; and their working arrangements arbitrary, slipshod and unscientific. For they regarded the dirt, the inefficiency and the inevitable waste, which their methods involved, as the natural companions to pot-making. . . .

One of the major problems was to ensure prompt and regular attendance at the works. As was usual Wedgwood set himself off from the rest of the potters by introducing the bell. One of his early factories was known in fact as the Bell Works, because the workmen were summoned by ringing a bell instead of blowing a horn as was the custom in the district. Moreover, Wedgwood laid down precise times when it should be rung—the first warning at 5.45 or a '¼ of an hour before (the men) can see to work', again at 8.30 for breakfast, at nine to recall them and so on until 'the last bell when they can no longer see. . . .

He also outlined a scheme which can be compared only with a primitive clocking-in system. He wrote,

> To save the trouble of the porters going round, tickets may perhaps be used, in the following manner—Let some sheets of pasteboard paper be printed with the names of all the work people, and the names cut off, about the size of half a card. Let each person take two of these tickets with him when he leaves work every evening; one of which he is to deliver into a box when he goes through the lodge in the morning, and the other when he returns from dinner. The porter then, instead of going round the works in the morning, looks over these tickets only; & if he finds any deficiency, goes to such places only where the deficiency appears. If the persons have neglected or refused to deliver their tickets on going through, they are to be admonished the first time, the second time to pay a small fine to the poors box. . . . It will be necessary to have divisions for the tickets in alphabetic order, for the greater facility of giving them out.

To speed up this process, he considered further refinements, and proposed a list of 'all the names in alphabetic order on a board hung up in the lodge' to be marked by the porter with different coloured chalks to record the time of arrivals. He reinforced this system and prevented its abuse by imposing a fine of 2*s*. on 'any workman scaling the walls or Gates', and a similar 'forfeit' on 'any workman forseing (*sic*) their way through the lodge after the time alowed (*sic*) by the Master'. Wedgwood's intentions were clear from his factory. The most prominent feature of Etruria was the bell, the next—the clock. To judge from his workers' wage-sheets, on which were marked the time of arrival, of departure and the time they had for meals, his methods were fully successful. His workmen even organized a pay demand outside the main gates, promptly at half-past six.

His attention to cleanliness and the avoidance of waste was equally scrupulous. Apart from his fines—'any workman leaveing a fire in there (*sic*) rooms at night forfits 2*s*. 6*d*' and 'any workman leaveing there scraps in there rooms so as to get dirty forfits 2*s*.'—he introduced scales and a clerk to operate them. . . .

This clerk's duties rapidly grew. After weighing the clay he was 'to lay it up with as much cleanness as if it was intended for food'. He had further to supervise the scraping, the sponging, the breaking open and the careful examining of the clay 'for red or yellow veins & other foulnesses'. Each process must take place on a separate bench, cleaned beforehand and set 'before a good light'. It is almost needless to say that the *utmost cleanness* should be observed thro' out the whole slip & clay house—the floors kept clean—& (even) *the avenues leading to the slip & clay houses sho*[d] *be kept clean likewise*. And in sumer (*sic*) time when it is dusty, watered likewise.' To defeat evasion the clerk was to pay 'check visits irregularly' and search for dirt when he was least expected.

Every aspect of the potter's trade was covered in his instructions: there were separate directions for the throwers, the handlers, the pressers, the finishers and the dishmakers; and there were equally detailed comments on every object made—from the need to check the number of holes in the grate of a teapot spout to the need for heavy bases for salt cellars. All had to be checked by the overseer or Clerk of the Manufactory. . . .

His rules, however, were not for his benefit alone. There were, for example, strict precautions 'taken to avoid the pernicious effects of lead' poisoning. And his rules for dippers show his usual attention to detail: 'The dipping rooms to be cleaned out with a mop *never* brushed', 'A pail of water with soap and a towel & a brush for the nails to be always at hand', 'No one to be allowed to eat in the dipping room', 'The men and the boys to have an upper dress to throw off when they leave the room, for instance a sort of smock frock with long sleeves & open behind would be convenient', 'Some ware (such) as tiles require to have part of the surface freed from the glaze, this should be done with a sponge, & *not as now* by brushing when dry'. Such prophylactics are still the basis of industrial hygiene in the potteries today. . . .

By his own persistence, by an unfailing attention to detail, by founding, if not creating, the traditions of a foreman class and equipping it with rules and regulations, he transformed a collection of what in 1765 he called, 'dilatory drunken, idle, worthless workmen', into what ten years later he allowed to be, 'a very good sett of hands'. He never fully achieved the reformation he had hoped for. He never made 'such *Machines* of the *Men* as cannot Err', but he certainly produced a team of workmen who were cleaner, soberer, healthier, more careful, more punctual, more skilled and less wasteful than any other potter had produced before.

CHAPTER 8

Scientific Medicine and Social Statistics

Introduction by Russell C. Maulitz

In the world of everyday habits and familiar things, few events affect us as dramatically as illness. As individuals we may move between sickness and robustness with alarming swiftness. *Illness* can change overnight. But *medicine*—that whole pattern of ideas and people and machines with which we confront illness—seems to move at a glacial pace. Nothing much seems to change from one month, or even one year, to the next. It helps us, therefore, to take a longer view and to look back, say, over a span of twenty-five years.

Ask someone who recalls how medicine was practiced a mere twenty-five years ago. The differences between then and now are stark. They leap out at us from the scanner rooms of hospitals, the sick rooms of AIDS patients, and the accounting rooms of the Health Care Financing Administration. If we want to investigate the growth of medicine and public health over the past two hundred years, then, it may help to take a few "snapshots" at quarter-century intervals. A good place to start is the time between the two great revolutions of the late eighteenth century. Let us begin in 1783, after the American Revolution and before the French Revolution.

But at what sorts of "samples" of medical history shall we point our snapshot camera? The classic triad of doctor, patient, and disease is probably as good a focal point as any. And let us bear in mind that each of these elements of the medical system is a moving target. The medical profession itself, for example, has changed its complexion in more ways than one. "Physicians" were once almost all white, male practitioners who did not admit even the surgeons into their ranks. And doctors have not been the only individuals to provide health care. Public health, nursing, and apothecary personnel—often cooperating with one another—are but a few of the professional groups who historically have acted as care-givers and providers in our health care system.

It is well to remember also that people become "patients" when they either label themselves to be "sick," or are labeled "patients" by the society in which they live. Even today, individuals considered "patients" in one culture might not be given that tag in another. (Someone suffering from AIDS would probably be called "sick" and a "patient" almost anywhere, but what about someone with a "neurosis" or "dyslexia"?) And let us recall, finally, that disease itself has not been static over time, but sometimes appears as epidemics.

Can we postulate the existence of "new diseases": was Legionnaire's disease "new" in 1976, or AIDS five years later? Ultimately doctors and patients face the common enemy. But what is it? Illness? "Disease"? Ultimately individuals suffer "illness" if their bodies or minds malfunction. They call themselves (or we call them) "sick." Must there be a disease behind every such illness? Doctor, patient, and disease are all relative terms, therefore, in that their definitions change over time. Yet we have to remain mindful that behind every "arbitrary" designation there is an underlying biological or social reality.

1783

By the 1770s, looking back at most nations of the Old and New worlds, we can already discern some of the rudiments of the modern era in medical theory and practice. Increasingly, hospitals were becoming centers for medical education. Though the hospital was most prominent in the European teaching centers of Holland, England, and Scotland, it was also at least beginning to be consequential in larger American cities such as Philadelphia and Boston. Institutions resembling hospitals, primarily religious orders' "hospices," had been around since the early medieval period. But in the late eighteenth century three new factors propelled the hospital into a new and much more central role.

The first of these factors was the secularizing of the general society. The United States *began* as a secular culture, without an established religion, while France in 1783 was on the verge, if not of disestablishing religion, then certainly of at least diluting the influence of church and monarchy. Hospitals came increasingly under the sway of boards of lay governors, still later to be dominated by the doctors themselves. But even under lay administration, hospitals formed splendid professional environments for the scientific and professional advancement of doctors.

The rise of hospital medicine was also a function of a second force growing from within the medical profession, namely, the remarkable increase in power, prestige, and knowledge enjoyed by the surgeons. Men such as Pierre Desault in Paris, John Hunter in London, and the father-and-son John Warrens of Boston represented a new breed of elite surgeons. Standing side by side with their physician (read "internist") colleagues, they could lay claim to equal (if not greater) knowledge of the human body, equal (if different) competencies in the therapeutic realm, and equal access to the patronage of aristocratic patients. And, at least in the major cities, they plied their craft in hospitals.

Indeed, the cities themselves, by fostering the crowded conditions that promoted diseases such as tuberculosis, played a critical part—the third of our factors in the growth of hospital medicine. In the great cities of a world on the brink of industrialization, hordes of patients were crammed, two, three, and four to a bed, into poorly plumbed and ventilated places, places that were rapidly becoming veritable warehouses of disease. Great for the doctors to learn from. Not so great for leaving the hospital.

And who were these patients? For anyone of means, illness was still an intensely private matter, to be addressed, indeed *ministered to,* with infinite solicitude by a physician or surgeon in the comfort and confines of home. *Sickness,* the result of the person's peculiar constitution, was the province of the rich. With the exception of a very few illnesses such as smallpox, sickness remained just that—something privileged and belonging uniquely to individuals born to the "higher orders." *Disease,* on the other hand, was something depersonalized and, increasingly, highly specific in the way people thought about it. This thing, disease, was most of all the province of the poor. The poor were the specimens in those warehouses of disease, hospitals. Here disease itself, like the anatomy of the human body, could be laid open and probed for its secrets. So in 1783, disease—or at least the knowledge of disease and its inner secrets—began, too, to be the special province of the doctors.

1808

In 1808, medicine had become a unified profession in at least two parts of the Western cultural world, if for rather different reasons in each case. In America, most of the new nation was still frontier, with most doctors' schooling limited to their apprenticeships. Here there was little room for the academic niceties of colleges of physicians clashing with colleges of surgeons, a struggle then still going on in most of Europe. Except for France. In Paris, and in a couple of other French cities, the Revolutionary powers had drawn the surgeons and physicians together under one roof.

Why should we care how doctors were trained? Almost everywhere, after all, treatment remained "heroic," that is, based on radical medical and surgical interventions like bleeding and purging. The answer lies in a point already made: Each group, physicians and surgeons, had useful but quite proprietary outlooks on the way the body functioned—and malfunctioned. Surgeons cut things out, and saw the human body in "local" terms. Physicians gave medicines, and saw the body in humoral or "holistic" terms. A fusion of the two outlooks, eased by the unification of medical education in France, made it possible to "see" the body in a new way, as a collection of tissues, and thereby to diagnose disease in new ways.

Philippe Pinel (1745–1826), Xavier Bichat (1771–1802), and René Laënnec (1781–1826) all promoted this view of the body—a new physiology (normal anatomy and function) and a new pathology (abnormal structure and function). One member of this trio, Laënnec, inventor of the stethoscope, went on in the 1810s to extend the new pathology to diagnosis through the simple act of placing his hollow "cylinder" on the patient's chest. What did he hear? Likelier than not, he mostly heard the whooshing and grating sounds that reflected the ravages of tuberculosis. But always he heard the "airs, waters, and places" flowing through the tissues of the (usually) dying patient's belly and chest.

And thus, three-quarters of a century before the specific bacterial cause of tuberculosis would be disclosed, it was possible for the first time to take the new empirical evidence afforded by the senses—those senses aided for the first time by the technology of the stethoscope—and account for what one found through the novel concept of *specific* disease. "Phthisis" (read tuberculosis) was the *same, specific* disease whether it appeared in the belly or (classically) in the chest.

The range of new disease concepts arriving from France—not to mention the availability of bodies to dissect and even, just beginning, laboratories of medical chemistry—formed a powerful attraction. It was a magnet for people interested in the new medicine, enough to move the "nerve center" from Scotland, where the medical worthies of colonial America had studied, to the continent, particularly for English and American doctors-in-training. Thus, in the early decades of the new century, many individuals, lay and medical alike, at least those who were well placed and well educated, looked increasingly to Paris for the best of medicine. And by the 1820s a few of them were even beginning to talk about how to make it "scientific."

1833

Twenty-five years later Germany, under increasingly strong central government administration, was beginning to move into the limelight as well, though France still held the center. Americans continued to flock to Paris in record numbers. One of them, Jacob Bigelow (1786–1879) of Boston, returned to write his landmark work on "self-limited diseases," one of the earliest pleas for moderation in the treatment of illnesses that could, he argued, often cure themselves through what many regarded as the "healing power of nature."

Bigelow's notion of milder treatment caught on with many practitioners in the 1830s and 1840s. But why did doctors abandon heroic treatments? The answer is, partly, that some did not. Often they stuck to bleeding and purging for exactly the same reason that others moved away from it: the patients. Some patients had begun to leave the regular doctors in favor of their new competition, the "irregulars," a broad group of new practitioners that included homeopaths, naturopaths, botanists, and others. What most of these eclectic or irregular practitioners had in common was their dependence on less-arduous forms of treatment for patients. Some of the patients doubtless were tired of being bled within an ounce of their lives. (Others, on the other hand, probably derived quite a thrill from such strenuous therapies!)

The regular physicians, meanwhile, reacted to unaccustomed competition in either of two predictable (but opposite) ways. Some clung more firmly than ever to the old therapies. Others, rationalizing their actions as they went, justified a change to milder treatments, if only, some said, because disease itself was changing. So the way in which doctors thought about disease depended not just on science, but also on the pressures of actual practice—on their patients.

But science was unquestionably changing the face of medicine as well. In 1833 the microscope had just been moved from the curio shelf to the serious biological laboratory bench, though certainly not yet to the doctor's office. Better optics meant the opening up of the world of the small. Bichat's tissues could now be broken down into the tiny living cells, little metabolic factories, functional building blocks of life now recognized by German investigators like Matthias Schleiden (1804–1881) and Theodor Schwann (1810–1882).

But Schleiden, Schwann, and almost all of their co-workers were more the forerunners of today's biomedical scientists, men and women of the laboratory, than of today's superspecialists of the high-tech hospital. The cell theorists compel our interest not because their ideas immediately explained disease. Indeed, quite the contrary. To the cell biologists of the 1830s and 1840s, the living cell seemed to develop in ways rather more, say, like crystals than like human tissues or germs. Cells revealed next to nothing about why people got sick. But the early cell biologists were exceedingly important because they were exemplars of the new German emphasis on scientific medicine, and because they set the stage for developments in pathology and bacteriology that would come along in the next generation.

And finally, Europe and America in 1832 suddenly faced one of the most serious challenges for generations of pathology, bacteriology, and public health experts. It was a disease that, while not new, hit with special force in 1832–1833: cholera.

1858

By the middle of the nineteenth century industrialization had changed the face of all three parts of our system: doctors, patients, and disease. The seemingly simple issue of sanitation, in merchant ships, bustling ports, and most especially in the swelling cities, became ever more inflamed as living conditions became more crowded and workers' poverty took its awesome toll in bodies and lives. Cholera attacked again and again, leaving behind a stunning legacy of disarray and death.

One result, in almost every industrial nation during the 1840s through the 1860s, was an extraordinary step-up in attention to the needs of public health. As the loss of sound bodies put their worker-based economies at risk, governments became acutely aware of the rough translation process in which health and wealth were almost interchangeable—a relationship to which the German "medical police" advocate, Johann Peter Frank, had pointed out almost a century earlier. Now lay experts like Edwin Chadwick (1800–1890) in England and Lemuel Shattuck (1793–1859) in the United States made common cause with doctors like Southwood Smith in seeking "sanitary reform."

These "sanitarians" were a new breed and a new vanguard. They argued for broad-based measures to clean up the environment and correct the conditions leading to the spread of cholera and other diseases. In some ways they were the precursors of today's environmental activists, urging basic reforms such as sewage control, quarantine, and public health monitoring. Interestingly, what they urged *against* in many cases was any sort of germ theory of disease.

The germ theory, after all, was old hat, having been around since the middle of the sixteenth century. Ironically it was by now, at least according to the sanitarians, soundly discredited. Most people, after all, had never seen a germ. And those who had knew that germs were everywhere, in healthy bodies and healthy environments as well as unhealthy ones. What was more (and perhaps most important), blaming disease on germs seemed to divert attention away from the real "causes" of disease—now you could say: just get rid of the germ!—and hence to undermine the broad program of sanitary reform.

Even the German proponents of the new scientific medicine were, in the 1850s and 1860s, at best lukewarm toward the idea that the etiology (scientific cause) of diseases like tuberculosis and cholera could be explained in terms of tiny "ferments" or microorganisms. For a long time Rudolf Virchow (1821–1902), probably the most eminent German medical scientist of the 1800s, was, for example, more of a sanitarian than a germ theorist. But that did not deter Virchow, or Max Von Pettenkofer (1818–1901), or many others, from advancing scientific medicine in significant directions. In Virchow's case that new direction lay also in another part of the world of the microscopically very small.

In the year of this snapshot, 1858, Virchow published his lectures on *Cellular Pathology,* advancing the idea that all cells grow, not like crystals, but from preexisting cells. The implications were dramatic. Diseased tissues could now be *explained* in ways that made a great deal of sense to physicians, if not to the sufferer—also an important implication, because this helped alienate doctor from patient. A cancer cell could be "explained" as some sort of deranged daughter cell that originally, ultimately split off from a normal parent cell.

Not only did cellular pathology make the cell the center of anatomical study, it also allowed for practical results in the university or hospital laboratory. Tissue could be studied to differentiate, say, a cancer from a benign growth or a foreign body reaction, and appropriate steps could then be taken to treat one or the other. At the professional and organizational level, cellular pathology had an effect somewhat analogous to that of the stethoscope. It put intellectual and technical power in the doctor's hands, demystified the causes of disease, and foretold the growing importance of laboratory technology in the ensuing century.

1883

By 1883 repeated outbreaks of cholera and other infectious diseases had made some people—as many of them laypersons as doctors—suspect that germs had something to do after all with cholera in particular and, by extension, with most infectious diseases. One layman who had moved over from his native field of chemistry to investigate this idea, first in agriculture and later in medicine, was Louis Pasteur (1822–1895). As early as our last "snapshot," 1857–1858, Pasteur had performed experiments that suggested the importance of the little yeasts and "ferments" he was able to grow in his test tubes and glass dishes. Working "in the field," he showed that eliminating the spores of microorganisms could lead to the elimination of the epidemic they were supposed to cause.

Now, in 1882 and 1883, a canny German physician, Robert Koch (1843–1910)—soon to be quite a rival of Pasteur's!—extended the Frenchman's investigations by establishing beyond doubt the bacterial etiology of both tuberculosis and cholera. He did so in a way that impressed doctors and patients alike, using elegant laboratory methods to demonstrate that the organisms were essential for diseases to emerge in the host. Sometimes the "host" was a laboratory animal—or even a gelatin culture in a glass container—removing the logic of diagnosis from the bedside to the laboratory bench. But even if the "host" was the patient, the new science of microbiology further reinforced the scientific part of medicine, and thereby further separated patient and doctor into two different worlds.

By now, the 1880s, in their quest for scientific medical educations, Americans by the thousands and perhaps tens of thousands were veering away from Paris and turning instead to the Germanic capitals, to the laboratories of Virchow, Koch, and many others. What they found in Berlin, Vienna, and Leipzig was a federation of university-based physicians for whom the white coat and microscope were but the outer manifestations of the inner trappings of science. Doctors in practice were often of two minds about this. They did not want "scientists" running their business. But they knew that science gave medicine, in the eyes of their colleagues and, most of all, in the eyes of their patients, a transcendence and a legitimacy greater than any it had enjoyed before. And, lest we forget, one final fact was also not without importance: It worked!

1908

Just how well it worked had perhaps not been entirely clear in 1883. The germ theory and bacteriology meant a disease might be more easily diagnosed, but it still could not be treated definitively. Moreover, for a decade and more, many doctors and many others stuck to the sanitarian position that they had grown comfortable with, and that they *knew* worked. But beginning in the mid-1890s, and culminating in our final snapshot of 1908, therapeutic results began to emerge. The first, perhaps, can be said to have been the dramatic cure one Christmas eve of a young boy in Vienna, dying from diphtheria, using the serum therapy of Emil von Behring (1854–1917). Von Behring, like most of the other early innovators in serotherapy, could trace his scientific lineage to the German laboratories of Robert Koch and that earlier generation of 1883. So, too, could serotherapy's direct offspring, chemotherapy, typified in 1908 by the use of the new arsenical drug, salvarsan, in treating the last (after cholera and tuberculosis) of the great trio of infections, syphilis.

In the nineteenth century American doctors had never been ones to shrink from making an extra dollar by teaching the ever-eager (and ever-expanding) supply of new medical students. But now, for at least a decade before our final watershed year of 1908, Americans returning from Europe sought, with a fair degree of success, to implant both the professional values, and the technical value, of the new medical science on native soil. By the time of the 1910 Carnegie Foundation report on medical education, authored by the philanthropic gadfly (and non-physician) Abraham Flexner (1866–1959), medical educational reform was well under way.

Flexner's report, however, had a galvanizing effect on both the profession (hence, the doctors) and the public (hence, the patients). The resulting faith in scientific medicine was, in retrospect, even more remarkable in that at this point, before the First World War, outside of serotherapy and chemotherapy, it had not really yet *delivered*. Only much later, indeed probably not until after the *Second* World War, did the promise finally seem fulfilled. It was this *promise* of medical science that seems to have gripped the imagination of so many Progressive Era reformers, physicians, and lay people.

What may make the Flexner Report even more important for us, however, than its genesis, is the impact it has had ever since. We are still locked into virtually the same system for training doctors that Flexner's friends, fresh from Europe, put in place at the beginning of the twentieth century.

Still, in 1908, one *could* truthfully argue that, for the first time in human history, if a patient went to a doctor the odds that he or she would actually *benefit* from the encounter were better than even. One response to this, of course, is: so what! Patients had always gone to doctors, even when the latter could do little to alter the natural outcome of the disease. And now that doctors *could* help and actually make a difference, in the most general public health sense, it is not at all clear, as Thomas McKeown shows us in his reading, whether improvements in health in the past two centuries had anything much to do with doctors in the first place.

And yet the drumbeat continues. As Leighton E. Cluff's article indicates, doctors and patients in the late twentieth century continue to face new challenges. Some, like AIDS, are dramatic and command both our scientific attention and our empathy. Others, like the aging of our population, are more insidious, but make no less a claim on our concern. Patienthood and doctorhood are also changing, with one common denominator throughout being the increasing involvement of government in medicine.

In the United States, events in almost every decade since the 1930s have foretold these changes: the transformation of the Hygiene Laboratory into the National Institutes of Health (1930), the Second World War (1939–1945), Medicare (1965), and Diagnosis Related Groups (mid-1980s). Government involvement in research has been joined to involvement in the frontline provision of health care. Only one thing, in fact, seems certain: The medicine of twenty-five and fifty years from now will appear as different from that of today as today's does from that of a century, or even a quarter-century, ago.

Philippe Pinel
The Clinical Training of Doctors

Philippe Pinel (1745–1826) is best known for his efforts in treating mentally ill patients in Paris at the turn of the nineteenth century. But he was also a pioneer in proposing important reforms in French medical education. His 1793 essay, of which the following is an excerpt, presenting his ideas on everything from proper hospital design to patient care and medical education, was actually lost for many years. But it foresaw many of the key changes that other reformers put into practice within the next few years.

Medicine must be taught in the hospitals – The healing art should be taught only in hospitals: this assertion needs no proof. Only in the hospital can one follow the evolution and progress of several illnesses at the same time and study variations of the same disease in a number of patients. This is the only way to understand the true history of diseases. . . .

General hospitals inadequate for medical teaching – Are the requisite facilities available in general hospitals? These are filled with cases that cannot be precisely diagnosed and that are often examined in a cursory and incomplete manner. Even an exact diagnosis may lead nowhere, because of the superficiality and negligence with which the nature of diseases is established, on the basis of trivial and often misleading signs. In less than an hour's visit, the physician examines hundreds of general hospital patients who present vague and confusing data: he seems to parody rather than to practice medicine. . . .

Need for an appropriate teaching hospital – This is why all modern medical schools have emphasized observation as a characteristic aspect of medicine when it is viewed as a major branch of the natural sciences. Leyden, Edinburgh, Vienna, Pavia, etc., have stressed the need of selecting a small number of patients for didactic purposes and grouping them on teaching wards. This offers the advantage of focusing the students' attention on a small number of well-defined cases that they must watch with greatest care, without neglecting any aspect of cleanliness or health. . . .

Project for a teaching hospital: Proposal by the Society of Medicine – At a time when education is being restored in France and public instruction organized, the Society of Medicine turns its attention to a matter of supreme importance, the creation of teaching hospitals. Only clinical teaching can spread knowledge of the healing art in a uniform manner and restore the rigorous, oft-neglected principles of observation. All other pubic teaching of medicine by the lecture method is pointless and unproductive. . . .

Topography of the hospital site – Hippocrates always keenly felt the need for an exact topographic description of the place where a physician intends to practice medicine. In his treatise on *Airs, Waters, Places,* the Father of Medicine insists that the physician ascertain whether the inhabitants obtain their water from a stagnant pool, a running stream or a source, whether the land is hilly and the climate very hot, or at a great elevation and naturally cold. Hippocrates advocates a careful look at the food, the occupation, and the usual way of life of the patient. Thus enlightened, the physician will be well prepared to understand the true nature of endemic and prevalent seasonal diseases.

Main aspects of this topography – Progress in the natural sciences of course provides modern physicians with resources for topographic research that the Father of Medicine lacked. They can gather exact information on the elevation of a terrain above sea level, its latitude and longitude, the air currents and their direction, the rivers, the nature of the soil, the hot or cold mineral waters, the mines, quarries, swamps, lakes, woods and mountains, their location, elevation, and their productivity in the three realms of nature. . . .

Nosologic meteorology – Meteorologic observations were imperfect in Hippocrates' time. Nevertheless, this great physician never fails to use them in beginning his descriptions of epidemics. Sydenham pays little heed to the measurable qualities of the air such as pressure, temperature, or humidity. He insists only on those alterations that instruments cannot record, but that the nature of prevailing sicknesses may reveal. Other authors . . . give an exact account of atmospheric variations at the beginning of their nosologic observations, even though there is usually no relationship between the state of the air and the nature or course of disease. . . . It is in fact undeniable that sudden changes in atmospheric pressure, temperature, and humidity sometimes produce very noticeable effects and these are easier to study on several patients suffering from different diseases.

Meteorologic instruments for teaching wards – In order to render meteorologic observations as complete as possible, I propose that the physician take care to provide teaching wards with 1) a barometer; 2) a double thermometer, one outside, one inside; 3) a comparative hygrometer, for example that of M. Saussure; 4) an outdoor raingauge, to know the number of aqueous meteors during a given time; 5) an electrometer, since it is undeniable that atmospheric electricity acts on the animal economy; 6) an anemometer to record the direction of the winds; 7) a magnetic needle, to observe its variations. . . .

Influence of the air on physical health – The third book of Hippocrates' *Aphorisms* is almost entirely devoted to the different illnesses caused by seasonal variations. The illnesses do not always correspond to the same seasons. . . . It is clear that this part of medicine requires keen observation in a teaching hospital, great sagacity in distinguishing slight details, a ban on one's own imagination, and total reliance on sensory evidence.

Principles of contagion: are they found in the air? – In order to elucidate and confirm Sydenham's distinction between stationary and intercurrent fevers, one should compare the condition of patients, such as those assembled in a teaching hospital, with patients on the outside, subject to the same diseases. But, whatever the authority of a great name, it is difficult to believe that the principle of stationary fever resides in miasmata scattered in the air. On the contrary, all the facts concerning the propagation of plague, of smallpox, etc. indicate that their causative principles, far from being disseminated in the air, on the contrary adhere to furniture and clothing and are easily propagated by all kinds of threads, of cotton or wool. . . .

Questions to ask upon admitting a patient – But it will nevertheless be necessary, at the admission of each patient to the teaching ward, to elicit in an orderly manner information regarding all previous and present aspects of the patient's condition by a series of discriminating and methodical questions. The *Collegium Casuale* of Edinburgh proceeds in the following way: one records 1) information regarding details about the patient's age, sex, temperament, and profession; 2) a description of the symptoms at the moment treatment will begin, including i) manifest signs; ii) the external and internal symptoms enumerated by the patient; iii) the state of his principal functions such as pulse, temperature, respiration, and excretions; 3) recollections regarding the onset of illness and its course, including i) circumstances surrounding its beginning; ii) early symptoms; iii) their duration; 4) an evaluation of long-range causes that might have influenced the illness. That is to say one asks about i) the conjectures of the patient regarding such causes; ii) factors to which the patient may have been exposed before onset of the illness; iii) his previous state of health; iv) sickness among his parents or the persons with whom he habitually lives; 5) an account of the medication already administered, recording i) remedies used; ii) their effects; iii) the regimen and condition of the patient since the onset of illness; iv) the results to date.

Arrangements conducive to precise observations – It is obvious that medical observations can be precise and conclusive only if the evidence is reduced to the smallest possible number of facts and to the plainest data. This procedure is followed in all other branches of the natural sciences. To facilitate observation, nothing should interfere with the course of the illness nor the progress of nature. The arrangement of the clinical wards, the distribution of patients in their beds, diligent attention to cleanliness and salubrity, the administration of simple remedies whether plant or mineral, a meticulous choice of food and drink, an untiring supervision of students and their attempts to comfort the patients, all the details of the domestic service, all these must be supervised by the physician with the strictest circumspection and greatest enlightenment without hindering the healing process.

Structure of a teaching hospital: General views – The large number of plans published in recent years regarding the construction and internal disposition of hospitals obviate the need to enter into details on that subject. . . . It would suffice to add all the requirements for the public teaching of medicine to the general dispositions for service and treatment of patients. An assembly hall would be needed for the professor's regular lectures and for the discussion of specific aspects of therapy after medical rounds. In addition, space would be required for chemical analyses, anatomical dissections, and research in experimental physics applied to medicine. It is perhaps superfluous to discuss architectural details regarding the distribution of these different rooms in a teaching hospital, since these will always have to depend upon the general hospital building where the teaching wards are located and be modified by a variety of circumstances. . . .

Urgent need for teaching hospitals in France – We hope that the French nation, having just reconquered its most inalienable rights, will create a clinical school promptly. Such an establishment is an urgent and indispensable priority since one of the worst ills that can afflict humanity is ignorance and lack of principles on the part of physicians. We must therefore profit from the experience of other nations and also put the progress made in the other sciences to use in our new establishment. Good judgment and precision will render medical observations useful and conclusive.

Should teaching wards be located in a large or small hospital? – A fundamental question is whether teaching wards should be located in a large or small hospital. This question is solved for surgery: experience indicates that this part of medicine has flourished and progressed only in large hospitals. . . Indeed, small or medium-sized hospitals cannot provide a sufficiently large number of clearly characterized surgical cases such as serious and compound fractures, numerous varieties of hernia operations, kidney stones, aneurysms, lachrymal fistulas, etc. Students [at a small hospital] therefore cannot survey the total subject matter during their academic studies nor can they progress in surgery. Therefore I believe that a surgical teaching ward must be part of a large hospital in order to be an effective center of instruction.

But for medicine a medium-sized hospital . . . can be sufficient since it can always provide twenty or thirty well-defined illnesses. Also, the acute illnesses characteristic of each season are available there in sufficient variety to offer students correct and precise data. It is worth mentioning that the basic goal of a teaching hospital is to give students a good grasp of what is already known in medicine. The description of illnesses and all that concerns diagnosis and prognosis offers little ground for new research, but relies on repetitious references to the medical literature. . . .

Grouping of patients on teaching wards – However one might group and distribute illnesses, one cannot avoid being somewhat arbitrary. . . . I believe that nature suggests a distribution according to age and sex. Every age has, so to speak, its own way of life and sickness, and demands fundamentally different therapy for the same disease. This holds equally true for the two sexes. One should first establish two general divisions, one for men, the other for women. The first would be subdivided into 1) boys up to puberty; 2) adults to about fifty years of age; 3) men from the climacteric to senescence. The women's section would be subdivided in a similar way and would comprise 1) childhood up to menstruation; 2) the whole period of fertility, that is, from onset to the end of menstruation; 3) from menopause to what is called *femina effeta*. Each of these subdivisions might hold three or four patients, a total of eighteen to twenty-four, a maximum, if one wishes to avoid hasty judgment caused by an excess of work.

Subdivision of teaching wards to distribute patients – One must never lose sight of an important aim in teaching hospitals, which is that patients should be cared for as if they were at home. This applies to regular, clean, and prompt service, and to the isolation of patients so that nothing complicates nor deflects the course of the illness by causing new symptoms to appear. It is therefore important that not only the whole section of the teaching hospital reserved for women be separate from that for men, but also that everything related to kitchen, laundry, and pharmacy be entirely independent so that it can be subjected to strict supervision. I go even further: the three subdivisions of women cared for in the hospital should be indicated by three small rooms with three to four beds each, and I would further ask that each room be divided into compartments by partitions so that each bed is isolated and that the state of each patient cannot be aggravated either by the sight of infirmities that might be upsetting, or by odors and fetid smells, or by contagious miasmata. . . .

The best method to teach medicine and pitfalls to avoid – The true method to teach medicine is the one appropriate to all the natural sciences: focus the students' attention on concrete situations, impart high standards of accuracy for their perceptions and observations, warn them against hasty judgments and fanciful reasoning; choose readings that confirm their taste for rigor, and impose an orderly progression on their studies—in a word, train their judgment rather than their memory and inspire them with that noble enthusiasm for the healing art that masters all difficulties.

The clinical professor should not consider teaching of secondary importance, as so many doctors do in order to become fashionable and wealthy. He should devote himself to teaching as one of the most sacred and noble tasks to be fulfilled in society. The professor should have unlimited power of inspection over all patient needs and provide continuous and tireless supervision together with the hospital administrators, who alone manage the budget. His duties are far reaching since he must also, alone or with others, function as physician in the general hospital of which the teaching ward is a part, like in the Vienna or Edinburgh schools.

His detachment from any other commitment and the need constantly to follow the cold process of observation will help him avoid many errors that have plagued all teaching staffs in the past and have prevented the spread of sound ideas. Among these errors is the subservience to the opinions of famous men who are passionately dogmatic, filled with the ambition to lead a sect, and skilled at courting the admiration of the younger generation, so susceptible to blind devotion. . . .

Another source of corruption for medical teaching, from which we are now delivered, has been the power mediocre physicians have often derived from important posts and from the favor of kings. The servile disciple of Boerhaave, the commentator Van Swieten, first physician to the late Empress-Queen, wielded a brazen sceptre over all those who aspired to practice the healing art in the lands ruled by the House of Austria. He imposed a blind conformity to the Leyden school. The first physician Chirac exerted a powerful and unfortunate influence on French medicine: his writings contain the most dangerous principles, quite contrary to the sane maxims followed by observers of all ages. While he was still alive, adulation proclaimed him a legislator of medicine.

General drawbacks of the lecture method – A professor who seeks fame is flattered to see many disciples crowd around him, eager to collect all his precepts on the healing art like so many oracles. Therefore he rarely resists the sweet pleasure of lengthy discourse, and often tries to instill in his students a feeling of admiration that his conceit always turns to his own advantage. In contrast, the professor who is impervious to all the frivolous seductions of vanity seeks only to excite the devotion of the young and to enlighten them without any other reward than the knowledge of having done right. He will rarely indulge in a vain display of his knowledge, but rather point the students toward observation, teach them to perceive facts objectively rather than crush them under the weight of his superiority. He will give his lessons at the sickbed, emphasize the signs and symptoms that characterize the illness, indicate the elements of a correct diagnosis, their order and succession until the illness terminates. But these truths require some explanations, so that teaching can be attuned to the students' knowledge.

First year of academic studies – The first year of academic studies should be used chiefly for anatomy in the amphitheatres and for the study of some elementary works on physiology and hygiene. Indeed, the human body is basic to the doctor's occupation. He must know it well, its structure, the vital functions in the healthy state, and the general means of preserving health. These three functions seem to me inseparable. We know that anatomy is only learned well through the study and dissection of cadavers and by comparisons with zoology.

Physiology should not concern itself with vain hypotheses, but rather with the knowledge of structure as it relates to all parts of the human body. These should be compared with analogous parts in all kinds of animals. The animal economy must become very familiar and also the varied phenomena that characterize each function in the healthy state according to age, sex, and the special make-up of each individual. In order to acquire exact and precise knowledge of digestion, respiration, circulation, reproduction, etc., in the healthy state, one mentally reviews a sort of history of each function, without forcing the data into any preestablished system.

Physiology must not be divorced from hygiene: as one studies a specific function, one should think of the many ways in which the "things non-natural" might affect it. For instance, if one wishes to know the number of pulsations per minute in the radial artery of various individuals, one should study the variations that can be produced by hard liquor, violent exercise, sleep, or strong passion. Or, if one studies the phenomenon of digestion, one should examine the special characteristics of foodstuffs locally available and their propitious or damaging influence on various individuals. Observation would thus constantly complement reading. Training in the study of varied normal phenomena would aid in the subsequent observation of the pathologic process. Indeed, sound judgment is often based on comparison. Students would only be admitted to participate in medical rounds upon proof that they had successfully completed the duties of the first year.

Main studies in the second year – I think that second-year students should not be admitted to clinical lessons right away. They have not yet mastered the general principles of practical medicine nor acquired the habit of grasping pathologic symptoms. They have not yet reached the level that diagnosis and prognosis require. I would therefore propose that these students spend a great deal of time during their second year attending medical rounds on the general hospital wards. At the same time they should concentrate on studying the best elementary text of practical medicine, for example that of Cullen. One should also try to rid them as much as possible from the habit of seeing through someone else's eyes and trusting someone else's judgment. In order to train them to see for themselves, one might encourage them to follow three or four patients closely. They could write a history of these illnesses while profiting from the remarks of the physician on rounds or from those of some more advanced student. Thus they would learn to recognize external and manifest signs that indicate the nature and severity of illnesses. They would carefully record their observations. In order to stimulate rivalry in a sort of competition, one might encourage two or three students to write the history of the same case, individually and without comparing notes, as soon as their competence permits.

The reading of these case reports would take up part of the professor's class in practical medicine. This course is essential in a teaching hospital. The professor's presentation of the subject matter should be as brief as possible since the students can consult the required textbooks. The other, longer part of the class should be used for the reading and critique of the students' individual work. This would encourage strict habits and teach them to master difficulties. Such classes, or rather, such informal groups, could meet without the paraphernalia of pedantry and give burgeoning talent free play. The professor could enjoy the moving experience of watching passionate and well-motivated young men make rapid progress, unrestricted by sterile efforts at memorization.

Another way to further their progress is for the physician carefully to describe pathologic symptoms during his clinical rounds in the hospital. He would not only point out the various genera, but also similar species of illnesses, according to the seriousness of the case and the danger to the patient's life. He would, for example, train the students to compare the febrile and the healthy pulse; to count heartbeats per minute; to establish the correspondence between increasing and decreasing pulse and animal heat measured by the thermometer; to determine the relationship between variations in respiration and pulse, etc.; to recognize changes in hardness, weakness, intermittence of pulse that cannot be ascertained by instruments and that depend on the cultivation and delicacy of the sense of touch. In this manner the students will familiarize themselves with pathologic symptoms in the second year. They will not yet know how all diseases evolve. But they will begin to acquire the elementary notions basic to more continuous and difficult work. And that will be the scrupulous examination and rigorous observation of illnesses treated on the teaching wards.

Third year studies: structure of sickbed lessons – Students who have mastered the first two years of medical school will have the maturity needed for clinical lessons. They are worthy of learning from the clinical professor according to their ability and knowledge. He will be able to identify the particularly talented students who can assist him in the strict supervision and care of patients. For, of course, each patient shall have his student supervisor, like at the Vienna clinical school, or his assistant supervisor, like at Pavia. It is here especially that an important principle must be applied: always confront the student with difficulties that make him practice under the watchful eye of the professor who will rectify mistakes, fill the lacunae, and preside without fail over the prescription of remedies.

The most outstanding students in the first two years will thus be entrusted with the special care of one patient each. They will render an exact account of the origins and course of the illness since its onset, beginning with a picture that derives from the series of questions customary in the *Collegium Casuale* of Edinburgh. . . . Finally, they will proceed to a methodical examination of the patient in his present state. . . .

Means of training the students' judgment rather than their memory – The physician's visit to the clinical ward thus essentially consists of listening to the student in charge report on the illness of each patient. In the third year of medical school especially, the progress of the students' knowledge and the acquired habit of observation relieves the professor from going into a host of details. He can be content to remedy omissions, rectify errors, and offer new thoughts on difficult cases. Thus patient care can be prompt, attentive, and enlightened, and at the same time the students can gain experience, a solid foundation for further progress. Every case they observe will be fresh in their minds because it is grounded in concrete observations. It adds to their knowledge and illustrates nature's resources for the cure of diseases and, in certain cases, her errors and impotence. The students' attention is livelier and deeper when concentrated on one object at a time or at least only on similar cases found in the literature or elsewhere in the hospital. . . .

Some opinions regarding the structure of a teaching hospital – Much has been written these past few years about the building of hospitals, from the point of view of health and the efficiency of the service. I shall therefore limit myself to a few remarks regarding the establishment of a clinical school according to the preceding principles. If a new clinical hospital were to be built, I would propose a three-storey pavilion like those provided for a general hospital. . . . The ground floor would comprise a convalescent ward, a meeting room for the professor's lectures and for the sessions of the previously mentioned medical society, a well-chosen library of practical medicine, a chemical laboratory for simple pharmaceutical preparations, for the analysis of plant and animal substances, and apparatus for physics applicable to medicine. The third floor would serve as living quarters for the personnel and as storage rooms. There would be appropriate space for drying frequently used plants. A carefully kept and labelled herbarium would contain the most common plants, as well as those that need study to ascertain their genus and species. If found useful, they would be dried. Of equal importance are areas for the storage of foodstuffs such as apples, pears, grapes, etc., as well as dried fruit and starches such as rice, noodles, potatoes, various cereals, prunes, figs, etc. These provisions would diversify the patients' diet and even fulfill medical purposes. . . .

The second floor would be reserved entirely for patients. It would be divided into a men's and women's section, separated by a wall with one door to permit passage. These two divisions would have 12–15 rooms each, their doors opening onto a kind of well-aired dormitory. . . . In each division there would always be two or three empty rooms, in excess of patient requirements. These could be occupied whenever one wished to vacate a room for a few days to purify it, in order to lessen the risk of contagion.

The present sketch of a teaching hospital seems to me based on principles prized by medical observers throughout history. Their implementation will hasten progress in medicine along the lines followed by all the other natural sciences. Medicine has lost this advantage in France because it has lacked a solid basis for teaching. Also, permission to practice has been granted with shameful ease, after a few pedantic efforts of memorization. It is a crime against humanity that the most difficult practical science, constantly toying with human lives, should have become a totally venal profession, so numerous among us that the title of doctor is almost ridiculous. Only in a teaching hospital, under the eyes of a dedicated and enlightened professor, can eager young students pursue the exact observation of pathologic phenomena and investigate the complex factors that exert a more or less marked influence on the animal economy. What exquisite judgment, cultivated talents, and deep knowledge of ancient and modern medicine such a professor must possess to be equal to his task! I dare say that, only once in a hundred years, will such a supremely difficult position be filled in a worthy manner.

Lemuel Shattuck

A General Plan for the Promotion of Public and Personal Health

The "sanitarians" of the nineteenth century in Europe and the United States were individuals both inside and outside of the medical profession. They gathered statistics on epidemic diseases such as cholera, and they attempted to link the incidence of disease and death, especially among the working population, to the horrid environmental conditions prevailing in the rapidly growing industrial cities. Lemuel Shattuck (1793–1859), a layman and one of the founders of the American Statistical Association, published his General Plan *for a sanitary survey of Massachusetts in 1850. It is a good example of the fact-finding and statistical thinking that characterized the sanitarians' preventive efforts.*

We recommend that provision be made for obtaining observations of the atmospheric phenomena, on a systematic and uniform plan, at different stations in the Commonwealth.

The atmosphere or air which surrounds the earth is essential to all living things. Life and health depend upon it; and neither could exist without it. Its character is modified in various ways; but especially by temperature, weight, and composition; and each of these modifications have an important sanitary influence. . . .

The atmosphere is corrupted in various ways. Man himself cannot breathe the same air twice with impunity. Every minute of every day he appropriates to the vitalization of his blood 24 cubic inches of oxygen, and supplies its place with 24 inches of carbonic acid gas. When present in large quantities, from whatever cause produced, carbonic acid gas is destructive of life. Charcoal burned in a close room is an illustration. Some other gases are also very destructive. The experiments of Thenard and Dupuytren proved that birds perish when the vapors of sulphuretted hydrogen and ammonia exist in the atmosphere to the extent of a fifteenth thousandth part; that dogs are deprived of life when the air contains a thousandth part; and that man cannot live when the air he breathes is impregnated with a three-hundredth part; and suffers in corresponding degree when a less proportion of these poisonous gases exists. Persons frequently fall dead when entering a well, vault, tomb, sewer or other place, filled with these gases, or with stagnated air in which are diffused emanations from decomposing animal, vegetable or mineral substances.

Such are a few only of the facts which illustrate the important agency of the atmosphere in the animal economy. What that peculiar condition is which produces a specific disease, or what changes produce different diseases, are as yet unknown; it has not been ascertained, "because meteorological science, as connected with the propagation and spread of disease, is as yet in its infancy. We have, indeed, some knowledge of the influence of two of the obvious conditions, namely, those of heat and moisture; but of the action of the subtler agents, such as electricity and magnetism, the present state of science affords us little information. Still there are unequivocal indications that there is a relation between the conditions of the atmosphere and the outbreak and progress of epidemic diseases, though we are as yet ignorant of the nature of that relation."

"The earth, it is well known," says the Registrar General, "is surrounded by an atmosphere of organic matter, as well as of oxygen, nitrogen, carbonic acid, and watery vapor. This matter varies and is constantly undergoing transformations from organic into inorganic elements: it can neither be seen, weighed, nor measured. The chemists cannot yet test its qualities. Liebig, with all the appliances of the Giessen laboratory, cannot yet detect any difference between the pure air of the Alps, and the air through which the hound can tell a hare, a fox, or a man has passed; or the air which observation shows will produce small-pox, measles, scarlatina, hooping cough, dysentery, cholera, influenza, typhus, plague. These matters may either be in a state of vapor, that is elastic, or inelastic; or like water, they may exist in both states. They are most probably in the state of suspension; hang, like the smoke in cities, over the places in which they are produced, but are spread and driven about like vesicular water in clouds. A stream of aqueous vapour of the same elasticity from the Atlantic, passing over England, is, in one place, perfectly transparent; in another, mist; in another, rain: so clouds of epidemic matter may fleet over the country, and in one place pass harmless by, in another destroy thousands of lives. The emanations from the living, the graves, the slaughter-houses, the heaps of filth rotting, the Thames—into which the sewers still empty,—raise over London a canopy which is constantly pervaded by zymotic matters; in one season this, in another that, preponderating.". . .

After the above was written, the legislature passed the following "Resolve relating to meteorological observations."

"*Resolved,* That his excellency the governor be authorized and requested to fix upon suitable stations, not exceeding twelve in number, in which shall be included the three Normal Schools and the three Colleges in this Commonwealth, where shall be deposited the instruments necessary for making systematic observations in meteorology, according to the plan recommended by the Smithsonian Institute, at an expense not exceeding one hundred dollars for each station, to be defrayed from the school fund, and that he be authorized to draw his warrant therefor accordingly."

If suitable agents are appointed under this resolve, our recommendation can be fully carried out without further legislation.

We recommend that, as far as practicable, there be used in all sanitary investigations and regulations, a uniform nomenclature for the causes of death, and for the causes of disease.

In making a survey of different places, or different articles, it is proper that uniform names should be given to measures and weights; and that uniform instruments should be used. In a sanitary survey the causes of death and the causes of disease will be the principal objects of investigation; and it is expedient, and even necessary, that such names should be given to each as have a definite meaning and can be universally applied. . . . Hence the reason for the above recommendation in a plan for a sanitary survey of the State will be apparent.

A report containing a nomenclature and classification of the *causes of death* was drawn up, and adopted by the National Medical Convention in 1847. Extracts from a revised copy, approved by the Massachusetts Medical Society, are inserted in the appendix. We hope that the directions and suggestions they contain will be carefully observed by all physicians, and others concerned in carrying the sanitary laws of the State into effect.

The *causes of disease,* in all sanitary inquiries, deserve equal, if not greater attention, than disease itself. They have been differently classified and named by different authors. By some they have been divided into *external* or *extrinsic,* and *internal* or *intrinsic;* by others, into *principal* and *accessory;* and into *remote* and *proximate;* and in other ways. Copland, (Diction. Vol. I, page 645) divides them into four classes,—*predisposing, exciting, specific,* and *determining* or *consecutive* causes; and makes several sub-classes under each. Bigelow and Holmes (Marshall Hall's Practice of Medicine, Am. Ed. pp. 67–83) divide them into *general* and *specific* causes; and subdivide the former into *predisposing* and *exciting,* and the latter into *contagious* and *non-contagious*. Williams (Principles of Medicine, p. 23, Am. Ed.) divides them into *predisposing* and *exciting* causes; and makes a subdivision of the second into *cognisable* and *non-cognisable agents*. None of these classifications, however well they may be adapted for professional use, seem well designed for general sanitary purposes. They are not sufficiently clear to be generally understood and practically useful. Bigelow and Holmes say, this classification "must be considered convenient rather than strictly philosophical." Even Williams himself says that "these divisions of causes are rather conventional and convenient than natural and philosophical;" and every one who may examine them will probably come to the same conclusion. It is easy to perceive that one may be a predisposing cause in one case and an exciting cause in another; and vice versa, according to circumstances. . . .

We recommend that, in laying out new towns and villages, and in extending those already laid out, ample provision be made for a supply, in purity and abundance, of light, air, and water; for drainage and sewerage, for paving, and for cleanliness.

It is a remarkable fact, that nearly the whole increase of the population of Massachusetts, during the last twenty years, is to be found in cities and villages, and not in the rural districts. The tendency of our people seems to be towards social concentration. And it is well to inquire what will probably be the consequences of these central tendencies; and how, if evils are likely to arise from this cause, they may be avoided. It has been ascertained that the inhabitants of densely populated places generally deteriorate in vitality; and that, in the course of years, families frequently become extinct, unless recruited by a union with others from the country, or with other blood of greater vital force. This is a significant fact, which should be generally known. Cities are not necessarily unhealthy, but circumstances are permitted to exist, which make them so.

"Every population throws off insensibly an atmosphere of organic matter, excessively rare in country and town, but less rare in dense than in open districts; and this atmosphere hangs over cities like a light cloud, slowly spreading—driven about—falling—dispersed by the winds—washed down by showers. It is matter which *has lived,* is dead, has left the body, and is undergoing by oxidation decomposition into simpler than organic elements. The exhalations from sewers, churchyards, vaults, slaughterhouses, cesspools, commingle in this atmosphere, as polluted waters enter the Thames; and, notwithstanding the wonderful provisions of nature for the speedy oxydation of organic matter in water and air, accumulate, and the density of the poison (for in the transition of decay it is a poison) is sufficient to impress its destructive action on the living—to receive and impart the processes of zymotic principles—to connect by a subtle, sickly, deadly medium, the people agglomerated in narrow streets and courts, down which no wind blows, and upon which the sun seldom shines.

"It is to this cause that the high mortality of towns is to be ascribed; the people live in an atmosphere charged with decomposing matter, of vegetable and animal origin; in the open country it is diluted, scattered by the winds, oxydized in the sun; vegetation incorporates its elements; so that, though it were formed, proportionally to the population, in greater quantities than in towns, it would have comparatively less effect. The means of removing impurities in towns exist partially, and have produced admirable effects; but the most casual observation must convince any one that our streets were built by persons ignorant as well of the nature of the atmosphere, as of the mortality which has been proved to exist, and is referable to causes which, though invisible, are sufficiently evident.

"The occupations of men in towns are mostly carried on in-doors, often in crowded workshops, while the agricultural laborer spends the greater part of the daytime in the open air. From the nature of the particles of animal matter thrown into the atmosphere, it is impossible to place the artisan in circumstances as favorable as the laborer; the sun and wind destroy and waft away the breath as soon as it is formed; but in the workshops of towns the men are shut from the sun, and no streams of the surrounding air carry off the steaming breath and perspiration, so that the mortality of workingmen in the metropolis is much greater than the mortality of women at the corresponding ages."

Rudolph Virchow

On the New Advances in Pathology, with Reference to Public Health and Etiology

Rudolf Virchow (1821–1902) was one of the most influential scientists of the nineteenth century. In the late 1840s he wrote a definitive report on the devastating typhus epidemic in Silesia (Germany), gaining him prominence in the field of public health. He went on in the 1850s to apply the new cell theory to the science of microscopic pathology. In this lecture, given in 1867, Virchow described how new developments in scientific pathology were important for progress in combating such diseases as typhus and cholera, and for improving public health.

I have undertaken today to speak on the progress made in pathology. . . . The subject I propose to treat is perhaps the most difficult . . . because all other branches of the natural sciences stand in somewhat closer relation to one another than they do to medicine and specifically to pathology. All other sciences are a little easier to understand, all others have a larger number of common premises, a greater fund of common knowledge. Pathology lies in a much more remote sphere. Therefore, I must consider an attempt to present a review on . . . any various positive gains which pathology has attained in the last years as impossible from the outset, as it would take hours to clarify all those postulates. . . . During the general development of culture there occurred a strange phenomenon: whereas, not so very long ago medicine included, so to speak, all other natural sciences and was the carrier of the natural sciences, the healer being in fact a "physikus" of which the name only now bears witness, at the present time physics and all the more physical aspects of natural science assume conversely an attitude of superiority and look down on pathology and medicine in general. They contribute not a little to nourish in the layman prejudices which are not justified by the actual state of our science. This is, in part, due to a certain carelessness the members of the other disciplines permit themselves in their views on pathology, while they, on the other hand, place the highest demands on the medical man, whom they expect to be completely at home in all other spheres of natural manifestations and not to make a single mistake. Most of the natural scientists of other branches have not as clear an idea of the increasingly complex conditions of research in our field as they, on their part, demand from us on subjects in their field of study. If this were otherwise, medicine would not have lacked the essential support that it needs so as to develop in a manner corresponding to the other branches of science. . . .

Our aim in pathology, and especially mine, goes beyond the somewhat limited ends that Prof. Wundt has just discussed, i.e., the cell. True enough, ten years or more ago, I demanded that physicians should not only make the whole body the object of their considerations, but also its individual parts, in particular those which appear to be its ultimate independent units. But even then I never stated that this was to be the limit of our thinking and never meant to exclude from pathology the essentially physical mode of thinking. Quite the opposite. When I directed attention to the cell, I wished to coerce research workers to discover exactly the processes inside the cell, the happenings inside the smallest elementary organism, and it was self-evident that continued further investigations could have no other aim than to determine the physical and chemical foundations on which the manifestations of life and the functions of the cell are based.

At the present time, with research delving more deeply in other fields, we too feel the urge of thinking on physical lines, according to well understood principles of mechanics. Nevertheless, I must even now declare that the postulate of linking such thinking to the cell, the smallest elementary organism of which the human body is composed, will always hold true in pathology. However far man may advance in his understanding of the physical and chemical processes occurring inside these elementary organisms, no research will, to my mind, ever lead us beyond considering the cell as the real and essential basis of our medical understanding. For it is in the cell that we find a uniform expression of the processes of life, and the cell therefore seems to be the carrier of the uniform functions of life.

Let us compare these conditions with another example of wider application that Mr. von Pettenkofer has recommended to your attention. . . . Public health must necessarily be based to an increasing extent upon physical, chemical and on generally exact investigations. Not without cause has Mr. von Pettenkofer raised the question of whether the combination of public health with public pharmacology (though in itself a most natural relation) would not prejudice the public health services and hinder their development. This is the very question that is to be posed in respect to medicine in general. I can certainly agree in principle with Mr. von Pettenkofer's sentiments in appreciation of the fact that three Bavarian universities have placed public health under professors of chemistry, so that instead of the former partnership with public pharmaceutics, an alliance is now established with chemistry. But I cannot agree with him when he thinks it desirable that public health should always be dependent on the chairs for chemistry. On the contrary, I consider it useful and suitable that it should remain allied with pharmacology. It is natural and in order that the physicians appointed by the state to look after the public interest should be trained in both subjects. However, I cannot deny that nowadays the professors of chemistry are better informed on what is good and useful for public health than many a professor of pharmacology—this I will not hide—and from this I conclude that the latter must learn more chemistry and physics. I conclude in regard to them exactly what I do in regard to all physicians. I require, in fact, that all physicians should learn more on these two subjects. I ask that, from the outset, physics and chemistry, these two most important foundations not only of medicine but of all understanding, of all thought, of all human knowledge, should be taught more thoroughly in our universities. . . .

. . . I believe no one will be able to deny that no matter how much physics, chemistry, geology, meteorology or whatever else we wish to introduce into the question of health, it will ultimately always be one of human beings, of individuals. These are the essential factors. Physics may enable us to trace exterior processes right into the interior of human beings. It may be possible to investigate how a certain harmful influence that attacks the population on a large scale acts on the individual. But the health worker must always tell himself: I am dealing with individuals, the individual citizen is the object of my investigations. In exactly the same manner should the physician tackle, within the individual human body, the individual citizens belonging to that state, i.e. the single cells. The cell is not a whit less individual citizen, a legitimate representative of individual existence, than everyone of us claims to be in the body of human society, in our state such as it is. . . .

However, even if we agree on the principles of research, on a general point of view, pathology still presents another exceedingly great difficulty. Any area of natural knowledge that is on the verge of turning into a science usually begins by arranging the factual material already available in a way we call scientific. We make a classification, we make certain general divisions which are further subdivided at a later time. This, understandably, has also been the case with diseases. For a long time one of the serious tasks of the physician was to classify known material into scientific categories in as complete a manner as possible. For nearly the whole of the last century medicine spent itself in an attempt to make a classification which could be regarded as a durable basis. Even in this century some of the most prominent men, those who have exerted the greatest influence on their students, have invested all of their pride in creating systems of pathology. They have progressed from artificial to natural systems. It was attempted to establish in pathology a classification the equal of that existing in the other natural sciences. But no one has ever succeeded in creating a truly satisfactory system. With such a history it might be said that pathology has not even reached the level that the other natural sciences reached long ago, i.e. that of having an accepted scientific system on a generally understandable basis. This I must concede, but I can even confess with a certain feeling of pride that we have emancipated ourselves from these systems, that we have ceased to consider attempts at general classification as essential or profitable. We are done with all this and have thus thrown off the last remnants of the constraining shackles that were transmitted to us from previous generations, i.e. the dogmatic nature of knowledge, mere tradition. We have realized that some of the concepts which in an earlier time were the expression of honest conviction have in later times become nothing more than an obstruction to thought. The essence of dogma has always and everywhere been that at the time of its inception it is the highest expression of conviction and after a few centuries it has become the most burdensome of fetters for any further development. No other science has felt this more than has pathology and in none did it have longer after-effects. . . .

In our attempts to establish a system, we have stripped away the last of the dogmatic chains that fettered us, and now we stand as free in the field of research as all other natural sciences. I am proud to point out that, time and again, even now we have men in medicine capable of assisting other branches of the natural sciences and who are counted among the most eminent authorities in these other branches too. It is certainly to our honor that in our generation physics in particular should have received the most vigorous support from among medical master minds inasmuch as first class physicists have arisen from the ranks of medicine. The limits between pure science and applied medicine are not as sharp as perhaps many of you may think. Good will only is needed to make them gradually disappear. I would like to praise as one of the finest advantages of medicine the circumstance that a number of eminent men took the road back to applied science from physics, from physiology, and by this step gained the greatest benefits in both practical medicine and in their specialized field. May I only remind you of the incredible achievements in ophthalmology during the past 15 years which elevated it to the rank of one of the most prominent sciences that now exist. These advances were in part attained by the fact that representatives of physical physiology happened to be working in practical ophthalmology and did not disdain to aid and assist individual patients, nor did they cut short their activities at the very point where they began to be useful. This step taken with such success by the ophthalmologists, this action on their part, has lent truth to the old Baconian postulate which, as is well known, requires science to be useful. We do know, of course, that no kind of knowledge is useless, and that ultimately every piece of real knowledge can become of use. But there is a big difference indeed, whether the benefit due to a given scientific advance takes effect only after decades or perhaps even generations, or whether it is already felt within a few months; whether the men of science who discover a new law leave it to an eventual third party to use this law, or whether they themselves take the usually much lesser trouble of reflecting whether the matter could perhaps be of general benefit. This is what actually occurred in ophthalmology, where advances in knowledge were made in a few years such as, in my experience, have not occurred in any other field. This is mainly due to the circumstance that men of pure science have not disdained to also be men of applied and useful science.

If I now again return to the cell, it is . . . to show that there is no other field of pathology in which the cell has become the center of all thinking to the extent that it has in ophthalmology. In fact, we have here arrived at the point where we can directly observe the process in the various cells, where we can observe changes in single diseased cells in the eye itself during life, i.e. in the dark background of the eye, and can assess the number of diseased elements which induce the pathological processes, of those elements from the impairment of which there results what we later designate by the collective term "disease."

With this change in conception, by tracing disease to cell life, we have relinquished an important point which held sway over the science of former days and which most clearly necessitated classification. This is the idea of the "unity of disease," the concept that disease has to a certain extent a being of its own, a form of existence, which has invaded the body as something foreign and simultaneously independent and is perceived as something separate from the parts of the body. This concept has been abandoned gradually. Nobody nowadays conceives of disease and life as having a separate existence side by side. On the contrary, we now know that the concept of the corporeal nature of disease, of its material reality, could only arise from the circumstance that the various diseased parts, i.e. the diseased elements (cells) were not yet known and could not be demonstrated. The merely philosophic concept of the material existence of disease was developed from ontology by virtue of generalized abstraction, while we now demonstrate the actual material existence of disease in the living component elements of the body. . . .

. . . Pathological-anatomical studies are not the only means of knowledge. Let us examine how investigation has to be conducted. Scientific considerations must always be based on anatomy, no matter whether judgments are made at the bedside or at the autopsy table, because processes, functions, also relate to an anatomical basis. Every kind of change that we perceive occurs in a certain given part of the body; it is not present in the body everywhere in a generalized way, but at some spot, in a definite location. There it has its site, and from there it spreads. May the changes be ever so physiological and living, we yet have the obligation to refer them to a certain part of the body, to localize them in a certain focus of effectiveness. In this sense, I believe, the postulate of an anatomical basis does not mean that the sole basis for medical knowledge should be founded on results of pathological-anatomical studies. The clinician, too, must base himself on an anatomical foundation if he wants to think physiologically. However, he frequently does this only contemplatively, only in the mind, but not in reality, as does the pathological anatomist. We may require that processes in the anatomical sector should be studied physically or chemically, but the first requirement will remain that every physician should think anatomically and that he should investigate where the disease is localized.

However, we now see that there occur many morbid changes which not only induce changes at any one site but do so at many sites, in which four, five or six parts of the body are affected. But even in this case the disease is localized; never is the whole of the body affected. If we find that four, five or even more organs are diseased, we can no longer give a short anatomical designation. Such complicated situations cannot be expressed by anatomical means. Should we attempt in every case of disease to enumerate all the organs that are affected, the designations available would not suffice and the terms would have to be increased inordinately. That is why it has always been a natural need to name the simple diseases that are limited to single sites in the body after these sites, on the one hand, and, on the other hand, to choose another designation for the compound affections involving several organs at the same time. In doing this, a point of view that is most important in other fields too, i.e. the genetic point of view, would seem most appropriate. For everyone asks right away: how did this complicated process take place? What has caused the different organs to be changed in such and such a way? And this takes us to the subject of the causation of disease. It would no doubt be extremely useful if we could say in every case: this series of changes that we have before us is induced by such and such causes. In fact, whenever we clearly recognize the cause of a morbid condition we do not hesitate for a moment to name the whole complex of disturbances, no matter what their nature, after its cause. In this connection, I am able to quote one of the greatest recent advances in our knowledge, i.e. the recognition of the mode of action of many poisons, those poisons, in fact, the nature of whose action had been most doubtful and in respect to which a certain uncertainty of interpretation was most widespread.

The poison from which the modern trend toward anatomy received its impetus was phosphorus. Up to now cases of phosphorus poisoning, which are becoming so frequent, were thought to be due to a sort of cauterization affecting the stomach and the organs near to it. Now, however, the conviction is taking hold that cases of cautery are exceedingly rare and exceptional, and that phosphorus acts in an entirely different manner, mainly through minute changes in the cell. Many important organs, which may be widely separated, and not only the stomach which is directly affected, but also the liver, kidneys, heart and the muscles, show most characteristic changes, so characteristic indeed that mere anatomical inspection will entail suspicion of poisoning. The moment it had become clear that a poison which had previously been thought to act locally caused changes in a large number of organs, the moment these changes were found to be gross anatomical changes that could be referred with greatest ease and certainty to cellular elements, a new attitude prevailed in our comprehension of a number of poisons and of related substances which the blood carries and distributes to the various organs, where they induce an effect similar to that first recognized and clearly demonstrated for phosphorus. We thus recognize groups of changes that can be traced to certain causes. Beside the anatomical groups we have certain etiological groups, groups in which the alterations in the different organs may even fall into quite different spheres and need not even agree in nature, but may vary according to the nature of the organs affected, but the common origin of which, nevertheless, gives us the right to place the whole complex in one etiological group and to call it by one name.

In many other, partly very serious diseases, we have for a long time practised such grouping, without actually knowing their causes. Of cholera, typhus, smallpox, scarlet fever and a number of other most serious morbid conditions it was previously believed that they were of simple nature. It was thought that scarlet fever and smallpox were merely skin diseases, and that cholera and typhus were just intestinal diseases, while we now have become convinced that apart from the skin in scarlet fever and smallpox and apart from the intestine in cholera and typhus, a large number of organs are affected, not only in the later stages of the disease, but also in the very early ones. It follows that we are not dealing with the disease of a single organ but with a complex, with an etiological group of affections. Unfortunately, we still fall short in our understanding of their actual causes. We place cholera in a complex, not because we know exactly what the cause of cholera might be but because we observe that this complex is always reproduced in the same way, and always happens in the same manner. The proof we adduce is therefore purely empirical. We conclude that because it has always been so, it will so be in the future, without being able to say with certainty: it is this or that substance which is spreading and acting. Nothing is further from my mind than to attack on this occasion those who believe that the cholera agent has already been found. I personally have not had the opportunity to examine recent data. I abstain from judgment. But I cannot say that I have been convinced by the mere assertion that we have now caught hold of this causative agent. I only assert that the whole experience of time past, the history of categorizing our knowledge, has resulted in considering cholera as one unit and typhus as another unit, no matter what convictions we have regarding the causative agents of cholera or typhus.

But this sort of grouping, the assignment of a particular name to certain modifications found to be interconnected empirically is only provisional. Such grouping will only turn into an incontestable reality, will obtain importance, when we proceed from mere empirical grouping to the building of truly scientific grouping by advancing proofs of causative connection. The mere observation that a series of changes occurs simultaneously is not the confirmation of a single etiological thread running through these changes. For this reason I called provisional the classification of disease in groups, without proof of their etiological connections. This is an imperfection of medicine, of pathology. Nobody will deny it. Nevertheless, I do not think that any other branch of the natural sciences may derive from it the right to reproach pathology or medicine in general. On the contrary, I would say that this same method is applied in every other natural science. I mean the method of assembling certain observations in an empirical group and of giving it a name, reserving for a later time the search for the exact cause explaining these empirically found facts. It is indeed a huge and truly colossal advance just to have reached the point where we can make such empirical groups. When, as at the present time, we set up every year a certain number of new groups of this sort, and as I believe, set them up with full justification; when not a single year passes without our advance in pathology becoming manifest, in that we can tell the world: Here is a new disease!; when we are able to isolate some particular facts from the formless mass of facts, and are able to group them in a special framework of experience—as long, I say, as we are able to do this, it must be conceded that we are on as direct a path to our goals as anyone else and it must be admitted that we are increasingly posing those questions which will ultimately be solved by our continued observations and those of the coming generation. We do what is also done in other branches of natural science not by trying to erect a system in an effort to reconcile all of our conflicting information, but rather by analyzing our old knowledge and by establishing new areas of experience.

Max von Pettenkofer

Cholera: How to Prevent and Resist It

By the 1880s Louis Pasteur and others had begun to demonstrate in their chemical and bacteriological laboratories that infectious diseases were linked etiologically to tiny self-replicating microorganisms that were previously considered to be "innocent bystanders." Many scientists, however, continued to adhere at least in part to the ideas of the earlier sanitarian movement. Among them was the German hygienist Max von Pettenkofer (1818–1901), who continued to believe that several interacting factors were responsible for epidemic diseases. Von Pettenkofer's essay on cholera appeared in 1883, the same year the bacteriologist Robert Koch isolated the microorganism responsible for cholera.

Mode of Propagation of Cholera – Although there are many points relating to the causes of Cholera which are still involved in obscurity, there are others which have been ascertained with a certainty which leaves no doubt. The disease has existed in certain parts of India time out of mind, like Typhoid Fever among ourselves. When the Portuguese first arrived in India, at the commencement of the sixteenth century, after the discovery of the passage round the Cape of Good Hope, they found the disease there, and soon had an unfortunate experience of its epidemic virulence. From certain localities in India, at the present day, it spreads at times over a greater or less extent of the country. Europe was first visited by Cholera during the early part of the present century, a visitation generally and correctly regarded as a result of the increase, and more particularly of the more rapid intercourse between Europe and India.

Cholera was at first thought to be contagious, because on its first invasion of Europe it advanced, both by land and sea, along the principal lines of traffic. More accurate observation, however, soon showed that in Europe, just as in its native India, Cholera only flourished at particular times and in particular places. It was observed that not only, under exactly similar conditions of traffic, did certain localities suffer most unequally, and some not at all, but also that places susceptible to Cholera are so only at certain times; and, further, that some places are attacked more and others less frequently, although this difference could not be attributed to a difference in the traffic.

Hence it has been concluded that, in addition to the specific Cholera-germ, which originates in India, and is connected in some way or other with human traffic, there must be some other element, not existing within the human body, but connected with the geographical situation, which spreads the poison, and which is not present at all times nor in all places. This element it is which acts as the local stimulant in the development of the specific Cholera-germ after it has been transported to a locality through the agency of human traffic. . . .

The occurrence of Cholera, and its frequency, depend therefore essentially on the simultaneous co-operation of several, but chiefly of three, causes, viz., the traffic, the local and temporal disposition, and the individual disposition. If one of these factors be wanting, no matter which, there can be no outbreak of Cholera. In order, therefore, to protect ourselves against Cholera, we can work in each of these three directions. The result of our exertions will depend partly on our knowledge, and partly on the influence which we can bring to bear on these three factors. But these three factors must form the centre of every system devised to protect us against the disease.

Let us therefore get a clear idea of what

has been established with certainty in each of these three directions, and of what it is in our power to accomplish.

Traffic – This, the first point for consideration, is the most difficult. Free communication between place and place, and man and man, is such an advantage that we could not deprive ourselves of it even to be protected from Cholera and many other diseases. A restriction of traffic to such an extent that Cholera could not be spread by it, would be a far greater calamity than the disease itself, and the bloodiest wars would be waged to remove the restriction if once imposed. . . .

The fact that Cholera radiates more from infecting neighbourhoods than from infected individuals, is of the greatest practical importance. It is the most powerful incentive to a fearless attendance on the sick. No one has any grounds for fearing a Cholera-patient who may be under the same roof, and we may unharmed render him any service. If the house has already become a source of infection, the unaffected inmates will not gain the least protection no matter how carefully they may avoid the patient; on the other hand, if the house be not a centre of infection, and the patient has caught the disease somewhere else, the patient himself cannot be regarded as a source of infection in the house. . . .

Dirty Linen a Carrier of the Poison – It may be assumed as a fact that a dangerous quantity of the specific poison of Cholera can be conveyed to a place, previously free from the disease, by means of wearing-apparel, especially if moist and soiled, which has been for some time in a town or house or other place infected with Cholera. The disease thus conveyed may attack those who come into immediate contact with the imported articles,—not, however, because they have belonged to a Cholera-patient, but because they have come from a Cholera-district. . . .

Great care must therefore be taken to avoid sending such articles, while dirty and not disinfected, from houses in which Cholera has been; and those people should also be very careful who have to receive them. Parcels of such things, for instance, should not be opened in the house, but in the open air. . . .

Other Carriers of the Poison – There are many cases on record which point almost decisively to the conclusion that damp, very watery, and slimy articles of food which have been in a Cholera-house, or other centre of infection, are especially likely to take up enough of the poison produced in the infected place to cause the disease in an unaffected place, if they are consumed without being thoroughly purified or re-cooked. . . .

The Excrements as Carriers of the Poison – In considering the influence of traffic on the spread of the disease, attention has hitherto been generally, if not exclusively, confined to the excretions of the patients, and more especially to the intestinal excretions. When it had been established as a fact that Cholera is spread through the agency of traffic, it became necessary to regard that agency as fixed and localized in something. It was considered the most logical course to assume that the specific poison of the disease is localized in the intestinal excretions, which are so prominent among the phenomena of the disease. Besides, facts gradually induced the belief that the fresh excretions of Cholera-patients are not dangerous, but only such as have become decomposed. The latest investigations, especially in India, the home of Cholera, have not strengthened the belief in the correctness of localizing the poison in the excretions, but, on the contrary, have considerably diminished that belief. Science is at present investigating with increased attention the modes of propagation of Cholera, and is less concerned to establish the various localizations of the poison on theoretical grounds than on the evidence of facts. This state of our knowledge does not, however, in the least justify us in disregarding the intestinal secretions, as being of no importance with regard to Cholera-epidemics; on the contrary, experience has afforded many reasons for attributing to careless management of the excretions an injurious influence over the local and temporal disposition to Cholera. . . .

Local and Temporal Disposition – Since Cholera has been watched in its course over large tracts of country, another influence on its epidemic extension in addition to that of traffic (importation) has become equally noticeable, and often even more so, viz., that of time and place. Hence many have been induced to overlook or even deny the influence, or at least the necessity, of traffic from place to place for the spread of Cholera. Not long after the first invasion of Europe by Cholera, which lasted from 1831 to 1837, there was a period when the great majority of medical men denied the transportability of Cholera from one locality to another (which then as now was confounded with contagiousness), because it was supposed that if importation were admitted, contagiousness must necessarily be implied. After Cholera had been at first regarded and treated as essentially a contagious disease, owing to the evidence of its importability which was noticed during its advance from Asia through Russia, along the lines of traffic, a great reaction occurred in public opinion in the opposite direction. This reaction occurred of necessity after all the costly regulations founded on the contagionistic theory, such as military cordons and quarantine, had proved useless. It was not until the reappearance of Cholera in Europe in 1848 that people presumed again by degrees to believe in the influence of traffic. But they soon again fell into the former extreme of belief in contagiousness, and again overlooked for a period the essential influence of time and place. Now at length science has opened up a new path, which does not start from *à priori* theories, and pass by all those facts which are irreconcilable with it, but one which proceeds from ascertained facts, assigns to each its due weight and leads step by step to the goal.

That the extension of Cholera does not depend on traffic alone, but also on local causes, is seen in every country subjected to epidemics of the disease. If we inspect a good map on which those places where deaths from Cholera, and especially where epidemics have occurred, are separately and distinctly marked, we shall always and everywhere be struck with the fact that the places attacked by epidemics are *not* grouped around the principal lines of traffic, but according to geographical position; *e.g.*, that the epidemics of Cholera are most unequally distributed along lines of railway. Hence it appears most evidently that traffic alone is insufficient for the propagation of the disease. In every country the localities subject to epidemic Cholera are grouped far more according to river and drainage districts, than according to the chief lines of traffic. . . .

Ground-Water – Just as in certain places the existence of typhoid fever exhibits a certain temporal dependence on the variations of the moisture of the soil, of the so-called "Groundwater," in like manner it is probable that Cholera is similarly dependent, though, owing to the fortunately less frequent prevalence of Cholera, the relationship cannot be so regularly and satisfactorily proved as in the case of Typhoid Fever. Since 1856, observations of the variations of the Ground-water have been regularly made in Munich, and are ready for comparison during the whole of this period of sixteen years. [See Chart.] . . .

It is not true that *every* place where Typhoid Fever occurs is equally susceptible to Cholera; although it is a fact well known that those places which are liable to epidemics of Typhoid Fever are especially liable to outbreaks of Cholera. The more contaminated, the more impregnated the soil of a place is, the more does it favour outbreaks of Typhoid Fever and of Cholera. . . .

The Cholera-Commission (Bavarian) of 1854 ascertained certain facts from their investigation of the circumstances of the epidemic which occurred in that year, and they are well worthy of attention now. Houses and localities situated in hollows, and especially those at the lowest point of hollows, or on terraces, or steppe-like formations close to the foot of declivities, showed in the great majority of cases a far greater disposition for Cholera than houses and localities situated on the summit of a ridge between two hollows, or further from the foot of declivities and sloping ground. . . .

Surveillance of the Cases of Diarrhœa – The transudation of large quantities of water from the various organs of the body, through the lining membrane of the stomach and intestines, is one of the principal symptoms of Cholera. Hence it may be laid down as the first preventative measure to be adopted, to avoid everything which experience has shown to be liable to cause vomiting or diarrhœa. It is a

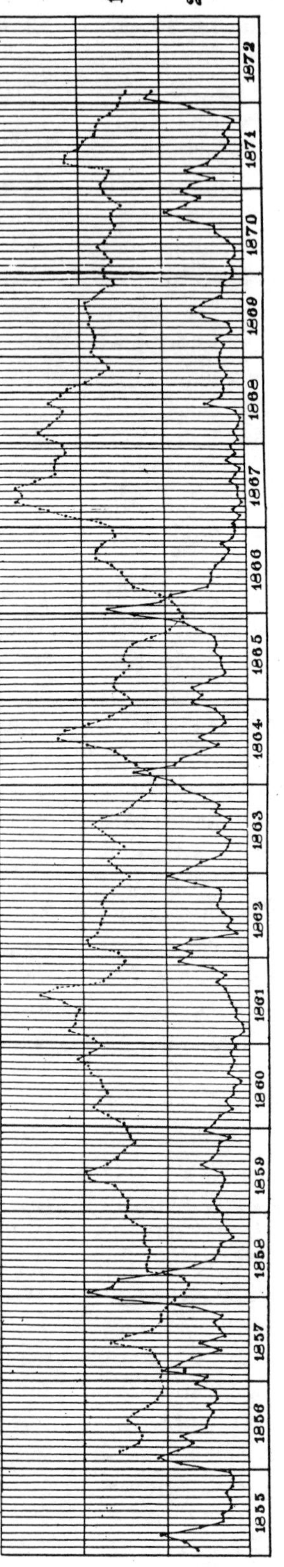

A CHART
SHOWING THE VARIATION IN THE
GROUND WATER AND IN THE MORTALITY
FROM TYPHOID FEVER IN
MUNICH.

1.—VARIATION IN THE GROUND WATER.
2.—MORTALITY FROM TYPHOID FEVER

matter of experience that almost every case of Cholera is preceded by a more or less severe attack of what is called choleraic diarrhœa, or cholerine, which lasts a shorter or longer time, and suddenly turns into severe Cholera. At the time of an epidemic, undoubtedly a large number of cases of choleraic diarrhœa occur which never turn into Cholera, and which cease without leaving any bad effects, and this even without medical treatment. But it is equally certain that those cases of diarrhœa which are subjected in time to proper medical attention scarcely ever become Cholera, whereas most cases of Cholera are the result of neglected diarrhœa. This is a fact which is independent of any theory.

Thomas McKeown

Infectious Diseases

Thomas McKeown is an English medical scientist whose specialty is social medicine and epidemiology. He has made broad investigations of the historical occurrence of infectious diseases. In this selection McKeown stresses the notion that the incidence of several such diseases began to decline before medical science intervened to prevent their spread. He points out that the isolation of the microorganism responsible for the disease is of importance, but equally so is the relationship among the host (individual), the parasite (organism), and the environment. Further, he emphasizes the central role of nutrition in the history of disease.

I have concluded that the fall of mortality since the end of the seventeenth century was due predominantly to a reduction of deaths from infectious diseases. . . . Why then did the infections decline from about the time of the modern agricultural and industrial revolutions which led to the aggregation of still larger and more densely packed populations? The answer to this paradox must be sought in the character of micro-organisms, the conditions under which they spread and the response of the human host, inherited or acquired.

However, for an understanding of the infections it is unsatisfactory to consider separately an organism and its host. They are living things which interact and adapt to each other by natural selection. The virulence of an organism is not, therefore, a distinct character like its size or shape: it is an expression of an interaction between a particular organism and a particular host. For example, a measles virus, whose effects on children in a developed country are relatively benign, may have devastating effects when encountered by a population for the first time. When assessing the major influences on the infections it will therefore be necessary to distinguish between the following:

(*a*) Interaction between organism and host. When exposed to micro-organisms over a period of time, the hosts gain through natural selection an intrinsic resistance which is genetically determined. In addition to this intrinsic resistance immunity may also be acquired, by transmission from the mother or in response to a post-natal infection. These types of immunity, inherited and acquired, are not due to either medical intervention or, as a rule, to identifiable environmental influences.

(*b*) Immunization and therapy. Immunity may also result from successful immunization, and the outcome of an established infection may be modified by therapy.

(*c*) Modes of spread. These are very different for different micro-organisms, and the feasibility of control by preventing contact with an organism is determined largely by the way it is transmitted. In a developed country it is relatively easy to stop the spread of cholera by purification of water; it is more difficult to control salmonella infection by supervision of food-handling; and at present it is impossible to eliminate an airborne infection such as the common cold by preventing exposure to the virus.

(*d*) The nutrition of the host. The results of an encounter with a micro-organism are influenced not only by the inherited or acquired immunity of the host, but also by his general state of health determined particularly, it will be suggested, by nutrition.

This classification provides a basis for an analysis of reasons for the decline of infectious diseases. It is against the background of an understanding of the interaction between organism and host that we must consider the possibility that the decline was due substantially to a change in the character of the diseases, essentially independent of both medical intervention and identifiable environmental (including nutritional) improvements. It is in relation to immunization and therapy that we must assess the contribution of specific medical measures. A judgement on the significance of reduction of exposure to infection must rest on understanding of the modes of spread of micro-organisms. And an estimate of the importance of an increase in food supplies requires appraisal of the association between malnutrition and infection.

Was the decline of the infections during the past few centuries associated with a change in the character of the diseases, that is, with modification of the relation between the micro-organisms and their hosts? Such a change is not independent of the environment; indeed it is determined largely by an ecological relationship to the environment. It is, however, of a kind which must have occurred continuously during man's history.

It has been suggested that a change of this type was important, and even that it was the main reason for the decline of mortality and improvement in health. Greenwood, for example, emphasized the importance of the 'ever-varying state of the immunological constitution of the herd' [population] and . . . Magill wrote [in 1955]: 'It would seem to be a more logical conclusion that during recent years, quite regardless of our therapeutic efforts, a state of relative equilibrium has established itself between the microbes and the "ever-varying state of the immunological constitution of the herd"—a relative equilibrium which will continue, perhaps, just as long as it is not disturbed, unduly, by biological events.' According to this interpretation, the trend of mortality from infectious diseases was essentially independent of both medical intervention and the vast economic and social developments of the past three centuries.

The grounds on which it was possible to reach so radical a conclusion are important. Magill based his views on the ineffectiveness and dangers of vaccination against rabies, the decline of tuberculosis long before effective treatment, the behaviour of diphtheria in the nineteenth century (it increased in prevalence and malignancy in the middle of the century and declined before the introduction of antitoxin), and the rapid reduction of pneumonia death-rates in New York State before the 'miracle' drugs were known, followed by an arrest of the decline from about the time when antibiotics were introduced. Moreover, these examples could be extended: the cholera vaccine required until recently by international regulations is almost useless; the reduction of mortality from diphtheria in the 1940s did not everywhere coincide with the introduction of immunization; and scarlet fever has had a variable history which appears to have been independent of medical and other influences.

Nevertheless, although specific measures had little effect on the trend of many infections, the question concerning the significance of changes in the character of the diseases is complex. It will therefore be desirable to examine the implications of the suggestion that the decline of mortality was due substantially to a favourable change in the 'ever-varying state of the immunological constitution of the herd'.

The immunological constitution of a generation is influenced largely by the mortality experience of those which precede it. This was particularly true in the past, when the majority of live-born people died from infectious diseases without reproducing. Under such conditions there was rigorous natural selection in respect of immunity to infection. The proposal that the decline of mortality resulted from a change in the immunological constitution of the population therefore implies that there was heavy mortality at an earlier period which led to the birth of individuals who were genetically less susceptible. According to this interpretation, the substantial and prolonged decline of infectious deaths was due largely, not to improvements since the eighteenth century, but to high mortality which must have preceded it. . . .

. . . Mortality from all the diseases was declining before, and in most cases long before, effective [immunization] procedures became available. It is doubtful whether a reliable estimate can be made of the effect of medical intervention on the whole class of airborne diseases, but it is probably safe to conclude that they were not the main influence on the trend of mortality even from the time when immunization or treatment was introduced, except in the case of tuberculosis and diphtheria. . . .

It is unlikely that treatment had any appreciable effect on the outcome of the [diarrheal] diseases before the use of intravenous therapy in the 1930s, by which time 95 per cent of the improvement had occurred. For the main explanation of the fall of mortality we must look to the hygienic measures which reduced exposure. . . .

Clearly, at least part of the decline of mortality from infectious diseases was due to reduced contact with micro-organisms. In developed countries an individual no longer meets the cholera vibrio, he is rarely exposed to the typhoid organism and he is infected by the tubercle bacillus much less often than in the past. But so far as can be judged there has been no considerable change in frequency of exposure to the streptococcus or the measles virus, and we must look elsewhere for an explanation of the decline of deaths from scarlet fever and measles.

The possibility of control of transmission of micro-organisms is determined largely by the ways they are spread. It is relatively easy (in developed countries) to prevent exposure to water-borne diseases; it is more difficult to control those spread by food, personal contact, and animal vectors; and it is usually impossible to prevent transmission of airborne infections. . . .

If the decline of mortality from infectious diseases was not due to a change in their character, and owed little to reduced exposure to micro-organisms before the second half of the nineteenth century or to immunization and therapy before the twentieth, the possibility that remains is that the response to infections was modified by an advance in man's health brought about by improved nutrition.

It should be said at once that there is no direct evidence that nutrition improved in the eighteenth and early nineteenth centuries. Evidence which could be regarded as convincing would be an increase in *per capita* food consumption or clinical observations showing an improvement in clinical state. These data do not exist, and few historical questions of such complexity would ever be resolved if they could be settled only by contemporary evidence of this kind. The case for the significance of nutrition is circumstantial. In this it is like the case for the origin of species by natural selection, which so far lacks confirmation by experimental production of a new species. It is nevertheless a convincing hypothesis because it has stood the test of critical examination in a variety of circumstances over an extended period. It is to a similar test that the suggestion concerning nutrition should be submitted.

The grounds for regarding better nutrition as the first and main reason for the reduction of infectious deaths are three-fold: this explanation is consistent with present-day experience of the relationship between malnutrition and infection; it accounts for the fall of mortality and growth of populations in many countries at about the same time; and when extended to include improved hygiene and limitation of numbers, it attributes the decline of the infections to modification of the conditions which led to their predominance. . . .

The rapid growth of populations in a large number of countries which differed in economic and other conditions (the United States, England and Ireland are remarkable examples) has led some historians to conclude that no single explanation is likely to be adequate. The opposite conclusion seems more plausible: the widespread expansion of numbers in many countries at about the same time in spite of variation in circumstances suggests the possible operation of a common major change.

The increase in food supplies which resulted from advances in agriculture and transport in the eighteenth and nineteenth centuries was such a change. . . .

Perhaps the most important requirement in a credible explanation for the decline of infectious diseases is that it should take account of the reasons for their predominance; we cannot be satisfied with an interpretation which suggests that the conditions which made them the common causes of sickness and death for ten thousand years remained essentially unchanged. This would be the case if immunization and therapy were the major influences during the past three centuries; or if mortality had fallen largely because of a reduction of virulence of micro-organisms. Moreover, if these were the reasons for the decline of the infections, we could be anything but confident about their future control. For in the light of experience of drug resistance we cannot foresee the long-term consequences of immunization and therapy; and if infectious deaths decreased because of a fortuitous change in virulence, they could quite readily increase again for the same reason.

The interpretation [previously] outlined attributes the predominance of infectious diseases since the first agricultural revolution to the expansion and aggregation of populations, poor hygiene and insufficient food; and their decline to modification or removal of these influences, spread over more than two centuries. During industrialization, however, the aggregation of populations increased; and in the early stages the hygiene, to put it cautiously, did not improve. It is therefore impressive that a large increase in food supplies coincided with population growth in many countries which differed widely in economic and other conditions. Of course, it is arguable that the expanded populations consumed all the additional food, so that there was no *per capita* increase. But in view of the other circumstantial evidence, as well as the lack of substance in alternative explanations, it seems more likely that at a time when birth-rates and death-rates were high, the population expanded because better nutrition resulted in increased resistance to infectious diseases, particularly in infants and children.

Leighton E. Cluff

America's Romance with Medicine and Medical Science

Leighton E. Cluff (b. 1923) is a leading medical educator who currently serves as president of the Robert Wood Johnson Foundation, one of the leading philanthropic forces for exploring change in late twentieth-century medicine. In this article, mindful of the changing organization of modern scientific medicine, Cluff uses medicine's past successes to launch a candid discussion of its present risks and future promise.

Though the medical advances before the twentieth century seemed remarkable, they pale in significance compared with what has followed. The discovery of insulin in the 1920s made diabetes mellitus a controllable disease. Major discoveries in nutrition and the identification of vitamins after 1912 virtually eliminated scurvy, rickets, beri-beri, and pellagra. These were all seminal events, but the golden age of therapeutics began really with the discovery of penicillin and sulfonamides, industrially produced, whose effectiveness was fully demonstrated during World War II. The sulfonamides, introduced in the 1930s, were incontestably important, but penicillin had an even more profound effect, by safely controlling serious wound infections, syphilis and gonorrhea, meningitis, pneumonia, and other common and dangerous infections. These successes changed public attitudes enormously; they demonstrated the power of medical science, the importance of industrial research and development.

Much of what has happened since, including the greater understanding of heredity and the working of genes, together with major developments in immunology, molecular biology, and drug therapy—and the application of this new learning toward improving the health of whole communities—has resulted from growing private and public support of medical research. . . .

The evidence of the successes of medical science to prevent and reverse the course of disease cannot be questioned. This is not to say, however, that all Americans, rich, poor, young, and old, have access today to the health care they require. Over the past several decades, elected officials have flirted with the possibility of introducing compulsory national health insurance; the United States is today the only major industrialized country to have a largely private medical insurance system. Not until twenty years ago, through Medicaid and Medicare, did the nation enact a public insurance program to assist the poor and the aged. Today, despite the continuing public support of medical science and medical technology, some thirty-five million Americans lack the insurance that gives them ready and easy access to the benefits of medical care.

The unprecedented prolongation of life—the ability to save infants weighing 500 grams [a little over one pound] at birth, the power of medical technology to defer death—together with the daily mass media accounts of new breakthroughs in medical science and technology, have only fueled America's passion for medical research. Still, important new social, legal, ethical, and economic concerns have also arisen; they lead some to question what the overall benefits and long-term consequences of medical science and technology are likely to be. Medical care is provided in new ways; the organization of the medical system is changing; there is a seeming eclipse of samaritanism by new impersonal technologies, together with the appearance of what many perceive to be a growing avariciousness, with proprietary interests and profit motivations figuring in medical care as they never have before—all these developments challenge America's romance with medicine. These new attitudes could affect further advances in medical research adversely, having major repercussions also for medical care. . . .

Between 1901 and the 1950s, the support of medical research in the U.S. was largely provided by private philanthropy. In fact, medicine and public health were the greatest beneficiaries of philanthropy; without its support, the development of medicine might have been quite different. Philanthropic encouragement was indeed essential to the advance of medical science in the U.S., even though federal support of research became more and more substantial during and after World War II. . . .

. . . Today, however, private funding of medical research and innovations in medical care is modest in comparison with federal and state or public expenditures. Indeed, between 1950 and 1976, the percentage of all support for biomedical research provided by private funding decreased more than five-fold, and has come to represent less than 5 percent of the total.

Voluntary health organizations, as opposed to private philanthropy, reflect a growth of public interest in and commitment to supporting medical research and advances in medical care. The first such voluntary effort in America began in 1904 when the Association for the Study and Prevention of Tuberculosis was founded. Yet the leading example of public awareness of the need for research in disease control and for special care of those afflicted was the work of the National Foundation for Infantile Paralysis. This voluntary effort, established in 1938 with a "March of Dimes," solicited broad public support for the study and prevention of poliomyelitis. Its successful development and testing of a vaccine between 1952 and 1954 captured public attention and removed the annual fear of a polio epidemic and its aftermath. This major breakthrough increased the enthusiastic support of all Americans for medicine and its potential contributions to people's lives. There are now numerous voluntary health organizations, most of which support and serve as advocates for medical research, training, and services, with the expectation that physicians will further understand and control specific diseases.

Public support of medical research has often developed as a by-product of attempts to solve practical problems. Authorization for the fundamental research needed to solve problems of specific diseases has been given by Congressional acts since 1912. . . .

An early contribution of the federal government was the creation of a National Board of Health in 1879 to coordinate the control of epidemic diseases. The life of the Board was short and its functions limited; its activities were transferred to the Marine Hospital Service in 1893. There, a bacteriological research laboratory, located in Washington, DC, became known as the Hygiene Laboratory. In 1912, legislation was enacted to create the U.S. Public Health Service (USPHS) out of the Marine Hospital Service. Based upon the scientific advances that were developing under the auspices of private and volunteer efforts, the USPHS was charged to study and investigate diseases damaging the nation's health. This allowed the support of classic studies in pellagra and eventually, with the help of private efforts, the USPHS demonstrated that this disease was the result of dietary deficiency. The role that government, together with public interest, could play in medical research was beginning to emerge.

Amidst the American prosperity and enthusiasm after World War I, Senator Joseph E. Ransdell introduced a bill that, in 1930, changed the name of the Hygiene Laboratory to the National Institute of Health (NIH), and provided the USPHS with authority to support the training of young men and women for careers in medical research. . . .

In 1937, seven years after Senator Ransdell's bill was enacted creating the NIH, Senator Matthew M. Neely inspired support for the statutory creation of the National Cancer Institute.

[Until 1946] the NIH supported the research largely of its own scientists, but government scientists had developed close collaboration with scientists in academic centers and research institutes. The NIH, therefore, departed from its tradition, and assumed responsibility for supporting the work of scientists outside the NIH. It also sought Congressional support for projects relevant to peacetime medicine, as well as extramural grants for research and training that would combat civilian problems. From that time on, the NIH has been by far the predominant national institution that supports biomedical research.

Using the National Cancer Institute as a precedent, interested scientists and citizens persuaded Congress to establish, in 1948, a National Heart Institute. Since then, the organizing principle of the NIH has been to establish institutes around categories of disease. . . .

As concern about government expenditures and economic decline has grown in the past decade, federal support for medical research has not escaped scrutiny. For more than thirty years following World War II, public and elected officials' enthusiasm for the promises of medical research, combined with the nation's thriving economy, had encouraged America's romance with medical science. However, as economic conditions deteriorated, public support of medical research weakened. . . .

As the science and technology of medicine rapidly developed, patients and members of the profession came to expect medical schools and their teaching hospitals to be on the leading edge, preparing physicians for the future, and providing the most up-to-date practitioners and care. After World War II, academic medical centers gradually became the foci of the public's enthusiasm for advances in medicine and medical care.

The need for training beyond medical school was inevitable if physicians were to be equipped with the necessary experience and skill to apply developments in science and technology as was expected of them by patients. Graduate medical education, which included hospital-based residency or house-staff training and post-doctoral fellowships, was essential; it could take only one year of "rotating internship" to prepare general practitioners, or several years of training to qualify in one of a growing number of medical specialties.

Physicians tended to confine themselves more to narrow fields of specialty, especially after World War II, and people wanted access to these highly skilled specialists. General practitioners were, by comparison, less esteemed, and the proportion of general physicians declined. Specialized technology was often available only in large urban hospitals, and the decision of many young physicians to practice in these settings led to a maldistribution of physicians across the nation; fewer physicians were available who were willing or able to provide general medical care; in rural areas the supply was particularly inadequate. . . .

In response to [an] apparent demand for general physicians to treat common medical problems and to provide continuing care outside the hospital, government and philanthropy provided incentives in the 1970s for medical schools and hospitals to increase the proportion of general physicians. Programs were financed to train family practitioners, general internists, general pediatricians, and other generalist physicians; funds were provided to train nurses in what became known as primary care; building on experiences with "medical corpsmen" during World War II, physician assistants were trained so that primary care could be extended to even more people.

With a predicted surplus of physicians, the future of primary care nurses and physician assistants is now uncertain. In addition, efforts to control excessive government expenditures and the mounting federal deficit may have the effect of reducing support for programs that train generalist physicians. . . .

Before World War I, most physicians practiced alone. There was little that these physicians, most of whom were general practitioners, could not do, assisted by a nurse and with a general hospital available for surgical procedures and supportive care. [In] World War I, however, the experience of physicians in the medical corps impressed upon many young doctors the value of practicing in groups. The model for private group practice was the organization established by the Mayo brothers in Rochester, Minnesota. After a faltering beginning, many large multi-specialty groups and clinics, as well as smaller single or multi-specialty groups, have been formed. These organizations have provided economic and professional advantages for physicians; in some instances they have opened investment and tax-saving opportunities, and increased access to the capital that is required for expanding technologically based office or clinic practices. Some patients have found that group practices, especially those composed of a full array of health professionals and medical specialists, provide a substitute for a single general physician. In other instances, patients were shuttled from one specialist to another, particularly when they had several different clinical problems; this was avoided only when the group practice or clinic identified one physician as the manager, or orchestrator of patient care. . . .

In the mid-forties, some private physicians countered the movement towards large organizations that had their own clinics, hospitals, and doctors by developing prepaid benefits like those offered by Kaiser and Puget Sound, allowing patients to retain their private physicians. The participating doctors accepted reduced fees if the prepaid revenues were inadequate, and policed their membership to prevent abuse. These organizations were the forerunners of "Health Maintenance Organizations," or HMO's, and "Independent Practice Associations," or IPA's. After uncertain starts, these new systems for medical care—large group fee-for-service practices, HMO's, and IPA's—are providing a way of controlling unnecessary medical care expenditures and of assuring access to health services, and a high quality of care. Many Americans are enrolling in such programs in order to obtain both the general and specialized medical services they desire. Many others continue to prefer the freedom to choose their own physician and place of care.

As the excitement generated by medical

science and technology grew, so too did hospitals' need for aid; people expected their hospitals to provide the very latest medical technology. In response to America's romance with medicine, the federal government provided the necessary resources for construction programs, such as the Hill-Burton Program, to improve both Veterans Administration and community hospitals throughout the nation. Advocates of the Hill-Burton Program believed it would provide access to powerful new medicine for families and communities that otherwise could not obtain it. The funds were, however, disproportionately allocated to middle-income communities; because communities were initially required to raise two-thirds of the construction funds, and the hospitals had to show that they were financially viable. Hospital construction was largely financed, then as it is now, by payments for patient services, adding to the growing cost of hospital care.

The method of reimbursing hospitals, largely through private and public insurance, for the costs they incurred in providing services had no limits and was financially rewarding for hospitals. The great resources that became available to community hospitals stimulated each to compete with the other for the latest in technology, as a means to attract physicians and patients. Physicians were the consumers of hospital services, solely responsible for the hospitalization of patients; doctors dominated hospitals; the competition between hospitals for doctors and their patients resulted in the widespread duplication and expansion of services and medical technology within many communities. The imposition of federal and state planning and regulatory agencies had little impact on hospitals, until stringent, tightly controlled mechanisms were instituted for paying hospitals for services they performed.

The profitability of hospital care led to a rapid increase in proprietary or for-profit hospitals and hospital chains. Yet, in many ways, not-for-profit, voluntary community hospitals began to resemble proprietary or for-profit hospitals and hospital chains as they sought to compete with one another; their characterization as "charitable" institutions, worthy of tax-exempt status and state or community bonding authority, is open to question. . . .

In the decade after World War II, Americans grew increasingly wealthy, and a larger proportion of them, who recognized the new power of medical science and technology, sought some type of hospital insurance. Insurance that covered hospital and physician services both benefited doctors and hospitals and improved people's access to more effective medical care.

Most medical insurance programs provided payment on the basis of fee-for-service and cost or charge reimbursement. Those who paid for the insurance, and insurers themselves, had almost no control over the cost of services provided by physicians and hospitals. Direct-service prepayment plans, however, became more firmly established after World War II, largely in response to the demands of employers and unions. The growth of such prepaid systems, including HMO's and IPA's, has accelerated recently, more because of their apparent control of unacceptable increases in health-care costs, than because they improve access to medical care and medical technology.

The number of people covered by commercial health insurance grew rapidly after World War II, but the unemployed, the poor, and the aged who were economically disadvantaged were left uninsured. In order to remain competitive with commercial insurance, Blue Cross and Blue Shield avoided the burden of insuring the poor and the aged. Responsibility for these groups fell to the government. Instead of creating an insurance system that would relieve economic problems and improve access to powerful new medicine for working Americans, the nation had developed a system that improved the access of middle-class people to hospitals, and of hospitals to middle- and upper-class patients. This insurance system had evolved under the control of hospitals and doctors and brought them professional and economic benefits; but it left to the state the responsibility of satisfying the need of less fortunate Americans for medical care. For many people, their romantic notion of medical development was not fulfilled, and this was especially true of those who had the greatest burden of illness—the poor and the old. However, in spite of the inequities in their distribution, the very existence of private health insurance and prepaid programs for industrial employees depressed any popular agitation for national health insurance. . . .

In 1965, the federal Medicare program was enacted to finance hospital care (in Part A) and physicians' fees (in Part B) for older citizens. A second program, Medicaid, was introduced, which gave the federal and state governments shared responsibility for financing the care of the poor. . . .

Nevertheless, too many Americans, either because they are unemployed, not poor enough, or do not meet other eligibility requirements, find that financial assistance is not available to them, that they do not have access to the medicine and medical technology they want, and from which they could benefit. . . .

At the same time, runaway costs for medical care are depleting resources. America's romance with medicine, however, persists, though in a faded form. It seems very unlikely that we shall ever agree to deprive ourselves, individually or as a society, of the benefits that advances in medicine and medical science bring. But social, legal, ethical, and medical problems must be discussed in order to *rationalize* (not ration) the use of medicine and medical technology. Only when medicine can better predict the benefits or risks of interventions for chronic or critical illness at the beginning and end of life, and only when medicine and society together can agree upon ethically appropriate medical care, will it be possible to control health-care costs, to avoid inequities in access to medical care, to revive or continue America's enthusiasm and support for medicine and medical science, and to establish the appropriate function of medicine that is restoring, preserving, and assuring people's good health.

CHAPTER 9

Darwinism as Science and Ideology

Introduction by Mark B. Adams

Charles Darwin (1809–1882) is one of the great figures in the history of science, yet even today—more than a century after his death—his ideas and his legacy remain controversial. Part of this controversy stems from an ignorance of the roots of Darwin's work and part from a misunderstanding of how his work was later appropriated—and distorted—to serve a variety of contradictory social and political agendas. In recent decades, the work of many historians has clarified these matters.

Natural History and Natural Order

The Age of Exploration led to a vast increase in the number of known plants and animals. As travel, trade, and commerce expanded, accounts and samples of new kinds of plants and animals poured into Europe from Africa, India, the Orient, the East Indies, and the Americas. It has been estimated that from roughly 1500 through 1750, the number of known plants and animals approximately doubled every generation. These began to be studied not simply because they were useful, but because they were interesting in their own right. As a result, the new sciences of botany and zoology were born.

During the eighteenth-century Enlightenment there was a general and systematic effort to extend the power of reason to encompass all of man and nature. The age produced many diverse and contradictory views, but all evinced a preoccupation with timeless natural order, and the greatest science of the age was classification. Naturalists who sought to understand the place of plants and animals in the natural order generally took one of two distinct approaches.

The first was the *scala naturae* or the "great chain of being." It derived from a view that in his infinite goodness, the Creator would surely create everything possible. There would thus be no breaks or gaps in nature—hence "Natura non facit saltum" (nature makes no leaps). In its fullest version, the chain of being arranged all things in the universe along a single, continuous, vertical line, from the lowest rocks, through fossils, "lower" plants, "higher" plants, "lower" animals, "higher" animals, humans—and on into a continuous range of spiritual beings up to God. This ascending scale went from the least to the most complex, from the least to the most perfect. The human place on the scale was special, for we were the highest of the material creations and alone among them in possessing a spiritual nature or soul. In some extreme forms, this approach held that species were simply arbitrary divisions imposed by us, that only individuals were real and formed the continuous chain. Even today, when we refer to "the missing link," or to "higher" or "lower" plants and animals, we are using terms derived from the *scala naturae*.

The second approach also derived from conceptions of the Creator, but to different effect. The biblical Book of Genesis tells that God created all living creatures, "each according to its own kind," and brought each kind before Adam to be named. "Species" means "kind" or "type." Classifiers who took this approach often assumed that all species had been originally created in Eden by God and named by Adam and that knowledge of these species and their names had been lost after the Fall, so they tried to sort out these original species and their proper names. According to this view of the natural order, then, nature consisted of a discrete number of distinct species, each designed and perfectly adapted to its mode of life in the "economy of nature."

These two models provided ways for eighteenth-century classifiers to organize the plants and animals coming into Europe, but neither was completely satisfactory. For those who believed all species had been named by Adam and had remained unchanged since creation, there were far too many species. How could so many, all fitted to such diverse environments, have existed in one Garden of Eden? How could each original species, with its original name, be unambiguously distinguished? And given their obvious similarities and differences, what larger groupings did they naturally form? On the other hand, for those who sought a single continuous chain of life, there were far too few species. Given the vast number of animals and plants, why was there not even a single piece of that continuous chain intact? Where were all the missing links? In their absence, how could the single right, "natural" lineup be establilshed?

Each naturalist who sought to establish the natural system in the century before Darwin had to confront these difficulties. Undoubtedly the most influential naturalist of the eighteenth century was Carolus Linnaeus (1707–1778). His multivolume *Systema naturae* (1735–1758) became the central authority for the classification of animals and plants. The work established the species as the central unit of classification and brought into general use "binomial nomenclature"—giving each form of animal or plant two names, first the genus, then the species (for example *Homo sapiens*). Many of his names are still in use today. Linnaeus was best known for his work on plants and was called "the Newton of botany." The key to his botanical classification was the female (pistils) and male (stamens) part of flowers. By counting the number of pistils, he sorted many plants into different "classes"; within each, he created different "orders" based on the number of male parts. This provided a powerful classificatory tool, but he acknowledged that this classification was arbitrary, and throughout his life he sought a more natural system. He recognized that a single linear order could not work. Having observed in his garden the appearance of what he regarded as a new species, *Peloria*, he came to believe that the "kinds" God had created were the natural orders; that he had mixed them to create genera; and that nature had crossed various genera to produce the species.

Lamarck and Cuvier

The two leading naturalists in the generation just before Darwin were the French zoologists Jean Lamarck (1744–1829) and Georges Cuvier (1769–1832). Like Linneaus before them, they had to grapple with the problems of establishing the natural system of classification, but each handled these problems quite differently. Lamarck began with the "great chain of being" approach, Cuvier with the "species" approach, but each ended up introducing modifications to handle the problems and materials they confronted.

Lamarck held that species were arbitrary divisions, that only individuals were real, and that one could align the major animal groups along a single scale of perfection. In 1802, that scale went from the "lowest," the infusorians, up through polyps, radiates, "worms," insects, spiders, crustaceans, annelids, molluscs, fishes, reptiles, and birds, to the most perfect, mammals. This alignment was justified by Lamarck's study of the "perfection"—that is, the complexity and differentiation in structure—of each group's central organ systems (nervous, circulatory, respiratory, digestive, and so forth).

Lamarck's work led him to make two modifications of the chain of being. The first stemmed from the discoveries of fossilized animal forms that no longer existed. It seemed to him unlikely that these forms represented former animals that had become extinct. Instead, he believed that they had been gradually transformed into more perfect types. Thus, by 1809 when his *Zoological Philosophy* was first published, the static *scala naturae* had become an escalator: at the lowest end, vital heat acting on wetness generated the lowest living forms, and over long stretches of time, through the natural innate tendency of matter to become more complex, these forms gradually moved up the scale of perfection. Thus, all kinds of creatures living today had a separate origin, and humans (the most perfect) are the descendents of the earliest living creatures generated from slime. The "missing links" are merely stages through which creatures have moved and will move. Lamarck's second modification was that the single linear "chain" became branching. It seemed to him that some major groups adjacent on the scale of perfection (for example, birds and mammals) were too different to be put next to one another, so he put them on two different lines branching off from reptiles. However, he still regarded mammals as more perfect than birds; they were simply following different pathways up the same scale of perfection.

Lamarck also believed that in adjusting to their environments and their modes of life, certain organisms modified their structures through "habit," "use," and "disuse." For example, according to Lamarck, giraffes have long necks because their ancestors stretched to reach the higher leaves, and the more developed neck was inherited by their offspring; whereas snakes, because they must crawl through the grass, developed longer bodies and lost their extremities (though other reptiles have them). Although it is now called "Lamarckism," this postulated inheritance of such acquired characteristics was a common belief long before Lamarck, and he used it as a secondary explanation to account for anomalies in the chain of life.

Cuvier's view of the organic world was utterly different. Like Linnaeus, he believed in the reality of species. For him, each kind of organism was a machine whose parts were perfectly correlated and fitted to one another. An organism's structure was in turn perfectly suited to its function, and they both were perfectly suited to that organism's "conditions of existence." For example, hooves were designed for running and fleeing, not for ripping or grabbing or clawing; so all hooved animals were herbivores and had mouth and jaw parts and a digestive system adapted to eating plants. Birds of prey needed good eyesight, speed and agility, claws or talons and beaks for grabbing and tearing flesh. Feathers were for flying and entailed lightweight bones, and certain characteristic back and rib features, including the "wishbone." Using these principles, and by making comparisons with the bones of living creatures, Cuvier studied the fossils of quadrupeds excavated from the Paris basin and was able to sort them out by type and to reconstruct what the whole animal must have been like from only a few bones. For this reason, he is often regarded as the father of modern comparative anatomy and paleontology.

Cuvier's machine concept of the organism also underlay his concept of "natural" classification. In his system, animal forms came in one of four basic designs or plans: the Vertebrates, the Molluscs, the Articulates, and the Radiates. Within each of these four basic designs were variations. For example, within the vertebrate plan were mammals, birds, reptiles, and fishes; within the articulate plan—crustaceans, spiders, insects, and earthworms. Those parts of organisms that showed the least variation, that necessitated certain other parts and excluded still others, were the most important for classification; those characteristics that varied within a species (such as color) were the least vital to its functioning and therefore the least important for classification. Cuvier noted with some satisfaction that fossil forms represented minor variations on his four basic structural plans. He thought that they had once existed but had been wiped out in catastrophic floods that had swept the globe.

For Cuvier, all organic species were machines perfectly designed for their place in nature, so no one was more or less perfect than any other—there were no "higher" or "lower" animals, no chain of nature. The gaps or "missing links" were there of mechanical necessity: intermediate forms were "incoherent"—they would not work. Nor was it possible for any one species to be transformed into any different form, any more than it would seem likely to us that a refrigerator could gradually change into a television: the change in any one part would necessitate a whole series of correlated changes in other parts, and structural intermediates would not do anything properly.

An approach comparable to Cuvier's—but more religious—was embodied in the contemporary works of the British clergyman William Paley (1743–1805) and especially in his *Natural Theology* (1802). If we saw a rock by the road, he writes, we might assume it was the product of natural causes; but if we saw instead a watch, we would assume it had been made and dropped there. Why? Because the watch is a mechanism carefully designed so that all of its parts fit together into a whole that serves a purpose—telling time. It shows design—and that means it had a designer and a maker. For Paley, however, the watch is crude indeed compared to the perfection of design embodied in the human eye (perfectly designed for seeing) and in all the other parts of animals and plants, each of which is adapted to the other parts in a harmonious whole to serve the organism in its role in nature. For Paley, organic design proves the existence of a supreme designer—God—and shows his personality and goodness. This conception, which linked natural history and theology, led to the *Bridgewater Treatises* of the 1830s, a series of books on geology, chemistry, anatomy, and other scientific topics commissioned to show God's handiwork in nature.

Charles Darwin (1809–1882)

Charles Darwin was born the same day as Abraham Lincoln in 1809. His father, a country doctor, wished him to become either a physician or a clergyman; while at the University of Edinburgh medical college, he found he could not stand the sight of blood, so he transferred to Cambridge University to study for the clergy. There he went on geological field trips and was an avid beetle collector. His development as a scientist, however, is associated with his six-year around-the-world voyage (1831–1836) on H.M.S. *Beagle,* an admiralty ship commissioned to chart the coast of South America for the British navy. Originally, the twenty-two-year-old Darwin was to be the gentleman companion of the twenty-six-year-old Captain Fitzroy; soon, however, he became an outstanding naturalist.

Fitzroy gave Darwin the first volume of *Principles of Geology* (1831) by the great British geologist Charles Lyell (1797–1875) as a going-away present. Lyell argued that the crust of the earth had been shaped over a long period of time by natural forces no different in kind or degree from those currently acting (volcanoes, earthquakes, erosion, deposition in water) to produce stratification and the gradual raising and lowering of land—elevation and subsidence. Darwin received Lyell's second volume while on the voyage. In it, Lyell dealt with the problem of fossils; he noted that in the most recent, "Tertiary," deposits, there were the fossilized bones of many species that still existed on earth, and the more recent the deposit, the more such fossils there were.

Lyell's *Principles* became the basis for Darwin's scientific studies on the voyage. While in Argentina, Darwin uncovered in the Pampa the bones of the glyptodon (a giant extinct kind of armadillo) and the megatherium (the extinct giant sloth). The war against the Indians in Argentina was leading to their gradual extermination; why, Darwin reasoned, could not such gradual diminution in numbers also have been the reason for the extinction of these fossil forms? While the *Beagle* was off the coast of Chile, an earthquake caused a tidal wave in Concepcion; Darwin noted that it had led to the elevation of marine shell beds some 13 feet above the water. He also mapped the land in Chile around Portillo Pass, and further north he noted step terraces that gave evidence the area had periodically been under shallow water. All this convinced Darwin that Lyell had been correct.

After returning from the voyage, Darwin began sorting his collected specimens of animals and plants, and during the spring of 1837 he became convinced that his observations of South America were incompatible with creationist natural theology. The main evidence was the geographical distribution of plants, animals, and fossils. The Galapagos Islands are located on the equator 600 miles off the west coast of South America, and are of recent volcanic origin; so are the Cape de Verde Islands, which are located roughly the same distance off the west coast of Africa. The two island groups form almost identical environments—so according to creationist natural theology, they should have the same kinds of animals and plants. Instead, Darwin found the Galapagos species are similar to those found in nearby South America, whereas Cape de Verde species resembled those in Africa. On the Galapagos, there are no native mammals, but several reptiles (giant tortoises and the land and marine iguanas); there are not very many kinds of land birds, but very many different kinds of finches. And why should the fossil forms he found in the Pampa closely resemble species currently living in the same general area? None of these facts made sense according to natural theology, but they would if species became transformed over time, if the Galapagos were originally populated by migrants from South America that gradually became transformed into different species, whereas Cape de Verde animals and plants had descended from African forms.

These facts led Darwin to formulate his "theory of descent with modification": current species are the modified descendants of former species closely related to them in space and time. But he was not immediately sure how these transformations took place. Then, in 1838, he read Thomas Malthus's *An Essay on the Principle of Population,* which described the geometric growth of human populations until they exhaust their food supply, leading to famine and death. This led him to the realization that in nature most plants and animals produce many more seeds or offspring than can survive; in this "struggle for existence," perhaps minor variations give certain offspring an advantage, and if those variations could be inherited by its offspring, they too would have an advantage and eventually the species would be transformed. He immediately began checking his theory out, getting information from British breeders of dogs, horses, and pigeons. Their different breeds were extraordinarily diverse in many characteristics—the same ones used by naturalists to distinguish species. In Darwin's view, then, all these breeds or varieties were "incipient species" produced by the breeders' selecting which animals with which characteristics would be used for breeding. In nature, there were also many variations, and only selected animals and plants would breed the next generation, but instead of conscious selection by the breeder, there was a "natural selection."

Darwin wrote out a sketch of his theory in 1842 and a longer essay version in 1844. He was composing a large work to be entitled "Natural Selection" when he got word that Alfred Russel Wallace (1823–1913) had come up with the same theory and was planning to publish it. Wallace had worked for many years in the Brazilian Amazon, so he had seen the great natural variation within wild species. He was working in the Malay archipelago when he, too, happened upon Malthus, whose work triggered his own theory.

Darwin's friends arranged for a short publication of Wallace's paper, together with a short paper by Darwin, in 1858. Darwin polished up excerpts of his longer work into an abstract and published it in 1859 under the title *On the Origin of Species by Means of Natural Selection, or the Preservation of Favoured Races in the Struggle for Life*. It had three general themes: to explain how evolution was possible through natural selection; to deal with the objections that might be raised to his theory; and to mobilize the facts of biogeography, morphology, systematics, zoology, botany, and geology to demonstrate that they were consistent with his theory. The book is a classic in the history of science.

Darwin's book is in the grand tradition of natural history, yet at the same time it reshaped that tradition. Darwin established that the long-sought "natural system" of classification was genealogical, and species could be sorted in ever-larger groups by how long ago they had diverged from a common ancestor. There were indeed missing links connecting current species; but they were not hidden in today's world, but were to be found in the fossil record—the common ancestors of today's diverse and divergent groups. Much of the subsequent scientific resistance to Darwin's theory would come from students of Cuvier; much of the subsequent religious opposition would emphasize creationism and the argument from design. This criticism is not without irony, for Darwin himself began as a scientific creationist, and he was led to abandon the theory, in the face of overwhelming evidence, precisely because he had taken it seriously as science. The shape of his theory also shows the influence of Cuvier and Paley, for like them he focussed on adaptation; but he went one step further and was able to explain its natural origin. Like Cuvier and Paley, Darwin thought all organisms were perfectly adapted to their environments—but some were more perfectly adapted than others, by virtue of small variations. For Cuvier and Paley, those variations were the least vital to the organism's functioning, almost by definition, and therefore the least important for classification; but for Darwin, they were the key to evolution, for they provided the material for natural selection.

The Scientific Reception

In the mid-nineteenth century, philosophical debates were underway about scientific method, the nature of science, politics, and philosophy. As these debates developed in various countries after 1859, the meaning of "Darwinism" was shaped by local disputes and agendas, with each side in the various debates seeking to invoke Darwin and his ideas for its own purposes—sometimes to castigate opponents, sometimes to provide scientific legitimacy for its philosophical and social views. At different places and times, "Darwinism" was used to mean a belief in evolution by any means; in evolution by natural selection; in the animal origins of man; or in atheistic materialism.

In different countries, different aspects of his theory were problematic. The British accepted the notion of "the struggle for existence" but had difficulty with the idea that "Nature" somehow "selects." In Russia, "selection" did not seem nearly so troublesome as the idea that members of the same species struggle and compete with one another. In Germany, "Darwinismus" came to be associated with views on the emergence of forms and the development of embryos. Conservative thinkers had difficulty conceiving how the mechanical process of natural selection, acting on "chance" variations, could produce the intricate adaptation and design evident in living organisms. Radical thinkers welcomed the materialistic aspects of Darwin's theory because of its challenge to orthodox religious views of man and nature, but they objected to the Malthusian aspects as reflecting "reactionary" social theory.

One criticism of Darwin's theory proved especially important. Fleeming Jenkin had pointed out an apparent flaw in Darwin's mechanism. Darwin had postulated that small advantageous variations that were heritable would be gradually accumulated and spread over long periods of time through their differential survival. But Jenkin pointed out that such a variation would first occur in a single individual that would have to breed with another organism that lacked the advantageous variation. Assuming that each parent contributed equally to the offspring and the parental traits would be blended, the offspring would have only half of the advantageous trait; and in subsequent generations, the trait would become increasingly diluted. Thus, slow gradual selection would soon have nothing to act upon. In response to this and other criticisms, Darwin felt compelled to explain how advantageous variations could accumulate faster and evolution could proceed more swiftly. In his later years, he toyed with various theories of inheritance and put increasing emphasis on the evolutionary significance of use, disuse, and the inheritance of acquired characteristics; but many were not satisfied with this revision. The development of cytology in the 1880s had demonstrated that the germ cells producing the next generation came directly from the parents unchanged. This led August Weismann (1834–1914) to postulate the "continuity" and "immortality" of the germ plasm and to deny that acquired characteristics and the effects of use and disuse could be inherited. By the turn of the century many biologists believed, in the words of one contemporary, that "Darwin's theory explains the *survival* of the fittest, but not the *arrival* of the fittest."

From roughly 1890 to 1930, Darwinism was in eclipse and "alternative" theories of evolution abounded. Most were Lamarckian in some way. One group came to believe that evolution occurred by "orthogenesis"—a directional development of species and groups along certain established lines, and, possibly, toward a goal. Others held to "neo-Lamarckism"—the importance of use, disuse, and the effects of the environment on shaping evolution. Henri Bergson (1859–1941) advocated "creative evolution," suggesting a development to higher and higher levels of complexity with emergent properties, toward a higher consciousness. Many experimentalists favored the "mutation theory" of the Dutch botanist Hugo de Vries (1848–1935), according to which "progressive" mutations occur in large numbers every few years and can produce wholly new species in a single generation. Other biologists proposed still other mechanisms of evolution.

During the period 1930–1945, however, Darwinism was reborn in the form of the so-called "synthetic theory of evolution," and a consensus emerged among many biologists that Darwin's theory had been essentially correct. The rediscovery of Gregor Mendel's laws in 1900, and the development of Mendelism and the chromosomal theory of heredity in succeeding decades, satisfied many that finally a successful theory of inheritance had appeared, and it was quickly tested and deployed among animal and plant breeders. Initially, genetics seemed antagonistic to Darwinism. But in the 1920s, the Russian naturalist S. S. Chetverikov, the British statistician R. A. Fisher, the physiologist J. B. S. Haldane, and the American animal breeder and biometrician Sewall Wright pointed out that Mendelian heredity actually "saved" Darwin's original theory: if genes were stable, and could be modified by mutations, then advantageous variations would not be "blended" away and Fleeming Jenkin's objection to Darwin's theory was invalid. Fisher, Wright, and Haldane developed the mathematical and statistical theory of the genetics of populations, showing that evolution would indeed occur by natural selection in a relatively short time. Chetverikov in Russia, and later Theodosius Dobzhansky in America, began studying the genetics of wild populations of the fruit fly *Drosophila*. This new subject of "population genetics" became the core of the new Darwinism. In subsequent decades, Ernst Mayr, Julian Huxley, G. G. Simpson, and others deployed this theoretical matrix to account for the facts of the systematics, taxonomy, biogeography, and paleontology of plants and animals.

This synthetic perspective became a consensus by the 1930s among many leading biologists in Britain, the United States, and the Soviet Union. In addition, the teaching and public acceptance of Darwinism in various countries became strongly affected by ideological, social, and political development.

Darwinism and Society

In the late nineteenth century, psychology, anthropology, and sociology were developing into independent scientific disciplines, and "evolutionism" became one of their major themes. This evolutionism was occasionally called "social Darwinism," but, although its supporters sometimes drew on Darwin's name and authority, it had little to do with his ideas. Actually, it derived more directly from Herbert Spencer (1820–1903), a British philosopher whose views on evolution had been greatly influenced by Lamarck's. Spencer conceived of progress toward increasing perfection and complexity as inevitable in the development of the universe, life, and society and had expressed these views in print even before Darwin's theory appeared. For Spencer, all developmental processes proceed through the same basic stages. It was Spencer's version of evolution, not Darwin's, that became the "evolutionism" of the new sciences of anthropology and sociology. For example, the notion that today's "primitive peoples" are at the same stage of social development as our own ancestors were many thousands of years ago is the view, not of Darwin, but of Spencer.

Spencer was especially popular in the United States. In the late nineteenth century, a group of American philosophers, social thinkers, and entrepreneurs espoused views that led them to be called "social Darwinists" by their opponents and critics. Their central argument was that a person's position in society should rightfully reflect merit and achievement rather than class origins or inherited wealth. Sometimes they are caricatured as advocating the survival of the fittest in society—the triumph of the powerful and the exploitation of the weak—with no kindness or charity afforded to the sick, the poor, or the disadvantaged. Actually, they generally believed in altruism and charity, arguing, as had Spencer, that the pervasiveness of these traits in a society was a sign of its advanced evolutionary development. Although they alluded to Darwin favorably and may have used his scientific authority to help legitimate their own social views, those views had little to do with Darwin's.

Religious reactions to Darwin's theory were many and varied. Indeed, one of Darwin's earliest American supporters was the botanist and clergyman Asa Gray (1810–1888), who regarded Darwin's theory as supporting religion by showing God's method of creation. By the turn of the century, many Protestant, Catholic, and Jewish thinkers found Darwin's evolutionary theory perfectly compatible with their religious views. However, at that time fundamentalist opposition to Darwin intensified in the United States. Various state legislatures had passed laws forbidding the teaching of Darwinism in the public secondary schools, though such laws were rarely enforced. Some progressives sought to overturn the laws by engineering legal test cases. The famous Scopes Trial in Tennessee in 1924 began as such an episode. In his memoirs, John Scopes makes clear that he engineered his own trial, but the case soon attracted worldwide attention, pitting William Jennings Bryan, several-time Democratic Party presidential candidate and fundamentalist preacher, against the famous Progressive atheist lawyer Clarence Darrow. However, the strategy backfired: in the three or four decades following the trial, American textbook manufacturers sought to expand their markets and avoid controversy by leaving Darwinism and the theory of evolution out of their biology textbooks altogether. As a result, the teaching of evolutionary biology in American secondary schools lagged behind that in many other countries for many decades.

In Germany, Adolf Hitler's formulation of his credo in *Mein Kampf* in the 1920s and the rise to power of his National Socialist (Nazi) Party in the early 1930s brought their ideology into a central position in German society. That ideology emphasized the biological and moral superiority of the Aryan race, and it invoked the principle of the "survival of the fittest" to justify laws and actions against so-called mental defectives, gypsies, Jews, and other groups. During the Nazi period, many attempts were made to give this ideology scientific legitimacy by allusions to Darwin and Darwinism. Actually, however, Nazi racism had little basis in either contemporary genetics or Darwinian evolutionary theory; instead, it stemmed from certain German traditions of racial thought and certain teachings of physical anthropology that involved a sort of typological thinking Darwin had sought to eliminate from biology. Nonetheless, as taught in German schools during the Nazi period, "Darwinism" was blended with Nazi racist ideology.

Before the Bolshevik seizure of power in November 1917, "Darwinism" had been very popular among the Russian intelligentsia and closely linked with liberal thought, but it had been opposed as a materialist, atheist philosophy by many Orthodox theologians and conservative thinkers. After 1917, Darwinism gained widespread popularity as part of the new Communist government's campaign to spread literacy, popularize science, and undermine religion. Indeed, in the 1920s, some books had trouble with the censors because they were not "Darwinian" enough. During that decade Soviet evolutionists produced much important work. With the advent of Stalinism and the collectivization of agriculture (1929–1932), Party ideology became a dominant force in Soviet life. Around 1932, Darwinism became an official part of accepted Soviet ideology, and by 1939, departments of Darwinism were established at most major universities.

Although Darwinism had been officially accepted, its interpretation remained open to dispute. In the 1930s, many leading Soviet biologists were aware of population genetics and were busy synthesizing a theory of evolution very like that concurrently taking shape in the West. However, a different kind of Darwinism, so-called "creative Darwinism," was also gaining support. Formulated in the late 1930s by the agronomist Trofim Lysenko (1898–1976) and his philosophical ally Israel Prezent, this view drew upon Russian traditions to synthesize a more activist, practical, ideological variant of Darwinism. The contemporary Western synthesis emphasized Darwin's concept of intraspecific competition and natural selection, while substituting modern genetics for his "confused" views on inheritance. Lysenko's "creative Darwinism" took an opposite tack: it rejected Darwin's idea of intraspecific competition as a result of Malthus's unhealthy influence, and emphasized instead Darwin's belief in the plasticity of the organism under domestication, the power of the practical breeder, and the inheritance of acquired characteristics. In his most radical formulation around 1946–1952, Lysenko claimed that new species and even genera could be created by the breeder suddenly, in a single generation. Lysenko rejected modern genetics as "bourgeois" and "idealist" and of no practical benefit. In a famous 1948 meeting of the Soviet agricultural academy, Lysenko's theories were given the official approval of the Communist Party, and "creative Darwinism" was the version of evolutionary theory taught in almost all Soviet schools from 1948 through 1964. Lysenko's influence fell rapidly with the ouster of Khrushchev in October 1964. Since then, a Soviet variant of the evolutionary synthesis very similar to that held in the West has been largely reestablished.

Conclusion

In recent years, a number of new scientific techniques and findings have shed new light on evolution. Developments in molecular biology and biochemistry have made it possible to compare in detail the structure of the genetic material of various individuals and species and this has led to new ways of classifying organisms and tracing their genealogies. The development of plate tectonics in geology and the history of drifting land masses have clarified the interpretation of much fossil evidence and explained puzzling problems of geographical distribution and extinction. New fossil finds had led to both the reinterpretation of the dinosaurs and of the evolutionary history of humankind. Disagreements about rates of evolutionary change have led to such proposed modifications of the synthetic theory as the theory of so-called "punctuated equilibria."

In recent years new controversies concerning the human implications of Darwin's legacy have also emerged. The development of sociobiology has stimulated opposition among some social scientists, whose disciplines have tended to assume that virtually all human behavior is the result of learning; among some leftists, who hold that human nature is largely determined by social and economic relations and should not be "reduced" to biology; and among some religious groups, who hold that such views interfere with ideas about the freedom of moral choice and the human soul. In addition, religious fundamentalists have again become active in challenging the teaching of evolution in public schools and textbooks. Some have claimed the existence of "scientific creationism," but most of its arguments against evolution are more than a century old, deriving ultimately from Paley and Cuvier.

When viewed historically, such developments are hardly surprising. In the past, Darwin's legacy has been co-opted and adapted to serve various scientific, social, and political agendas, and our own age is no exception. But these current controversies should not blind us to a few basic facts of history. Whatever disagreements have existed and do exist among biologists concerning Darwin's views on the mechanism of evolution, the rates of evolutionary change, or the relative importance of natural selection, a myriad of scientific findings have only confirmed Darwin's theory of "descent with modification." Even during the eclipse of Darwinism at the turn of the century, virtually all the biologists who disagreed with Darwin's natural selection theory were nonetheless Darwinists in one fundamental sense: they all believed in evolution. For more than a century, biologists have regarded evolution as a fact. And it was Charles Darwin, through his scientific observations and his inspired reasoning, who established that fact and forced us to contemplate its implications.

Peter J. Bowler

The Origins of Darwinism

It is sometimes asserted that the principal precursor of Darwin was Jean Baptiste de Lamarck (1744–1829), who developed his theory of evolution "ahead of his time." Actually, in many respects Lamarck was "behind his time," and the theories of his French contemporary, Georges Cuvier (1769–1832), more directly addressed the concerns that would dominate Darwin's thinking. Lamarck's theory was resurrected and taken seriously only later in the nineteenth century, after Darwin, and was often misrepresented in the process. In these excerpts from his book Evolution: The History of an Idea, *historian Peter Bowler summarizes the vast literature on Lamarck, Cuvier, and the development of Darwin's theory.*

Erasmus Darwin and Lamarck

. . . At the end of the eighteenth century . . . there are two figures whose ideas seem to come much closer [than others] to the modern concept of organic development: Erasmus Darwin and Lamarck. Both avoided the temptation to see even complex forms of life as derived from spontaneous generation, and hence were forced to take more seriously the processes by which living things can actually change through time. Both have been hailed as founders of modern evolutionism. . . . Erasmus Darwin and especially Lamarck are important because they elaborated the most complex of the Enlightenment's efforts to deal with the problem of organic change; but we should not be misled by superficial similarities into assuming that they contributed directly to the Darwinian revolution.

Erasmus Darwin occupies a unique place in the history of evolutionism. He was a colorful personality, and perhaps the only thinker we shall encounter who put forward some of his ideas in the form of poetry. Works such as the *Botanic Garden* (1791) and the *Temple of Nature* (1803) were popular in their day, although Darwin's couplets are not adapted to modern taste. Erasmus was also the grandfather of Charles Darwin, and because his (nonpoetic) *Zoonomia* (1794–1796) proposed a theory of evolution, he naturally has been a target for those who wish to show that it was only by gleaning insights from his precursors that the younger Darwin was able to formulate his theory of natural selection. . . . But such anticipations always turn out to be superficial; for example, Erasmus' account of the "balance of nature" maintained *between* the species has been mistaken for the "struggle for existence" *within* each species used so effectively by his grandson.

Erasmus Darwin's views must be interpreted within their own context. He was a deist who believed that God had designed living things to be *self-improving* through time. In their constant efforts to meet the challenges of the external world, they developed new organs through the mechanism that the Lamarckians would make famous as the "inheritance of acquired characteristics." The results of the individual's efforts are inherited by his offspring, so that by accumulation over many generations a whole new organ can be formed. Darwin seems to have assumed that the overall results of this effort to adapt to the environment would be a gradual progress of life toward higher states of organization. . . .

Darwin was a physician rather than a naturalist, and the only consistent account of his theory is a single chapter in the *Zoonomia*. Jean Baptiste Pierre Antoine de Monet, chevalier de Lamarck, was a professional naturalist who wrote extensively on his own theory. Largely ignored in his own time, it is not surprising that he has received more attention from historians. . . .

Strictly speaking, Lamarck cannot be counted as an eighteenth-century evolutionist, because he did not come to accept the possibility of transmutation until just after 1800. He is one of those unusual figures who make a major shift in outlook at a comparatively late age. The fact that he was over fifty when he abandoned his original commitment to the fixity of species, allows us to see him as a product of the Enlightenment. His theory represents a combination of several themes characteristic of eighteenth century attempts to deal with the origin and development of life. There is the materialist belief in spontaneous generation, made more reasonable by limiting it only to the simplest forms of life. There is the steady ascent of a scale of organization, which may owe something to the temporalized chain of being. And finally, there is a process by which living things can change in response to new conditions, something accepted in various forms by many Enlightenment thinkers. Lamarck's tragedy was that he put all of these ideas together at a time when the Enlightenment's way of thought had gone out of fashion and he paid the price of being dismissed as a crank by many of his contemporaries.

In his early botanical works, Lamarck showed an interest in the hierarchical arrangement of classes, although from the first he realized that plants and animals would form two parallel hierarchies, not a continuous chain of being. At the same time, he was developing his unconventional theory of chemistry and a unique system of geology based on uniformitarian principles. In 1794 Lamarck was appointed to the Muséum d'histoire naturelle, now reorganized from the old Jardin du Roi by the revolutionary government and was given the task of classifying the invertebrates. He adapted so well to this that he is regarded as one of the founders of invertebrate taxonomy; but while developing this new skill he abandoned his original commitment to the fixity of species. His theory of organic development was first outlined in 1802, then reorganized as the basis of his best known work, the *Philosopohie zoologique* of 1809. . . .

Eventually Lamarck converted to the materialist belief that life is the product of matter via spontaneous generation. . . . Because only the simplest form of life could be produced directly, higher forms must have been derived from these simple ones by some kind of progressive development over many generations. For Lamarck, the species constituted a hierarchy of structures ranging up to the most complex, and this hierarchy represented the historical pattern along which life had advanced. The active powers of the nervous fluid carve out ever more complex channnels, and each generation advances slightly beyond the level of its parents. It has been argued that Lamarck's acceptance of this idea of continuous progress does not derive from the temporalized chain of being as advocated by Bonnet. Certainly, Lamarck did not believe in a unilinear sequence of forms and allowed two main branches instead of a single chain. Yet Bonnet too had accepted the possibility of a branching chain, and it is not unreasonable to suppose that Lamarck translated the unilinear chain into the more sophisticated notion of a hierarchy of organization. The element of overall linearity remained, because Lamarck insisted that in theory the progressive trend would produce an unambiguous sequence of forms. The hierarchy was not just an abstract scale defining degrees of organization, but a predetermined path by which life advanced. It is possible even that Lamarck saw the path of development as mapped out in advance by the Creator, because he was certainly not an atheist.

If the active power of nature compels life to mount steadily up the chain of being, how, he asks, can we still see the complete hierarchy today? Why have all living things not raised themselves to the same level as man? The answer to this is crucial for defining the difference between Lamarck's theory and the modern one. He did *not* suppose all forms alive to have evolved from a common ancestry. On the contrary, he believed that the organisms at each level of the scale today have progressed to that point separately; organisms at different levels are derived from different acts of spontaneous generation at different points in time. The direct formation of the simplest forms of life has gone on continuously throughout the earth's history. Today's highest forms have progressed over

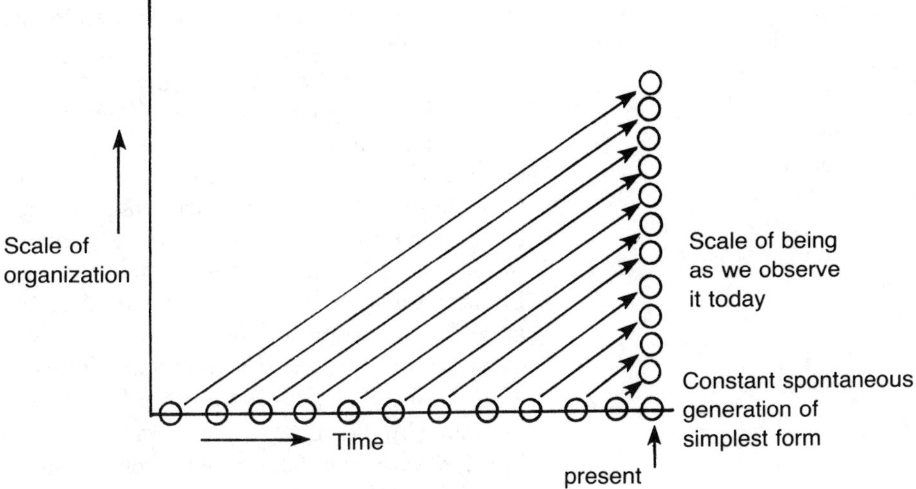

Figure 9–1 *Lamarck's Theory of Organic Progression.*
Each point on the scale of being we observe today has been derived by progression from a separate act of spontaneous generation. The lower down the scale the organism is today, the more recently its first ancestor was produced. Thus evolution is not a system of common descent but consists of separate lines progressing in parallel along the same hierarchy.

many generations from the earliest appearance of the simplest forms. Organisms now halfway up the scale have progressed from acts of spontaneous generation in the more recent past, while the simplest organisms we observe today just have been formed. The evolution of life thus consists of a whole series of lines each advancing separately along the scale of being, as shown in the diagram.

As a taxonomist, Lamark knew that he could not, in fact, observe a simple linear arrangement of forms in the animal kingdom. The chain of being as it exists for him has a number of major branches and many gaps, illustrated in a diagram added to the *Philosophie zoologique*. Branching has occurred because the mechanism that forces each line of development along the scale of being is not the only one involved. Lamarck knew from his geology that the surface of the earth is subject to constant, if very slow change, and fossils told him that living things also have changed through time. He refused to accept the possibility of extinction: the fossil species must have evolved into those of today, because nature is powerful enough to prevent any of her productions from being driven to extinction. There must be a mechanism by which life can adapt to changing conditions at the same time as it progresses up the chain of being. This mechanism was, of course, the inheritance of acquired characteristics—the only part of the theory still remembered and associated with its author's name. Yet for Lamarck himself it was only a secondary factor that disturbed the pure line of progress.

Lamarck believed that the animal's needs determine the organs its body will develop. This did not mean that the animal could grow itself a new organ by willpower alone, however. The needs determine how the animal will use its body, and the effect of exercise, of use and disuse, cause some parts to develop while others wither away. The environment creates the animal's needs, which in turn determine how it will use its body. Those parts that are strongly exercised will attract more of the nervous fluid; this fluid will tend to carve out more complex passages in the tissue and increase the size of the organ. Disused organs will receive less fluid and will degenerate. Lamarck gave no detailed theory of inheritance but assumed that the characteristics acquired as the result of effort would be transmitted to the offspring thereby enabling the effect to become cumulative. To give a famous example: the short-necked ancestors of

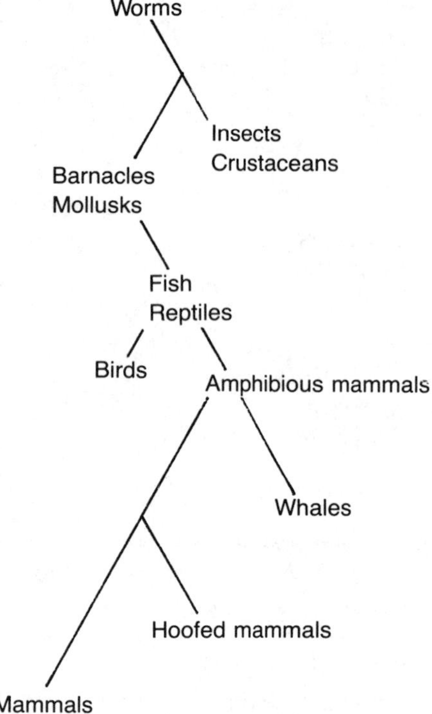

Figure 9–2 *Lamarck's Branching Chain of Being. Adapted from a diagram in the* Philosophie zoologique, *this shows how Lamarck thought the theoretically linear scale of organization had been forced into various branches by the necessity of life adapting to changing conditions throughout the earth's history. Note that this is* not *meant to be a system of genealogical relationships: no one part of the chain is derived directly from any other. The diagram is merely a more realistic representation of the chain produced in the previous diagram.*

the modern giraffe were at some point in their history forced to begin feeding from trees. All the individuals stretched their necks upwards and as a result this part of the body grew in size. The next generation inherited the extra neck-length and stretched it even further, so that over a long period of time the giraffe gradually acquired the long neck we see today.

The inheritance of acquired characteristics can be seen as an alternative to Darwinian natural selection as a means of explaining how living things slowly adapt to their environment. In almost every respect, however, the system into which Lamarck built this mechanism differs from that accepted in the post-Darwinian era. Darwin was not interested in the ultimate origin of life, whereas for Lamarck spontaneous generation was an integral part of the materialist viewpoint. For Darwin and almost all later naturalists, evolution was a process of divergence by which all forms of life have branched out from a few ancestral forms due to the long-continued effects of geographical isolation and adaptation to new conditions. For Lamarck, a series of distinct lines of evolution moved independently along the same scale of organization, progressing inevitably upward. The chain of being would be forced into branches by the effects of different conditions, but this was a secondary phenomenon and Lamarck had no interest in the problems of geographical distribution that were so crucial for Darwin. Species still had a real existence in Darwin's theory, in the sense of distinct breeding populations, but Lamarck denied the existence of species altogether. He pictured both his evolutionary mechanisms, progressive and adaptive, as essentially continuous processes that would never produce sharp divisions in nature. Lamarck even predicted that the march of biological discovery would fill in the gaps between what were regarded as distinct "species." In almost every respect, his viewpoint harks back to the eighteenth century, not forward to the nineteenth.

Although in some respects alien to modern evolutionism, Lamarck's theory was worked out in more detail than any other Enlightenment account of the origin of life. He attracted no immediate followers, however, and died in obscurity. To a large extent his eclipse was engineered by his greatest rival, Georges Cuvier. Cuvier not only built up a rival system of biology but also rose to political power in France, an eminence that enabled him to ensure that Lamarck's views were dismissed as outdated speculations. Cuvier developed techniques of comparative anatomy that even Lamarck adopted, but Lamarck remained with the invertebrates and Cuvier was appointed to a more prestigious post at the Muséum classifying the vertebrates. Cuvier astounded the world by applying his techniques to the reconstruction of extinct species from their fossil remains, while Lamarck failed even to see the possibility of using the new paleontology to support the idea of an evolutionary progression.

Cuvier divided the animal kingdom into four basic "types" that could not be ranked into a hierarchical order as Lamarck and many of his predecessors had maintained. The real power of the Linnaean system based on parallel rather than hierarchical relationships now could be appreciated. Although the types were distinct, representing four viable plans upon which an organic form could be based, each was infinitely flexible in the way its external modifications could be adapted to the demands of the environment. The relationships between species now were seen to be based on the fundamental similarities of their internal structures, rather than an ordered ranking of their external characters. Foucault's thesis is that this approach made possible the Darwinian view of natural development, in which a single original form diversifies through adaptive radiation, its basic characteristics remaining to serve as the link by which we classify the descendants. . . . This thesis needs to be qualified in a number of cases, yet it remains true that to a surprisingly large extent eighteenth-century evolutionism was conceived as change taking place within an orderly framework.

Cuvier himself was no evolutionist, though— it was his classification that was modern, not his views on the mutability of species. The sense that he developed for understanding the way in which internal parts of the body fit together convinced him that species are fixed and distinct. The new comparative anatomy gave so detailed a picture of each animal's internal structure that it seemed impossible for such complex forms to be created by a natural process. Materialist speculations on the origin and mutability of life were dismissed as incompatible with the new depth of biological knowledge. The Enlightenment's more speculative approach was swept away by the increasingly conservative tastes of Napoleonic France. In Britain, an even stronger reaction was creating a new interest in natural theology and making it impossible for Hutton's antibiblical geology to be taken seriously. Lamarck founded no school because he developed his theory twenty years too late, when the materialist spirit he had turned to was being swept away by political and cultural revolution. Evolutionism now would have to make a new start, taking into account the new sciences and the new thought processes of the nineteenth century. In many respects, Darwinism would be shaped more by these ostensibly conservative developments than by the speculations of Enlightenment evolutionists. . . .

The Crucial Years: 1836–1839

The *Beagle* reached England in October 1836. In March of the following year Darwin took lodgings in London and began what he later described as the two most active years of his life. He read papers to the Geological Society and engaged in the scientific life of the capital and at the same time he pondered the implications of his discoveries on the voyage, especially those of the Galápagos islands. It is suggested occasionally that Darwin had come to accept the transmutation of species during the final stages of the voyage itself. The majority of modern historians, however, believe that the conversion took place after his return, probably between March and July of 1837. . . . In September of 1838 his reading of Malthus on population [see below] finally enabled him to put the basic idea of natural selection together, and from this point on he was committed to a comprehensive reinterpretation of natural history in the light of his new theory. . . .

We begin with an outline of the basic argument upon which natural selection is based. Despite the apparently rambling character of Darwin's later writings, the theory indeed has a logical structure resting upon certain verifiable assumptions about nature. Perhaps its most important foundation is a new interpretation about exactly what a biological species is. Darwin pioneered what [zoologist Ernst] Mayr calls "population thinking" to replace the old typological view of species. This is a willingness to treat the species as a population of unique individuals united only by the fact that they are potentially capable of breeding together. The species *is* the population, whatever the amount of variation in physical structure between the individuals concerned. It is not defined by some ideal type upon which the individuals are supposed to be modeled. Variation is not a trivial disturbance of the ideal form but an essential character of the population and hence of the species. If an external factor favors certain individuals in the population at the expense of others, then the average nature of the population will change and, by definition, so will the species.

The second major factor is a suitable concept of heredity. The individual must be able to pass on his unique characteristics to his offspring. Darwin's early interest in Lamarckism showed a willingness to accept what is now known colloquially as "soft" heredity: the belief that what the parent transmits to its offspring is subject to modification by external causes. But if the environment can control heredity, there should be little individual variation, because all members of a population will have absorbed similar influences. To make a selection theory plausible, Darwin had to abandon "soft" in favor of "hard" heredity: the belief that what the parent transmits to its offspring cannot be affected by outside influences and depends only upon what *he* inherited from his parents. This gives a deeper significance to individual variation, allowing Darwin to appreciate how selection could change a population by picking out a certain kind of variation that then will be preserved by heredity.

In his efforts to understand the variability of animals and plants, Darwin began to collect information from breeders and horticulturists. His interest was natural. . . . It was inevitable that to study the past development of life he would turn to the one area where organic change could be observed today. Animal and plant breeding offered an experimental way of studying the effects of variation. But it is no accident that the significance of both variability and hard heredity were recognized first by breeders, who knew that they owed their success to manipulation of these factors by selection. Darwin always presented natural selection through an analogy with the artificial form. The breeder picks out those individuals in his group that possess something of the characteristic he seeks and breeds his next generation solely from these. He thus isolates the desired characteristic and by selecting further variations in the same direction can improve it in later generations.

The only problem with this analogy is that some force in nature must be found that can substitute for the element of choice exercised by the breeder. Darwin eventually was satisfied that the struggle for existence would play this role, by selecting out those individuals best adapted to the environment. This would provide a natural explanation of adaptive evolution, with no implication of divine supervision. Darwin inferred the selective power of struggle from Malthus' principle of population, coupled with the observed fact that the population of any species must remain more or less constant. Malthus' principle showed that numbers of any species potentially can increase at an exponential rate. Yet both observation and common sense tell us that the population of a wild species cannot increase significantly from year to year, because of the limited nature of the food supply. Many individuals born in each generation must die before their time, because available resources simply cannot support the potential increase in population. From this Darwin deduced that there must be a constant struggle for existence in nature, as individuals compete to see who will get enough of the limited food supply to stay alive and breed.

By putting all these points together, the argument for natural selection emerges. Given a degree of variation among individuals, it seems obvious that some will do better than others in the struggle for existence and will pass on their advantageous characteristics through breeding. We can develop the argument by reworking Lamarck's famous example of the giraffe evolving its long neck to feed off trees. In the original population of grass-feeders, some individuals would by chance have longer than average necks, others shorter. When the grass began to disappear, those with longer necks would be able to reach leaves on trees more easily; because they could exploit the alternative source of food more effectively, they would be healthier and able to breed more readily; their offspring would be more numerous and inherit the extra length of neck. Conversely, those animals with shorter necks would get less food and not breed so easily; in the extreme case they would die of starvation, although a difference in rate of reproduction is all the mechanism requires. It then follows that in the next generation more individuals will come from long-necked parents than from short; the average length of neck in the population will have increased because of the preservation of this characteristic by heredity. If there is continued variation in future generations, the selection process will continue, resulting eventually in the major change in the species, such as the modern giraffe's long neck. . . .

. . . It is important to note that by the late 1830s Darwin had broken with the interpretation of natural theology still accepted by the majority of his contemporaries. Even if he did believe that selection could be reconciled with design, his mechanistic interpretation meant that one no longer needed to use the Creator's intentions as an explanatory tool. Natural selection would work whether or not one believed that the Deity continued to keep an eye on things, because it depended solely upon the operations of the everyday, deterministic laws of nature. To find a theory with this characteristic, in fact, had been the driving force of Darwin's investigation. At an early stage in his research, he had decided that the origin of species must be treated as a purely scientific problem that could be answered without appealing directly to the Creator's guiding hand. The search that ended with the discovery of natural selection was inspired by his desire to find a theory that would provide a scientifically acceptable solution to this problem. If this sounds today like a natural extension of Lyell's uniformitarian method into the organic world, it was an extension that Lyell himself deliberately refused to allow. Indeed the majority of naturalists at the time believed that some form of higher power *did* interfere with the normal operations of nature to generate new species, putting the phenomenon outside the range of scientific investigation.

Looking in more detail at Darwin's discovery, one must be aware that he was acutely conscious of the need to follow a method of investigation that would be seen as fully scientific. Precisely because he was extending science into an area that his contemporaries thought unsuitable, he was determined to minimize the risk of being criticized on grounds of inadequate methodology. His later emphasis on the fact-gathering side of his work was intended to show that he was not a mere speculator, someone who rushed into wild theorizing with an insufficient basis in hard facts. The methodological debates of the time clearly acknowledged a role for theorizing in science, and [it] has [been] argued that the writings of Sir J. F. W. Herschel and William Whewell were particularly important to Darwin. Herschel emphasized the need to balance theory and experimental work. Whewell pointed out that a powerful scientific theory achieved its status through its ability to link diverse areas of study, as in the case of Newton's theory of gravitation. This made Darwin aware of the need to expand his own theory's explanatory power, especially once he realized that he had no adequate understanding of the causes of individual variation. . . .

Darwin never gave up the belief that variations are caused when external conditions disturb the reproductive system, but he now realized that the effect would produce random, rather than adaptive changes. He now was committed to searching for a mechanism that could pick out those variations that were by chance of some use to the organism. Forced to admit that his study of the cause of variation was blocked, he simply accepted the existence of random individual differences as an observable fact that legitimately could be built into a theory. By speaking of variation as "random" he signified not only that it occurred in all directions (useless as well as useful) but also that its causes were beyond the scope of immediate investigation. Darwin thus pioneered a new approach to scientific explanation that would become characteristic of the later nineteenth century. Use of the populational approach required acceptance of an explanation based on factors that only could be described statistically, bypassing the need for reducing everything to absolutely fixed laws. As with the kinetic theory of gases, the Darwinian theory required that the laws used by the scientist must be seen as merely the average effect of a vast number of individual events, each of which is subject to causation, but at a level impossible to describe. . . .

By the summer of 1838, Darwin's biological investigations had brought him a long way toward the theory of natural selection. He knew that transmutation must occur through changing the proportion of varying individuals in an isolated population, presumably through the environment picking out those with a useful character. It was at this point, according to the autobiography, that he read Malthus and suddenly realized that the pressure of population must create a struggle for existence in which only the fittest individuals would survive and breed. This leads us to the hottest debate over the development of Darwin's ideas. Did Malthus merely serve to drive home the effectiveness of a selection mechansim that Darwin had already put together on the basis of purely scientific influences? Or does the population principle symbolize the ideological content of the theory, showing how he translated the competitive ethos of Victorian capitalism into a principle of nature?

The need for a balanced account of the factors influencing Darwin is stressed even by those modern accounts that lay most emphasis on "external" forces. The notebooks confirm just how far he had come toward natural selection on the strength of his biological investigations. His readings in *laissez-faire* economics merely reinforced the population thinking already derived from animal breeders. Malthus did not come as a sudden insight that revealed the extent to which struggle was inevitable within society, and hence within nature. Yet those who have suggested a more significant role for Malthus have a case that does not rest on the notebooks alone, because Malthus symbolizes what is perceived as a more pervasive ideological influence on Darwin's thought. Even if only a catalyst, the concept of individual struggle seems to have gained great strength in Darwin's mind as a result of reading Malthus. Whatever its origins, we still may ask why the struggle metaphor appealed so strongly to him and whether its appeal owed anything to a sense that his own society itself was governed by individual competition. . . .

The modern study of Darwin's notebooks has revealed the enormous breadth of the influences that affected his thinking, but it will not resolve the debate over the ideological origins of natural selection. What some historians consider to be the most important external influence may have been transmitted in so indirect a manner that we may never be able to confirm its presence. The link between Darwin's theory and capitalism is too pervasive for it to be detected in the notes recording his conscious thoughts. It rests instead upon the general congruity between the Darwinian image of struggle and the realities of nineteenth-century society. The historian must choose whether he will accept the logic of this indirect argument or concentrate on the wealth of scientific insights revealed by the notebooks.

Development of the Theory: 1840–1859

After putting together the outline of a naturalistic theory of species transformation, Darwin spent the next twenty years exploring its details and expanding its explanatory powers. In 1842 he wrote out a short sketch, followed two years later by a substantial essay that was meant to be published if he should die. He had no intention of publishing yet, however, especially as the reaction to Chambers' *Vestiges* had revealed the strength of public and scientific opposition to transmutation. Darwin only gradually let a few friends into the secret, including Lyell and the botanists Joseph Hooker and Asa Gray. Informally, he was trying to build up a community of scientists who would speak the new language of evolution. Only in the later 1850s did he begin to write a large-scale work with the aim of eventual publication. In 1858 his hand was forced by the arrival of Wallace's paper on natural selection, and he began to write the shorter account that appeared as the *Origin of Species*.

This work was carried out under conditions quite unlike those of the earlier years. Darwin married in 1839 and moved to Down House, in the countryside of Kent. Soon he began to develop the debilitating illness that was to remain with him for the rest of his life. The nature of this illness is uncertain. At one time it was thought that Darwin had picked up a nervous ailment transmitted by a South American insect. More recently it has been suggested that he was poisoning himself with patent medicine or that the symptoms were the result of psychological stress. The illness limited Darwin to a few hours work a day, and prevented him from entering public life except within his local community.

Modern historians have focused attention on two issues in the development of the theory during this period. One concerns the relevance of numerous projects in natural history that Darwin undertook, especially his major study of the barnacles. We now can see that these projects were not merely sidelines, taken up so that Darwin could establish his reputation as a competent biologist. On the contrary, they were used directly to test the theory of evolution and work out its consequences for general biology. Even more crucial were developments taking place within the theory itself. As expounded in the 1844 "Essay" it was *not* complete, and much attention recently has begun to focus on the additions he was forced to make. Darwin began to realize that he would have to relate natural selection to broader trends in the history of life, particularly the constant branching and specialization revealed by many trends in the fossil record. Only when he had explained this through his "principle of divergence" did he feel confident enough to begin his big book on the species question. In the meantime, his thinking on the role of factors such as geographical isolation had changed dramatically.

Darwin's interest in the barnacles was stimulated by the discovery of some unusual specimens while on the *Beagle* voyage. He now saw that the preparation of a complete description of this little-studied subclass would be an ideal way of testing his views on evolution in practice when applied to morphology and classification. His intentions were not made obvious in the published monograph, but in effect Darwin was exploring the effect of evolution on biologists' understanding of natural relationships. . . .

Turning to the developments that occurred within the theory of natural selection, we must first note a disagreement over the state of Darwin's thoughts in the early 1840s. It normally is assumed that by this time he already had come to see natural selection as a unrelenting force, the struggle for existence constantly promoting the fit at the expense of the unfit, even in a stable environment. But [the late historian of science Dov] Ospovat has argued that in its early form, the theory had not yet outgrown its origins in Paley's natural theology. Even in the 1844 Essay, Darwin still implied that species normally exist in a state of perfect adaptation, with little or no individual variation, where the struggle for existence is unnecessary. Only when the environment changed did individual variation appear, providing natural selection with the raw material it needed to change the species until it was once again in a state of perfect adaptation to the new conditions. According to this interpretation, Darwin only gradually over the next decade came to appreciate the full significance of population pressure, which would generate a struggle for existence whatever the external conditions or the degree of individual variation. He thus moved slowly toward the mature version of his theory, in which evolution is not an episodic process separating periods of stability, but a constantly active force that will not only adapt species to new conditions but wherever possible also will increase the level of adaptation even in a stable environment. Only then did he begin to realize the difficulty of reconciling a mechanism based on never-ending struggle with the existence of a benevolent Creator.

Whether or not we accept Ospovat's view of Darwin's original theory, it is certainly true that he became aware increasingly of the extent to which natural selection produced a steady trend toward increased specialization of function. He always had known that evolution was a branching process, an inevitable consequence of the discovery that new species arise when small populations derived from an existing form become geographically isolated under new conditions. At first this point simply was taken for granted, but a problem emerged when Darwin began to realize that after the initial branching there was a continued tendency for the branches to diverge ever farther away from one another. His own research must have helped him to recognize this point, but Ospovat argues that the main stimulus was the work on divergence and specialization by other naturalists such as W. B. Carpenter, Richard Owen, and Henri

Milne-Edwards. Modeling their concept of development upon K. E. von Baer's theory of embryological growth, these biologists argued that not all species are adapted perfectly to their own way of life. On the contrary, they show differing degrees of specialized divergence from the common archetypical form of their group. Carpenter and Owen also showed that the degree of specialization within a group increased in the course of geological time, as shown by the sequence of fossils. This forced Darwin to see that evolution was not a process for merely maintaining adaptation. Instead, it would have to be considered a developmental force that acted upon the once very generalized types from which each class began, producing first a number of separate branches and then progressively specializing each branch for its own particular way of life.

By reinterpreting adaptive evolution as a process of constant specialization, Darwin was able to link his theory with the work of more conventional naturalists. In the *Origin of Species* he even could refer to Owen's paleontological studies on this theme as evidence in favor of his own theory. At the detailed level, Darwin's opposition to teleology forced him to disagree with the interpretation that most of his contemporaries put on the trends they had discovered. Yet in its general outline, his theory was now able to mesh quite well with the latest developments made by those who adopted a more conventional approach to morphology, paleontology, and embryology. His theory would be even easier to promote now that it required only a reinterpretation of existing knowledge, rather than setting up of a whole new picture of the history of life. . . .

The concept of branching evolution inevitably undermined the traditional view that man is the preordained goal of organic progress, but did it necessarily destroy the whole idea of progress? In a branching scheme there can be no unambiguous hierachy of forms each with its allotted position. Instead, it might be possible to set up an abstract scale of organic complexity for ranking entirely different kinds of living structures. Darwin realized that this would be very difficult in practice, but he admitted that most naturalists have an instinctive feeling that some organisms are "higher" than others. In this case, it would be possible to say that evolution was progressive in the sense that it pushed each form toward a higher level of organization within the context of its own peculiar kind of structure. Darwin was tempted to believe that increasing specialization was indeed a form of progress, because it meant that the descendants were better prepared than their ancestors to cope with a particular way of life. This had the advantage of allowing him to retain his old faith in an overall purpose for natural selection, a comfort in his effort to retain a link with the argument from design. He was forced to admit, however, that some kinds of specialization—parasites, for example—result in actual degeneration. Obviously, evolutionary breakthroughs that lead to the establishment of entirely new classes do *not* arise from the more highly specialized members of the previous class. Only a more generalized form can undergo so drastic a change of structure, while specialization easily can become a trap that prevents a species from adapting to rapid changes in its environment and thus leads to extinction. Darwin continued to believe that natural selection could give rise to a form of progress, but he had to concede that it was at best a slow and irregular by-product of the mechanism's chief function of adaptation. . . .

When Darwin put the pieces of this puzzle together in 1856, he regarded his theory as complete. With divergence seen as a necessary consequence of natural selection, he could explain a far greater range of biological phenomena than with the original form of the theory. Only then did he feel confident enough to begin writing his projected "big book" on the species question, a project that would soon be interrupted by his decision to write the *Origin of Species*.

Charles Darwin

On the Origin of Species

In the 1850s, Charles Darwin (1809–1882) was writing his "big book," a long theoretical work to be called Natural Selection. *When he learned in 1858 that the young Alfred Russel Wallace (1823–1913) had independently arrived at the same theory and would soon publish it, Darwin abridged his long draft into a shorter "Abstract" and published it as* On the Origin of Species *in 1859. It is one of the great books in the history of science, and one of the most readable. These excerpts from its introduction, fourth chapter ("Natural Selection"), and conclusion express Darwin's theory in his own words. Today, we call it Darwin's "theory of evolution," but that is not what he called it; then, "evolution" meant development. Instead, he refers to his "theory of descent with modification": that all current species are the modified descendants of other species closely related to them in space and time.*

When on board H.M.S. 'Beagle,' as naturalist, I was much struck with certain facts in the distribution of the inhabitants of South America, and in the geological relations of the present to the past inhabitants of that continent. These facts seemed to me to throw some light on the origin of species—that mystery of mysteries, as it has been called by one of our greatest philosophers. On my return home, it occurred to me, in 1837, that something might perhaps be made out on this question by patiently accumulating and reflecting on all sorts of facts which could possibly have any bearing on it. After five years' work I allowed myself to speculate on the subject, and drew up some short notes; these I enlarged in 1844 into a sketch of the conclusions, which then seemed to me probable: from that period to the present day I have steadily pursued the same object. I hope that I may be excused for entering on these personal details, as I give them to show that I have not been hasty in coming to a decision.

My work is now nearly finished; but as it will take me two or three more years to complete it, and as my health is far from strong, I have been urged to publish this Abstract. I have more especially been induced to do this, as Mr. Wallace, who is now studying the natural history of the Malay archipelago, has arrived at almost exactly the same general conclusions that I have on the origin of species. Last year he sent to me a memoir on this subject, with a request that I would forward it to Sir Charles Lyell, who sent it to the Linnean Society, and it is published in the third volume of the Journal of that Society. Sir C. Lyell and Dr. Hooker, who both knew of my work—the latter having read my sketch of 1844—honoured me by thinking it advisable to publish, with Mr. Wallace's excellent memoir, some brief extracts from my manuscripts. . . .

In considering the Origin of Species, it is quite conceivable that a naturalist, reflecting on the mutual affinities of organic beings, on their embryological relations, their geographical distribution, geological succession, and other such facts, might come to the conclusion that each species had not been independently created, but had descended, like varieties, from other species. Nevertheless, such a conclusion, even if well founded, would be unsatisfactory, until it could be shown how the innumerable species inhabiting this world have been modified, so as to acquire that perfection of structure and coadaptation which most justly excites our admiration. Naturalists continually refer to external conditions, such as climate, food, &c., as the only possible cause of variation. In one very limited sense, as we shall hereafter see, this may be true; but it is preposterous to attribute to mere external conditions, the structure, for instance, of the woodpecker, with its feet, tail, beak, and tongue, so admirably adapted to catch insects under the bark of trees. In the case of the misseltoe, which draws its nourishment from certain trees, which has seeds that must be transported by certain birds, and which has flowers with separate sexes absolutely requiring the agency of certain insects to bring pollen from one flower to the other, it is equally preposterous to account for the structure of this parasite, with its relations to several distinct organic beings, by the effects of external conditions, or of habit, or of the volition of the plant itself. . . .

It is, therefore, of the highest importance to gain a clear insight into the means of modification and coadaptation. At the commencement of my observations it seemed to me probable that a careful study of domesticated animals and of cultivated plants would offer the best chance of making out this obscure problem. Nor have I been disappointed; in this and in all other perplexing cases I have invariably found that our knowledge, imperfect though it be, of variation under domestication, afforded the best and safest clue. I may venture to express my conviction of the high value of such studies, although they have been very commonly neglected by naturalists.

From these considerations, I shall devote the first chapter of this Abstract to Variation under Domestication. We shall thus see that a large amount of hereditary modification is at least possible; and, what is equally or more important, we shall see how great is the power of man in accumulating by his Selection successive slight variations. I will then pass on to the variability of species in a state of nature; but I shall, unfortunately, be compelled to treat this subject far too briefly, as it can be treated properly only by giving long catalogues of facts. We shall, however, be enabled to discuss what circumstances are most favourable to variation. In the next chapter the Struggle for Existence amongst all organic beings throughout the world, which inevitably follows from their high geometrical powers of increase, will be treated of. This is the doctrine of Malthus, applied to the whole animal and vegetable kingdoms. As many more individuals of each species are born than can possibly survive; and as, consequently, there is a frequently recurring struggle for existence, it follows that any being, if it vary however slightly in any manner profitable to itself, under the complex and sometimes varying conditions of life, will have a better chance of surviving, and thus be *naturally selected*. From the strong principle of inheritance, any selected variety will tend to propagate its new and modified form.

This fundamental subject of Natural Selection will be treated at some length in the fourth chapter; and we shall then see how Natural Selection almost inevitably causes much Extinction of the less improved forms of life, and induces what I have called Divergence of Character. In the next chapter I shall discuss the complex and little known laws of variation and of correlation of growth. In the four succeeding chapters, the most apparent and gravest difficulties on the theory will be given: namely, first, the difficulties of transitions, or in understanding how a simple being or a simple organ can be changed and perfected into a highly developed being or elaborately constructed organ; secondly, the subject of Instinct, or the mental powers of animals; thirdly, Hybridism, or the infertility of species and the fertility of varieties when intercrossed; and fourthly, the imperfection of the Geological Record. In the next chapter I shall consider the geological succession of organic beings throughout time; in the eleventh and twelfth, their geographical distribution throughout space; in the thirteenth, their classification or mutual affinities, both when mature and in an embryonic condition. In the last chapter I shall give a brief recapitulation of the whole work, and a few concluding remarks.

No one ought to feel surprise at much remaining as yet unexplained in regard to the origin of species and varieties, if he makes due allowance for our profound ignorance in regard to the mutual relations of all the beings which live around us. Who can explain why one species ranges widely and is very numerous, and why another allied species has a narrow range and is rare? Yet these relations are of the highest importance, for they determine the present welfare, and, as I believe, the future success and modification of every inhabitant of this world. Still less do we know of the mutual relations of the innumerable inhabitants of the world during the many past geological epochs in its history. Although much remains obscure, and will long remain obscure, I can entertain no doubt, after the most deliberate study and dispassionate judgment of which I am capable, that the view which most naturalists entertain, and which I formerly entertained—namely, that each species has been independently created—is erroneous. I am fully convinced that species are not immutable; but that those belonging to what are called the same genera are lineal descendants of some other and generally extinct species, in the same manner as the acknowledged varieties of any one species are the descendants of that species. Furthermore, I am convinced that Natural Selection has been the main but not exclusive means of modification. . . .

If during the long course of ages and under varying conditions of life, organic beings vary at all in the several parts of their organisation, and I think this cannot be disputed; if there be, owing to the high geometrical powers of increase of each species, at some age, season, or year, a severe struggle for life, and this certainly cannot be disputed; then, considering the infinite complexity of the relations of all organic beings to each other and to their conditions of existence, causing an infinite diversity in structure, constitution, and habits, to be advantageous to them, I think it would be a most extraordinary fact if no variation ever had occurred useful to each being's own welfare, in the same way as so many variations have occurred useful to man. But if variations useful to any organic being do occur, assuredly individuals thus characterised will have the best chance of being preserved in the struggle for life; and from the strong principle of inheritance they will tend to produce offspring similarly characterised. This principle of preservation, I have called, for the sake of brevity, Natural Selection. Natural selection, on the principle of qualities being inherited at corresponding ages, can modify the egg, seed, or young, as easily as the adult. Amongst many animals, sexual selection will give its aid to ordinary selection, by assuring to the most vigorous and best adapted males the greatest number of offspring. Sexual selection will also give characters useful to the males alone, in their struggles with other males.

Whether natural selection has really thus acted in nature, in modifying and adapting the various forms of life to their several conditions and stations, must be judged of by the general tenour and balance of evidence given in the following chapters. But we already see how it entails extinction; and how largely extinction has acted in the world's history, geology plainly declares. Natural selection, also, leads to divergence of character; for more living beings can be supported on the same area the more they diverge in structure, habits, and constitution, of which we see proof by looking at the inhabitants of any small spot or at naturalised productions. Therefore during the modification of the descendants of any one species, and during the incessant struggle of all species to increase in numbers, the more diversified these descendants become, the better will be their chance of succeeding in the battle of life. Thus the small differences distinguishing varieties of the same species, will steadily tend to increase till they come to equal the greater differences between species of the same genus, or even of distinct genera.

We have seen that it is the common, the widely-diffused, and widely-ranging species, belonging to the larger genera, which vary most; and these will tend to transmit to their modified offspring that superiority which now makes them dominant in their own countries. Natural selection, as has just been remarked, leads to divergence of character and to much extinction of the less improved and intermediate forms of life. On these principles, I believe, the nature of the affinities of all organic beings may be explained. It is a truly wonderful fact—the wonder of which we are apt to overlook from familiarity—that all animals and all plants throughout all time and space should be related to each other in group subordinate to group, in the manner which we everywhere behold—namely, varieties of the same species most closely related together, species of the same genus less closely and unequally related together, forming sections and sub-genera, species of distinct genera much less closely related, and genera related in different degrees, forming sub-families, families, orders, sub-classes, and classes. The several subordinate groups in any class cannot be ranked in a single file, but seem rather to be clustered round points, and these round other points, and so on in almost endless cycles. On the view that each species has been independently created, I can see no explanation of this great fact in the classification of all organic beings; but, to the best of my judgment, it is explained through inheritance and the complex action of natural selection, entailing extinction and divergence of character, as we have seen illustrated in the diagram.

The affinities of all the beings of the same class have sometimes been represented by a great tree. I believe this simile largely speaks the truth. The green and budding twigs may represent existing species; and those produced during each former year may represent the long succession of extinct species. At each period of growth all the growing twigs have tried to branch out on all sides, and to overtop and kill the surrounding twigs and branches, in the same manner as species and groups of species have tried to overmaster other species in the great battle for life. The limbs divided into great branches, and these into lesser and lesser banches, were themselves once, when the tree was small, budding twigs; and this connexion of the former and present buds by ramifying branches may well represent the classification of all extinct and living species in groups subordinate to groups. Of the many twigs which flourished when the tree was a mere bush, only two or three, now grown into great branches, yet survive and bear all the other branches; so with the species which lived during long-past geological periods, very few now have living and modified descendants. From the first growth of the tree, many a limb and branch has decayed and dropped off; and these lost branches of various sizes may represent those whole orders, families, and genera which have now no living representatives, and which are known to us only from having been found in a fossil state. . . . We here and there see a thin straggling branch springing from a fork low down in a tree, and which by some chance has been favoured and is still alive on its summit. . . . As buds give rise by growth to fresh buds, and these, if vigorous, branch out and overtop on all sides many a feebler branch, so by generation I believe it has been with the great Tree of Life, which fills with its dead and broken branches the crust of the earth, and covers the surface with its ever branching and beautiful ramifications. . . .

It is interesting to contemplate an entangled bank, clothed with many plants of many kinds, with birds singing on the bushes, with various insects flitting about, and with worms crawling through the damp earth, and to reflect that these elaborately constructed forms, so different from each other, and dependent on each other in so complex a manner, have all been produced by laws acting around us. These laws, taken in the largest sense, being Growth with Reproduction; Inheritance which is almost implied by reproduction; Variability from the indirect and direct action of the external conditions of life, and from use and disuse; a Ratio of Increase so high as to lead to a Struggle for Life, and as a consequence to Natural Selection, entailing Divergence of Character and the Extinction of less-improved forms. Thus, from the war of nature, from famine and death, the most exalted object which we are capable of conceiving, namely, the production of the higher animals, directly follows. There is grandeur in this view of life, with its several powers, having been originally breathed into a few forms or into one; and that, whilst this planet has gone cycling on according to the fixed law of gravity, from so simple a beginning endless forms most beautiful and most wonderful have been, and are being, evolved.

Herbert Spencer

Progress and the Social Organism

The British philosopher Herbert Spencer (1820–1903) was one of the most widely read thinkers of the nineteenth century. His various books and essays surveying biology, psychology, anthropology, sociology, and ethics profoundly influenced the social thought of his day. For Spencer, the law of universal history is progressive development, and all progress—cosmic, evolutionary, biological, social, economic, and cultural—derived from the same source, which he explains in the following excerpt from "Progress: Its Law and Cause." In the second excerpt, from "The Social Organism," Spencer uses his idea of progress to reexamine the old idea that society is organic, creating an evolutionary analogy that profoundly influenced the so-called "social Darwinists" of the late nineteenth century.

Progress: Its Law and Cause

The current conception of Progress is somewhat shifting and indefinite. Sometimes it comprehends little more than simple growth—as of a nation in the number of its members and the extent of territory over which it has spread. Sometimes it has reference to quantity of material products—as when the advance of agriculture and manufactures is the topic. Sometimes the superior quality of these products is contemplated: and sometimes the new or improved appliances by which they are produced. When, again, we speak of moral or intellectual progress, we refer to the state of the individual or people exhibiting it; while, when the progress of Knowledge, of Science, of Art, is commented upon, we have in view certain abstract results of human thought and action. Not only, however, is the current conception of Progress more or less vague, but it is in great measure erroneous. It takes in not so much the reality of Progress as its accompaniments—not so much the substance as the shadow. That progress in intelligence seen during the growth of the child into the man, or the savage into the philosopher, is commonly regarded as consisting in the greater number of facts known and laws understood: whereas the actual progress consists in those internal modifications of which this increased knowledge is the expression. Social progress is supposed to consist in the produce of a greater quantity and variety of the articles required for satisfying men's wants; in the increasing security of person and property; in widening freedom of action: whereas, rightly understood, social progress consists in those changes of structure in the social organism which have entailed these consequences. The current conception is a teleological one. The phenomena are contemplated solely as bearing on human happiness. Only those changes are held to constitute progress which directly or indirectly tend to heighten human happiness. And they are thought to constitute progress simply *because* they tend to heighten human happiness. But rightly to understand progress, we must inquire what is the nature of these changes, considered apart from our interests. . . .

In respect to that progress which individual organisms display in the course of their evolution, this question has been answered by the Germans. . . . Investigations . . . . have established the truth that the series of changes gone through during the development of a seed into a tree, or an ovum into an animal, constitute an advance from homogeneity of structure to heterogeneity of structure. In its primary stage, every germ consists of a substance that is uniform throughout, both in texture and chemical composition. The first step is the appearance of a difference between two parts of this substance; or, as the phenomenon is called in physiological language, a differentiation. Each of these differentiated divisions presently begins itself to exhibit some contrast of parts; and by and by these secondary differentiations become as definite as the original one. This process is continously repeated—is simultaneously going on in all parts of the growing embryo; and by endless such differentiations there is finally produced that complex combination of tissues and organs constituting the adult animal or plant. This is the history of all organisms whatever. It is settled beyond dispute that organic progress consists in a change from the homogeneous to the heterogeneous.

Now, we propose in the first place to show, that this law of organic progress is the law of all progress. Whether it be in the development of the Earth, in the development of Life upon its surface, in the development of Society, of Government, of Manufactures, of Commerce, of Language, Literature, Science, Art, this same evolution of the simple into the complex, through successive differentiations, holds throughout. From the earliest traceable cosmical changes down to the latest results of civilization, we shall find that the transformation of the homogeneous into the heterogeneous, is that in which Progress essentially consists. . . .

. . . From this uniformity of procedure, may we not infer some fundamental necessity whence it results? May we not rationally seek for some all-pervading principle which determines this all-pervading process of things? Does not the universality of the *law* imply a universal *cause?* . . .

. . . We pass at once to the statement of the law, which is this:—*Every active force produces more than one change—every cause produces more than one effect.*

Before this law can be duly comprehended, a few examples must be looked at. When one body is struck against another, that which we usually regard as the effect, is a change of position or motion in one or both bodies. But a moment's thought shows us that this is a careless and very incomplete view of the matter. Besides the visible mechanical result, sound is produced; or, to speak accurately, a vibration in one or both bodies, and in the surrounding air: and under some circumstances we call this the effect. Moreover, the air has not only been made to vibrate, but has had sundry currents caused in it by transit of the bodies. Further, there is a disarrangement of the particles of the two bodies in the neighbourhood of their point of collision; amounting in some cases to a visible condensation. Yet more, this condensation is accompanied by the disengagement of heat. In some cases a spark—that is, light—results, from the incandescence of a portion struck off; and sometimes this incandescence is associated with chemical combination.

Thus, by the original mechanical force expended in the collision, at least five, and often more, different kinds of changes have been produced. . . .

. . . From the law that every active force produces more than one change, it is an inevitable corollary that through all time there has been an evergrowing complication of things. Starting with the ultimate fact that every cause produces more than one effect, we may readily see that throughout creation there must have gone on, and must still go on, a never-ceasing transformation of the homogeneous into the heterogeneous. . . .

. . . Endless facts go to show that every kind of progress is from the homogeneous to the heterogeneous; and that it is so because each change is followed by many changes. And it is significant that where the facts are most accessible and abundant, there are these truths most manifest.

However, to avoid committing ourselves to more than is yet proved, we must be content with saying that such are the law and the cause of all progress that is known to us. Should the Nebular Hypothesis ever be established, then it will become manifest that the Universe at large, like every organism, was once homogeneous; that as a whole, and in every detail, it has unceasingly advanced towards greater heterogeneity; and that its heterogeneity is still increasing. It will be seen that as in each event of to-day, so from the beginning, the decomposition of every expended force into several forces has been perpetually producing a higher complication; that the increase of heterogeneity so brought about is still going on, and must continue to go on; and that thus Progress is not an accident, not a thing within human control, but a beneficent necessity. . . .

The Social Organism

. . . Universally in [the past], things were explained on the hypothesis of manufacture, rather than that of growth: as indeed they are, by the majority, in our own day. It was held that the planets were severally projected round the sun from the Creator's hand; with exactly the velocity required to balance the sun's attraction. The formation of the Earth, the separation of sea from land, the production of animals, were mechanical works from which God rested as a labourer rests. Man was supposed to be moulded after a manner somewhat akin to that in which a modeller makes a clay-figure. And of course, in harmony with such ideas, societies were tacitly assumed to be arranged thus or thus by direct interposition of Providence; or by the regulations of law-makers; or by both.

Yet that societies are not artifically put together, is a truth so manifest, that it seems wonderful men should have ever overlooked it. Perhaps nothing more clearly shows the small value of historical studies, as they have been commonly pursued. You need but to look at the changes going on around, or observe social organization in its leading peculiarities, to see that these are neither supernatural, nor are determined by the wills of individual men, as by implication historians commonly teach; but are consequent on general natural causes. . . .

. . . Those who regard the histories of societies as the histories of their great men, and think that these great men shape the fates of their societies, overlook the truth that such great men are the products of their societies. Without certain antecedents—without a certain average national character, they could neither have been generated nor could have had the culture which formed them. If their society is to some extent re-moulded by them, they were, both before and after birth, moulded by their society—were the results of all those influences which fostered the ancestral character they inherited, and gave their own early bias, their creed, morals, knowledge, aspirations. So that such social changes as are immediately traceable to individuals of unusual power, are still remotely traceable to the social causes which produced these individuals, and hence, from the highest point of view, such social changes also, are parts of the general developmental process. . . .

A perception that there exists some analogy between the body politic and a living individual body, was early reached; and from time to time re-appeared in literature. But this perception was necessarily vague and more or less fanciful. In the absence of physiological science, and especially of those comprehensive generalizations which it has but recently reached, it was impossible to discern the real parallelisms. . . .

Let us set out by succinctly stating the points of similarity and the points of difference. Societies agree with individual organisms in four conspicuous peculiarities:—

1. That commencing as small aggregations, they insensibly augment in mass: some of them eventually reaching ten thousand times what they originally were.

2. That while at first so simple in structure as to be considered structureless, they assume, in the course of their growth, a continually-increasing complexity of structure.

3. That though in their early, undeveloped states, there exists in them scarcely any mutual dependence of parts, their parts gradually acquire a mutual dependence; which becomes at last so great, that the activity and life of each part is made possible only by the activity and life of the rest.

4. That the life and development of a society is independent of, and far more prolonged than, the life and development of any of its component units; who are severally born, grow, work, reproduce, and die, while the body politic composed of them survives generation after generation, increasing in mass, completeness of structure, and functional activity.

These four parallelisms will appear the more significant the more we contemplate them. While the points specified, are points in which societies agree with individual organisms, they are points in which individual organisms agree with each other, and disagree with all things else. In the course of its existence, every plant and animal increases in mass, in a way not parallelled by inorganic objects: even such inorganic objects as crystals, which arise by growth, show us no such definite relation between growth and existence as organisms do. The orderly progress from simplicity to complexity, displayed by bodies politic in common with all living bodies, is a characteristic which distinguishes living bodies from the inanimate bodies amid which they move. That functional dependence of parts, which is scarcely more manifest in animals or plants than nations, has no counterpart elsewhere. And in no aggregate except an organic, or a social one, is there a perpetual removal and replacement of parts, joined with a continued integrity of the whole.

Moreover, societies and organisms are not only alike in these peculiarities, in which they are unlike all other things; but the highest societies, like the highest organisms, exhibit them in the greatest degree. We see that the lowest animals do not increase to anything like the sizes of the higher ones; and, similarly, we see that aboriginal societies are comparatively limited in their growths. In complexity, our large civilized nations as much exceed primitive savage tribes, as a vertebrate animal does a zoophyte. Simple communities, like simple creatures, have so little mutual dependence of parts, that subdivision or mutilation causes but little inconvenience; but from complex communities, as from complex creatures, you cannot remove any considerable organ without producing great disturbance or death of the rest. And in societies of low type, as in inferior animals, the life of the aggregate, often cut short by division or dissolution, exceeds in length the lives of the component units, very far less than in civilized communities and superior animals; which outlive many generations of their component units.

On the other hand, the leading differences between societies and individual organisms are these:—

1. That societies have no specific external forms. This, however, is a point of contrast which loses much of its importance, when we remember that throughout the vegetal kingdom, as well as in some lower divisions of the animal kingdom, the forms are often very indefinite—definiteness being rather the exception than the rule; and that they are manifestly in part determined by surrounding physical circumstances, as the forms of societies are. If, too, it should eventually be shown, as we believe it will, that the form of every species of organism has resulted from the average play of the external forces to which it has been subject during its evolution as a species; then, that the external forms of societies should depend, as they do, on surrounding conditions, will be a further point of community.

2. That though the living tissue whereof an individual organism consists, forms a continous mass, the living elements of a society do not form a continous mass; but are more or less widely dispersed over some portion of the Earth's surface. This, which at first sight appears to be a fundamental distinction, is one which yet to a great extent disappears when we contemplate all the facts. For, in the lower divisions of the animal and vegetal kingdoms, there are types of organization much more nearly allied, in this respect, to the organization of a society, than might be supposed—types in which the living units essentially composing the mass, are dispersed through an inert substance, that can scarcely be called living in the full sense of the word. . . .

Indeed, it may be contended that this is the primitive form of all organization; seeing that, even in the highest creatures, as in ourselves, every tissue developes out of what physiologists call a blastema—an unorganized though organizable substance, through which organic points are distributed. Now this is very much the case with a society. For we must remember that though the men who make up a society, are physically separate and even scattered; yet that the surface over which they are scattered is not one devoid of life, but is covered by life of a lower order which ministers to their life. The vegetation which clothes a country, makes possible the animal life in that country; and only through its animal and vegetal products can such a country support a human society. Hence the members of the body politic are not to be regarded as separated by intervals of dead space; but as diffused through a space occupied by life of a lower order. In our conception of a social organism, we must include all that lower organic existence on which human existence, and therefore social existence, depends. And when we do this, we see that the citizens who make up a community, may be considered as highly vitalized units surrounded by substances of lower vitality, from which they draw their nutriment: much as in the cases above instanced. Thus, when examined, this apparent distinction in great part disappears.

3. That while the ultimate living elements of an individual organism, are mostly fixed in their relative positions, those of the social organism are capable of moving from place to place, seems a marked disagreement. But here, too, the disagreement is much less than would be supposed. For while citizens are locomotive in their private capacities, they are fixed in their public capacities. As farmers, manufacturers, or traders, men carry on their business at the same spots, often throughout their whole lives; and if they go away occasionally, they leave behind others to discharge their functions in their absence. Each great centre of production, each manufacturing town or district, continues always in the same place; and many of the firms in such town or district, are for generations carried on either by the descendants or successors of those who founded them. Just as in a living body, the cells that make up some important organ, severally perform their functions for a time and then disappear, leaving others to supply their places; so, in each part of a society, the organ remains, though the persons who compose it change. Thus, in social life, as in the life of an animal, the units as well as the larger agencies formed of them, are in the main stationary as respects the places where they discharge their duties and obtain their sustenance. And hence the power of individual locomotion does not practically affect the analogy.

4. The last and perhaps the most important distincion, is, that while in the body of an animal, only a special tissue is endowed with feeling; in a society, all the members are endowed with feeling. Even this distinction, however, is by no means a complete one. For in some of the lowest animals, characterized by the absence of a nervous system, such sensitiveness as exists is possessed by all parts. It is only in the more organized forms that feeling is monopolized by one class of the vital elements. Moreover, we must remember that societies, too, are not without a certain differentiation of this kind. Though the units of a community are all sensitive, yet they are so in unequal degrees. The classes engaged in agriculture and laborious occupations in general, are much less susceptible, intellectually and emotionally, than the rest; and especially less so than the classes of highest mental culture. Still, we have here a tolerably decided contrast between bodies politic and individual bodies. And it is one which we should keep constantly in view. For it reminds us that while in individual bodies, the welfare of all other parts is rightly subservient to the welfare of the nervous system, whose pleasurable or painful activities make up the good or evil of life; in bodies politic, the same thing does not hold, or holds to but a very slight extent. It is well that the lives of all parts of an animal should be merged in the life of the whole; because the whole has a corporate consciousness capable of happiness or misery. But it is not so with a society; since its living units do not and cannot lose individual consciousness; and since the community as a whole has no corporate consciousness. And this is an everlasting reason why the welfare of citizens cannot rightly be sacrificed to some supposed benefit of the State; but why, on the other hand, the State is to be maintained solely for the benefit of citizens. The corporate life must here be subservient to the lives of the parts; instead of the lives of the parts being subservient to the corporate life.

Such, then, are the points of analogy and the points of difference. May we not say that the points of difference serve but to bring into clearer light the points of analogy. . . .The *principles* of organization are the same; and the differences are simply differences of application.

Trofim Lysenko
The Situation in Biology

In the West, the name Trofim Lysenko (1898–1976) is usually associated, not with Darwinism, but with "anti-Darwinism" and the repression of Soviet genetics. First rising to prominence in farming in the late 1920s, he was supported by Stalin beginning in 1935 and finally won all-out backing for his theories in 1948; thereafter, genetics was officially condemned as a bourgeois, capitalist science by the Communist Party and the Soviet government until 1965. In this excerpt from his triumphal 1948 presidential speech at the Lenin All-Union Academy of Agricultural Sciences, Lysenko presented a version of Darwinism (so-called "creative Darwinism") that dominated Soviet biology for more than a generation. Note that Lysenko's interpretation denies both intraspecific competition and the gradual emergence of new species—generally regarded in the West as the hallmarks of Darwin's theory—and emphasizes instead Darwin's belief in the inheritance of acquired characteristics, his reliance on the practical experience of breeders, and his denial of the origin of species through divine creation.

The History of Biology: A History of Ideological Battle

The appearance of Darwin's teaching, expounded in his book, *The Origin of Species,* marked the beginning of scientific biology.

The leading idea of Darwin's theory is the teaching on natural and artificial selection. Selection of variations favourable to the organism has produced, and continues to produce, the fitness which we observe in living nature; in the structure of organisms and their adaptation to their conditions of life. Darwin's theory of selection provided a rational explanation of the fitness observable in living nature. His idea of selection is scientific and true. In substance, his teaching on selection is a summation of the age-old practical experience of plant and animal breeders who, long before Darwin, produced varieties of plants and breeds of animals by the empirical method.

Darwin investigated the numerous facts obtained by naturalists in living nature and analyzed them through the prism of practical experience. Agricultural practice served Darwin as the material basis for the elaboration of his theory of evolution, which explained the natural causes of the purposiveness we see in the structure of the organic world. That was a great advance in the knowledge of living nature. . . .

The classics of Marxism, while fully appreciating the significance of the Darwinian theory, pointed out the errors of which Darwin was guilty. Darwin's theory, though unquestionably materialist in its main features, is not free from some serious errors. A major fault, for example, is the fact that, along with the materialist principle, Darwin introduced into his theory of evolution reactionary Malthusian ideas. In our days this major fault is being aggravated by reactionary biologists. . . .

Darwin himself, in his day, was unable to fight free of the theoretical errors of which he was guilty. . . . Today there is absolutely no justification for accepting the erroneous aspects of the Darwinian theory, those based on Malthus' theory of overpopulation with the inference of a struggle presumably going on within species. And it is all the more inadmissible to represent these erroneous aspects as the cornerstone of Darwinism. . . . Such an approach to Darwin's theory prejudices the creative development of its scientific core.

Even when Darwin's teaching first made its appearance, it became clear at once that its scientific, materialist core, the theory of the evolution of living nature, was antagonistic to the idealism that reigned in biology.

Progressively thinking biologists, both in

our country and abroad, saw in Darwinism the only right road to the further development of scientific biology. They took it upon themselves to defend Darwinism against the attacks of the reactionaries, with the Church at their head, and of obscurantists in science. . . .

In the post-Darwinian period the overwhelming majority of biologists—far from further developing Darwin's teaching—did all they could to debase Darwinism, to smother its scientific foundation. The most glaring manifestation of such debasement of Darwinism is to be found in the teachings of Weismann, Mendel, and Morgan, the founders of modern reactionary genetics. . . .

The Michurinists [the followers of Soviet biologist I. V. Michurin], in their investigations, take the Darwinian theory of evolution as their basis. But in itself Darwin's theory is absolutely insufficient for dealing with the practical problems of socialist agriculture. That is why the basis of contemporary Soviet agrobiology is Darwinism transformed in the light of the teachings of Michurin . . . and there by converted into Soviet creative Darwinism.

Many problems of Darwinism assume a different aspect as the result of the development of our Soviet agrobiological science, of the Michurin trend in agrobiology. Darwinism has not only been purified of its deficiencies and errors and raised to a higher level, but has undergone a considerable change in a number of its principles. From a science which primarily *explains* the past history of the organic world, it is becoming a creative, *effective means* of systematically mastering living nature, making it serve practical requirements.

Our Soviet Michurinist Darwinism is a creative Darwinism which poses and solves problems of the theory of evolution in a new way, in the light of Michurin's teaching.

I cannot in this report touch on many of the theoretical problems of great practical significance.

I shall dwell briefly on only one of them—namely, the question of intra- and interspecific relations in living nature.

The time has come to consider the question of speciation, approaching it from the angle of the transition of quantitative accumulation into qualitative specific distinctions.

We must realize that speciation is a transition—in the course of the historical process—from quantitative to qualitative variations. Such a leap is prepared by the vital activity of organic forms themselves, as the result of quantitative accumulations of responses to the action of the definite conditions of life, and that is something that can definitely be studied and directed.

Such an understandiing of speciation, an understanding of natural laws, places in the hands of biologists a powerful means of regulating the vital process itself and consequently speciation as well.

I think that in posing the question this way we may assume that what leads to the appearance of a new specific form, to the formation of a new species out of an old one, is not the accumulation of quantitative distinctions by which varieties within a species are usually recognized. The quantitative accumulations of variations which lead to the leap which changes an old form of species into a new form are variations of a *different order*. . . .

Species are not an abstraction, but actually existing links in the general biological chain.

Living nature is a biological chain broken up, as it were, into individual links or species. It is therefore wrong to say that a species does not retain the constancy of its qualitative definiteness as a species for any length of time. To insist on that would be to regard the evolution of living nature as proceeding as if along a plane, without any leaps.

I am confirmed in this opinion by the data of experiments for the conversion of hard wheat (durum) into soft (yulgare).

Let me note that all systematists admit that these are good, indisputable, independent species.

We know that there are no true winter forms among hard wheats, and that is why in all regions with a relatively severe winter hard wheat is cultivated only as a spring, not a winter, crop. Michurinists have mastered a good method of converting spring into winter wheat. It has already been mentioned that many spring wheats have been experimentally converted into winter wheat. But all of those belonged to the species of soft wheat. When experiments were started to convert hard wheat into winter wheat it was found that after two, three or four years of autumn planting (required to turn a spring into a winter crop) durum becomes vulgare, that is to say, one species is converted into another. Durum wheat with 28 chromosomes is converted into several varieties of soft 42-chromosome wheat, nor do we, in this case, find any transitional forms between the durum and vulgare species. *The conversion of one species into another takes place by a leap.*

We thus see that the formation of a new species is prepared by an alteration of vital activity under definite new conditions in a number of generations. In our case it is necessary to bring autumn and winter conditions to bear on hard wheat in the course of two, three or four generations. Then it can change by a leap into soft wheat without any transitional forms between the two species.

I think that it may be pertinent to note that what led me to study the essentially theoretical problems of species and of intraspecific and interspecific relations among individuals, was never mere curiosity or a fondness for abstract theorizing. I was and am led to study these questions of theory by my work in the course of which I have to find answers to purely practical problems. For a correct understanding of the relations among individuals within a species and between species it was necessary to have a clear idea of the qualitative distinctions of intraspecific and interspecific diversities of forms.

It thus became possible to find new solutions to such problems of practical importance as weed control in farming, or the choosing of ingredients for the sowing of grass mixtures, or the speedy and extensive afforestation of steppe areas, and many others.

That is what led me to make a new study of the problem of intra- and interspecific struggle and competition, and after a thorough and comprehensive investigation I have come to the conclusion that there exist no intraspecific struggle and mutual assistance among individuals within a species, and that there does exist interspecific struggle and competition and also mutual assistance between different species. I regret that I have so far done very little to elucidate the theoretical implications and practical significance of these questions in the press.

. . . Thus, Comrades, as regards the theoretical line in biology, Soviet biologists hold that the Michurin principles are the only scientific principles. The Weismannists and their followers, who deny the heritability of acquired characters, are not worth dwelling on at too great length. The future belongs to Michurin. (*Applause.*)

V. I. Lenin and *J. V. Stalin* discovered I. V. Michurin and made his teaching the possession of the Soviet people. By their great paternal attention to his work they saved for biology the remarkable Michurin teaching. The Party, the Government, and *J. V. Stalin* personally, have taken an unflagging interest in the further development of the Michurin teaching. There is no more honourable task for us Soviet biologists than creatively to develop Michurin's teaching and to follow in all our activities Michurin's style in the investigation of the nature of the development of living beings. . . .

Comrades, before I pass to my concluding remarks I consider it my duty to make the following statement.

The question is asked in one of the notes handed to me, What is the attitude of the Central Committee of the Party to my report? I answer: The Central Committee of the Party examined my report and approved it. (*Stormy applause. Ovation. All rise.*) . . .

The Michurin trend in biology is a materialist trend, because it does not separate heredity from the living body and the conditions of its life. There is no living body without heredity, and there is no heredity without a living body. The living body and its conditions of life are inseparable. Deprive an organism of its conditions of life and the living body will die. . . .

It is still not clear to some that heredity is inherent not only in the chromosomes, but in any particle of the living body. They therefore want to see with their own eyes cases of hereditary properties and characters transmitted from generation to generation without the transmission of chromosomes. . . .

. . . Experiments in vegetative hybridization provide unmistakable proof that any particle of a living body, even the plastic substances, even the sap exchanged between scion and stock, possesses hereditary qualities.

Does this detract from the role of the chromosomes? Not in the least. Is heredity transmitted through the chromosomes in the sexual process? Of course it is.

We recognize the chromosomes. We do not deny their existence. But we do not recognize the chromosome *theory* of heredity. . . .

The present session has demonstrated *the complete triumph of the Michurin trend over Morganism-Mendelism.* (*Applause.*)

It is truly a historic landmark in the development of biological science. (*Applause.*)

I think I shall not be wrong if I say that this session has been a great occasion for all workers in the sciences of biology and agriculture. (*Applause.*)

The Party and the Government are showing paternal concern for the strengthening and development of the Michurin trend in our science, for the removal of all obstacles to its further progress. This imposes upon us the duty to work still more extensively and profoundly to arm the state farms and collective farms with an advanced scientific theory. That is what the Soviet people expect of us.

We must effectively place science, theory, at the service of the people, so that crop yeilds and the productivity of stockbreeding may increase at a still more rapid pace, that labour on state farms and collective farms may be more efficient.

I call upon all Academicians, scientific workers, agronomists, and animal breeders to bend all their efforts and work in close unity with the foremost men and women in socialist farming to achieve these great and noble aims. (*Applause.*)

Progressive biological science owes it to the geniuses of mankind, *Lenin* and *Stalin,* that *the teaching of I. V. Michurin has been added to the treasure house of our knowledge, has become part of the gold fund of our science.* (*Applause.*)

Long live the Michurin teaching, which shows how to transform living nature for the benefit of the Soviet people! (*Applause.*)

Long live the Party of Lenin and Stalin, which discovered Michurin for the world (*applause*) and created all the conditions for the progress of advanced materialist biology in our country. (*Applause.*)

Glory to the great friend and protagonist of science, our leader and teacher, Comrade Stalin! (*All rise. Prolonged applause.*)

Peter Kropotkin

Mutual Aid

Famous as the "anarchist prince," Peter Kropotkin (1842–1921) grew from his Russian aristocratic origins into one of Europe's most prominent revolutionaries. In the early 1860s, Kropotkin traveled in Siberia and studied its natural history. He had read Darwin's Origin *and expected to see a harsh struggle for existence between members of the same species, but instead he was impressed by the ways in which members of wild herds and flocks cooperated against common enemies. This convinced him that the great Russian biologists of his day were right in criticizing Darwin's emphasis on competition and the struggle for existence. Only while in England, where this view was not generally held, was Kropotkin moved to write the book* Mutual Aid: A Factor of Evolution *(1902). The following excerpts are from the 1914 edition. Like many other synthetic thinkers of his day, Kropotkin did not make sharp distinctions between biological and social evolution, and he justified his political vision in biological terms: he advocated "anarchy"—society without government—because he believed that, like animals, humans in the natural state would cooperate with one another to overcome hardships.*

Two aspects of animal life impressed me most during the journeys which I made in my youth in Eastern Siberia and Northern Manchuria. One of them was the extreme severity of the struggle for existence which most species of animals have to carry on against an inclement Nature; the enormous destruction of life which periodically results from natural agencies; and the consequent paucity of life over the vast territory which fell under my observation. And the other was, that even in those few spots where animal life teemed in abundance, I failed to find—although I was eagerly looking for it—that bitter struggle for the means of existence, *among animals belonging to the same species,* which was considered by most Darwinists (though not always by Darwin himself) as the dominant characteristic of struggle for life, and the main factor of evolution.

The terrible snow-storms which sweep over the northern portion of Eurasia in the later part of the winter, and the glazed frost that often follows them; the frosts and the snow-storms which return every year in the second half of May, when the trees are already in full blossom and insect life swarms everywhere; the early frosts and, occasionally, the heavy snowfalls in July and August, which suddenly destroy myriads of insects, as well as the second broods of the birds in the prairies; the torrential rains, due to the monsoons, which fall in more temperate regions in August and September—resulting in inundations on a scale which is only known in America and in Eastern Asia, and swamping, on the plateaus, areas as wide as European States; and finally, the heavy snowfalls, early in October, which eventually render a territory as large as France and Germany, absolutely impracticable for ruminants, and destroy them by the thousand—these were the conditions under which I saw animal life struggling in Northern Asia. They made me realize at an early date the overwhelming importance in Nature of what Darwin described as "the natural checks to over-multiplication,"in comparison to the struggle between individuals of the same species for the means of subsistence, which may go on here and there, to some limited extent, but never attains the importance of the former. Paucity of life, underpopulation—not over-population—being the distinctive feature of that immense part of the globe which we name Northern Asia, I conceived since then serious doubts—which subsequent study has only confirmed—as to the reality of that fearful competition for food and life within each species, which was an article of faith with most Darwinists, and, consequently, as to the dominant part which this sort of competition was supposed to play in the evolution of new species.

On the other hand, wherever I saw animal life in abundance, as, for instance, on the lakes where scores of species and millions of individuals came together to rear their progeny; in the colonies of rodents; in the migrations of birds which took place at that time on a truly American scale along the Usuri; and especially in a migration of fallow-deer which I witnessed on the Amur, and during which scores of thousands of these intelligent animals came together from an immense territory, flying before the coming deep snow, in order to cross the Amur where it is narrowest—in all these scenes of animal life which passed before my eyes, I saw Mutual Aid and Mutual Support carried on to an extent which made me suspect in it a feature of the greatest importance for the maintenance of life, the preservation of each species, and its further evolution.

And finally, I saw among the semi-wild cattle and horses in Transbaikalia, among the wild ruminants everywhere, the squirrels, and so on, that when animals have to struggle against scarcity of food, in consequence of one of the above-mentioned causes, the whole of that portion of the species which is affected by the calamity, comes out of the ordeal so much impoverished in vigour and health, that *no progressive evolution of the species can be based upon such periods of keen competition.*

Consequently, when my attention was drawn, later on, to the relations between Darwinism and Sociology, I could agree with none of the works and pamphlets that had been written upon this important subject. They all endeavoured to prove that Man, owing to his higher intelligence and knowledge, *may* mitigate the harshness of the struggle for life between men; but they all recognized at the same time that the struggle for the means of existence, of every animal against all its congeners, and of every man against all other men, was "a law of Nature." This view, however, I could not accept, because I was persuaded that to admit a pitiless inner war for life within each species, and to see in that war a condition of progress, was to admit something which not only had not yet been proved, but also lacked confirmation from direct observation. . . .

. . . I consequently directed my chief attention to establishing first of all, the importance of the Mutual Aid factor of evolution, leaving to ulterior research the task of discovering the *origin* of the Mutual Aid instinct in Nature. . . .

. . . It is not love to my neighbour—whom I often do not know at all—which induces me to seize a pail of water and to rush towards his house when I see it on fire; it is a far wider, even though more vague feeling or instinct of human solidarity and sociability which moves me. So it is also with animals. It is not love, and not even sympathy (understood in its proper sense) which induces a herd of ruminants or of horses to form a ring in order to resist an attack of wolves; not love which induces wolves to form a pack for hunting; not love which induces kittens or lambs to play, or a dozen of species of young birds to spend their days together in the autumn; and it is neither love nor personal sympathy which induces many thousand fallow-deer scattered over a territory as large as France to form into a score of separate herds, all marching towards a given spot in order to cross there a river. It is a feeling infinitely wider than love or personal sympathy—an instinct that has been slowly developed among animals and men in the course of an extremely long evolution, and which has taught animals and men alike the force they can borrow from the practice of mutual aid and support, and the joys they can find in social life.

The importance of this distinction will be easily appreciated by the student of animal psychology, and the more so by the student of human ethics. Love, sympathy and self-sacrifice certainly play an immense part in the progressive development of our moral feelings. But it is not love and not even sympathy upon which Society is based in mankind. It is the conscience—be it only at the stage of an instinct—of human solidarity. It is the unconscious recognition of the force that is borrowed by each man from the practice of mutual aid; of the close dependency of every one's happiness upon the happiness of all; and of the sense of justice, or equity, which brings the individual to consider the rights of every other individual as equal to his own. Upon this broad and necessary foundation the still higher moral feelings are developed. . . .

It will probably be remarked that mutual aid, even though it may represent one of the factors of evolution, covers nevertheless one aspect only of human relations; that by the side of this current, powerful though it may be, there is, and always has been, the other current—the self-assertion of the individual, not only in its efforts to attain personal or caste superiority, economical, political, and spiritual, but also in its much more important although less evident function of breaking through the bonds, always prone to become crystallized, which the tribe, the village community, the city, and the State impose upon the individual. In other words, there is the self-assertion of the individual taken as a progressive element.

It is evident that no review of evolution can be complete, unless these two dominant currents are analyzed. . . .

To make even a rough estimate of their relative importance by any method more or less statistical, is evidently impossible. One single war—we all know—may be productive of more evil, immediate and subsequent, than hundreds of years of the unchecked action of the mutual-aid principle may be productive of good. But when we see that in the animal world, progressive development and mutual aid go hand in hand, while the inner struggle within the species is concomitant with retrogressive development; when we notice that with man, even success in struggle and war is proportionate to the development of mutual aid in each of the two conflicting nations, cities, parties, or tribes, and that in the process of evolution war itself (so far as it can go this way) has been made subservient to the ends of progress in mutual aid within the nation, the city or the clan—we already obtain a perception of the dominating influence of the mutual-aid factor as an element of progress. But we see also that the practice of mutual aid and its successive developments have created the very conditions of society life in which man was enabled to develop his arts, knowledge, and intelligence; and that the periods when institutions based on the mutual-aid tendency took their greatest development were also the periods of the greatest progress in arts, industry, and science. In fact, the study of the inner life of the mediaeval city and of the ancient Greek cities reveals the fact that the combination of mutual aid, as it was practised within the guild and the Greek clan, with a large initiative which was left to the individual and the group by means of the federative principle, gave to mankind the two greatest periods of its history—the ancient Greek city and the mediaeval city periods; while the ruin of the above institutions during the State periods of history, which followed, corresponded in both cases to a rapid decay. . . .

To attribute . . . the industrial progress of our century to the war of each against all which it has proclaimed, is to reason like the man who, knowing not the causes of rain, attributes it to the victim he has immolated before his clay idol. For industrial progress, as for each other conquest over nature, mutual aid and close intercourse certainly are, as they have been, much more advantageous than mutual struggle.

However, it is especially in the domain of ethics that the dominating importance of the mutual-aid principle appears in full. That mutual aid is the real foundation of our ethical conceptions seems evident enough. But whatever the opinions as to the first origin of the mutual-aid feeling or instinct may be—whether a biological or a supernatural cause is ascribed to it—we must trace its existence as far back as to the lowest stages of the animal world; and from these stages we can follow its uninterrupted evolution, in opposition to a number of contrary agencies, through all degrees of human development, up to the present times. Even the new religions which were born from time to time—always at epochs when the mutual-aid principle was falling into decay in the theocracies and despotic States of the East, or at the decline of the Roman Empire—even the new religions have only reaffirmed that same principle. They found their first supporters among the humble, in the lowest, down-trodden layers of society, where the mutual-aid principle is the necessary foundation of every-day life; and the new forms of union which were introduced in the earliest Buddhist and Christian communities, in the Moravian brotherhoods and so on, took the character of a return to the best aspects of mutual aid in early tribal life. . . .

. . . In the practice of mutual aid, which we can retrace to the earliest beginnings of evolution, we thus find the positive and undoubted origin of our ethical conceptions; and we can affirm that in the ethical progress of man, mutual support—not mutual struggle—has had the leading part. In its wide extension, even at the present time, we also see the best guarantee of a still loftier evolution of our race.

CHAPTER 10

Uncertainty in Physics and Society

Introduction by Richard Olson

As we saw in chapters 8 and 9, theories derived from the biological sciences had a substantial impact on the broader European culture during the nineteenth century. But just at a time when evolutionary thought seemed to overshadow other scientific theories in popular ideologies, new startling discoveries within physics occurred to recapture public interest and provoke bewilderment. These revolutionary developments—associated with the terms relativity and quantum theory—first occurred in connection with sophisticated mathematical theories that could be understood only by the most highly trained physicists, and they seemed far removed from ordinary human needs and experiences. But their impact soon expanded throughout the intellectual communities of Europe and America because each seemed to demand a fundamental rethinking of the nature and limits of human knowledge which was as radical as that demanded by Newtonian physics and Lockean psychology in the Enlightenment (see chapters 6 and 7). Together, the relativistic and quantum revolutions challenged traditional understandings of such widely used concepts as "time," "space," "matter," "cause"; and they even seemed to imply that at some fundamental level there is and can be no objective "real" and deterministic universe for humans to know.

In order to appreciate some of the major implications of the revolutions that created "modern" physics, we must briefly consider a few widely held principles associated with the "classical" physics of the eighteenth and nineteenth centuries as well as several closely related problems which absorbed much of the attention of nineteenth-century physicists. First, Newtonian science had effectively abandoned the mid-seventeenth century "mechanist" assumption that all of nature was constituted of perfectly hard particles moving in void space which interacted only through impact with one another. By the middle of the nineteenth century it was generally admitted that the ultimate nature of the smallest particles which aggregate to form observable bodies was not fully known; but it was clear that they had associated with them certain "forces" of attraction and repulsion, including gravitational, chemical, and electromagnetic forces, which could be mathematically described.

Any physical phenomenon thus became intelligible when it could be accounted for in terms of bodies moving in space and time under the influence of well-understood forces. Certain forces, including gravity, were assumed to be permanently attached to particles and to be extended through space without any consideration of time. Thus, many phenomena, including most of those associated with gases, were explicable in terms of the "kinetic theory" which modified the old mechanistic assumptions only by redefining the "impact" of particles in terms of forces extended in space rather than in terms of hard body collisions.

Additionally, by the late nineteenth century it was generally agreed that light, electricity, and magnetism were all associated with electromagnetic forces that could be propagated from one body to another in the form of waves which moved with a single velocity, the speed of light. With very few exceptions, physicists agreed that electromagnetic waves had to be propagated in some medium, as sound waves were propagated in the air and other materials. This medium was called the "luminiferous" or "electromagnetic" ether, and one of the great problems of classical physics was to discover its structure and character. Although physicists admitted that the nature of ether might be no more explicable than the nature of particles of matter, they generally agreed that it would be counted as understood if a model of it could be constructed.

In his selection on "ether and reality," Jonathan Powers discusses certain key features of electromagnetic phenomena which had to be explicable through any ether model—the production of magnetic effects by varying electric currents and the propagation of "transverse" waves with the velocity of light. Then he describes a mechanical ether model developed by James Clerk Maxwell during the mid-nineteenth century.

Not all nineteenth-century physicists sought the same kind of concrete, visualizable models that seemed to appeal to Victorian British physicists. As Pierre Duhem explains in the selection from *The Aim and Structure of Physical Theories*, French and German scientists were more drawn to abstract theories which emphasized what we might call mathematical models. But even those who attempted to develop the more austere and elegant mathematical models agreed that any acceptable model had to satisfy the requirements of Newtonian mechanics at the same time that it accounted for electromagnetic phenomena.

As the nineteenth century progressed, physicists became increasingly uneasy; for in spite of their collective ingenuity, after nearly a century of work they were still unable to provide an adequate mechanical or mathematical model of the ether. And new problems were beginning to arise in connection with the study of cathode-rays—rays that appeared when electrical discharges occurred in evacuated tubes—which stubbornly resisted attempts to classify them unambiguously as either waves or as collections of moving particles.

Beginning in 1881, moreover, when the young American naval officer Albert A. Michelson (1852–1931) made his first attempt to measure the velocity of the earth through the ether, a series of experimental discoveries began to undermine the foundations of the secure structure of classical physics. The very existence of the stationary ether, that "continuous substance filling all space, which can vibrate light, which can be shared into positive and negative electricity, which in whirls constitutes matter, and which transmits by continuity and not by impact every action and reaction of which matter is capable," was called into question by Michelson's experiment. And since much nineteenth-century physical theory depended on the ether's assumed existence, this uncertainty raised fundamental problems.

Again, in 1887, Henrich Hertz (1857–1894) made an odd and inexplicable discovery in the course of confirming some of the most impressive predictions of classical electromagnetic theory. He found that when ultraviolet light shone on the antenna of his apparatus it became a more sensitive detector of electromagnetic waves; but there was nothing in electromagnetic theory to explain why. Even more puzzling and disturbing were the discoveries that began in 1895 when Wilhelm Konrad Roentgen (1845–1923) of Munich detected a mysterious radiation—X-rays—emanating from electric discharge-tubes. During the next year, Henri Becquerel's (1852–1908) investigations showed that uranium gave off another kind of unknown radiation without any kind of external stimulation; and within the next few years three different forms of natural radioactivity, alpha-rays, beta-rays, and gamma-rays, had been isolated, if not understood. In 1900 the experimental work of Rubens and Kurlbaum seemed to show that even the heat radiation from ordinary materials had properties which were inconsistent with the most fundamental and general of classical theories, thermodynamics.

Throughout the later nineteenth century, one further set of experimental results also began to bother the physicists. As early as 1822 Joseph Fraunhofer (1787–1826) had discovered that a chemical element when heated to the gaseous phase emits light of certain frequencies that are characteristic of the element; that is, each element has a unique spectrum. By 1860 these characteristic spectra were being used to detect the presence of elements even in minute traces and to investigate the composition of the sun and other stellar bodies. It was natural that physicists should try to discover what it was in the structure of the elements that gave rise to these characteristic spectral emissions. Unfortunately, there were almost no clues to work from. In 1885 Johann J. Balmer (1825–1898), one of many searchers, discovered an empirical mathematical relation among some of the frequencies of light emitted in the spectrum of one element—hydrogen—but the quantitative relations discovered by Balmer and others seemed to complicate rather than simplify the attempt to discover the atomic structure from within classical physical theory.

These discoveries and many more made during the first decades of the twentieth century demanded dramatic changes in physical theories, changes which were not surprisingly paralleled by and to some degree mirrored in changes in social thought, literature, and art.

The two physical theories which seemed to call forth the greatest response among nonscientists were first the special theory of relativity, advanced by Albert Einstein (1879–1955) in 1905 in response to problems closely related to Michelson's experiments, and second, quantum mechanics, a body of theories developed in response to problems associated with the interpretation of Hertz's work, radioactivity, atomic spectra, and with the pattern of radiation emitted by hot bodies. Because of space limitations, the selections and comments in this chapter are confined to some of the semipopular responses to the theory of relativity and to the "indeterminacy" or "uncertainty" principle from quantum mechanics.

There can be no doubt that relativity and quantum theory have had profound effects on formal aspects of contemporary philosophy, or that modern science has provided dramatic analogs outside the limits of these two theories; but to most moderately well-educated individuals in Western society, "relativity" and "indeterminacy" are the most common and important terms which provide a link to the abstruse theories of men like Einstein, Max Planck, Niels Bohr, and Werner Heisenberg. And these terms or concepts have played a very important role in giving modern intellectuals a sense of their own limitations and a sense—whether it be real or illusory—of their own freedom to act and to control their destiny.

Einstein's theory clearly overthrew a variety of traditional beliefs. One physical implication, for example, was that the law of conservation of energy was not strictly valid and that mass and energy were somehow interconvertible. Similarly, the theory called into question our basic notions of space and of time. As Hermann Minkowski (1864–1909) wrote in 1908, for the relativistic physics, "space by itself and time by itself, are doomed to fade away into mere shadows, and only a kind of union of the two will preserve an independent reality." This interrelation between space and time undermined the commonsense notion of simultaneity; for in relativity theory, two events that are simultaneous in one frame of reference may occur at different times in another. In a like manner, the space-time relation overthrew the commonsense notion of length; for a body might have different spatial "lengths" in different frames of reference. Perhaps most fundamentally, Einstein's assumption of the invariable speed of light in a vacuum made it impossible to expect that any mechanistic structure of the ether might be developed to account for the propagation of electromagnetic waves. In fact, it precluded all possibility of a mechanical or traditional mathematical model and forced its adherents to accept the idea that a theory need be (perhaps can be) no more than a set of mathematical functions in a four-dimensional space along with rules for interpreting the values of those functions.

In order to provide a sense of how our commonsense notions come to be undermined by relativity theory, we include Bertrand Russell's (1872–1970) discussion of the issue of simultaneity from *The ABC of Relativity*, one of the clearest lay accounts of the theory and its implications.

In spite of the radical revisions which relativity theory calls for in traditional physical ideas, however, it does not in any way challenge the fundamental lawfulness of physical phenomena, nor does it justify a belief in the subjectivity of physical experience. The theory of relativity is premised on the assumption that no preferred frame of reference exists from which absolute motion or rest can be determined, but this does not mean that physical phenomena as seen from various vantage points in space-time are in any sense incompatible with one another or dependent on unspecifiable, personal aspects of the observer. In fact, in its most general form the theory explicitly demands that all physical laws have an identical form for all frames of reference and thus provides a way to discover precisely what the description of a physical event will be in any arbitrary frame of reference once it is described in one frame. Confusion on this issue reigned even among competent physicists during the early years of the twentieth century; and many popularizers and scientific critics provided the license which lay readers used to turn Einstein's theory on its head as a justification for a belief in the subjectivity of all knowledge.

A good example of the kind of distortion of the theory of relativity which reached the nonscientific public was L. T. More's statement:

> Both Professor Einstein's theory of relativity and Professor Planck's theory of Quanta are proclaimed somewhat noisily to be the greatest revolutions in scientific method since the time of Newton. That they are revolutionary there can be no doubt, insofar as they substitute mathematical symbols as the basis of science and deny that any concrete experience underlies these symbols, thus replacing an objective by a subjective universe.

More was correct insofar as he saw relativity as a challenge to traditional materialism, but the notion expressed here that Einstein's theory is less "objective" than mechanistic theories is absurd if one uses this term in its ordinary signification. The term objective usually means: based on experienced fact and free from personal biases, and this is precisely what relativity theory is—it is more in conformity with experienced fact than classical physics, and it by no means implies that personal attitudes should play a role in physical theory. Yet the implication that relativity theory was somehow subjective formed the basis for the lay reception.

One of the most important cultural uses of relativity theory was made by historians and other social scientists to reinforce a trend away from the positivistic and scientistic approaches to the study of historical and social phenomena, which had been popular during the nineteenth century. In "Charles A. Beard, the 'New Physics,' and Historical Relativity," Hugh I. Rodgers explains the background to this movement and provides a case study which shows how relativistic notions filtered down through scientific popularizations into important aspects of social thought.

The principle of uncertainty or indeterminancy enunciated by Werner Heisenberg (1901–1976) in 1927 as a fundamental principle of quantum mechanics has been no less important than Einstein's theory of relativity in calling into question important traditional beliefs and in giving new hope to those who see a deterministic and materialistic universe as unbearable and valueless. According to the most widely held interpretation of quantum mechanics, the uncertainty principle challenges the very notion of cause and effect which underlies amost all classical science and much traditional social and theological thought. In addition, by guaranteeing our inability to stipulate perfectly the state of any isolated part of the universe, the uncertainty principle weakens the argument of determinists in all fields and seems to give a new support for speculations about free will.

The basic contention of the uncertainty principle is simple. To any physical quantity Q which we might investigate, there corresponds another quantity P, such that we can never know the magnitudes of both P and Q exactly. In fact, the product of the uncertainties in the measurements of P and Q can never be less than the so-called quantum of action, a constant discovered by Max Planck (1858–1947) in 1901 in connection with his studies of radiant heat. Also, the principle implies that if we know the position of an atomic particle with perfect accuracy, we cannot predict its momentum.

Werner Heisenberg has himself been one of the most successful popular expositors of the philosophical implications of quantum theory. In the selection below, from a lecture delivered in 1959, he explores some of the challenges presented to traditional beliefs in "materialism" and "objectivity" and explains how some of the new developments point backward to notions developed by Plato and Aristotle but abandoned by "classical" scientists.

The uncertainty principle—and with it, all of quantum theory—can be interpreted in several different ways. According to one interpretation, statements from quantum theory can only be interpreted statistically. The statistical laws given by quantum theory may be perfectly determined, although the results of any single event cannot be known. Our uncertainty may only be the result of ignorance which is unavoidable because of the nature of all physical techniques of measurement. In order to measure or observe an event, the observer must disturb the situation he or she seeks to study; for example, in measuring the position of a particle the observer inevitably changes its velocity. This interpretation of quantum theory places an emphasis on the inseparability of observer and observed and reinforces the "subjectivist" notions of relativity. It makes the scientist play a more crucial role within scientific theory and is used to justify the interposition of personal values and biases within social theory. At the same time it leaves intact the basic belief in a deterministic universe.

This statistical interpretation of quantum mechanics, however, faces serious challenges on empirical grounds, and many scientists as well as laypersons see in quantum theory the justification for a belief in a fundamentally noncausal, indeterministic element in the structure of reality—an element which reinforces humanistic yearnings for a new legitimizing of the notion of free will. This second, more radical, interpretation of quantum theory has had the greatest impact on humanistic thought and is represented in the selection from John Lukacs's *The Historical Consciousness*. Lukacs provides a succinct summary of the major implications drawn from quantum mechanics by humanistic scholars—implications which seem to signal a possible reversal of the modern confidence in scientific objective methods in many fields of human endeavor.

Jonathan Powers

Aether and Reality

Jonathan Powers' recent attempt to explicate the philosophical assumptions and implications of modern physics discusses the use of mechanical analogs—what James Clerk Maxwell called "physical analogies"—in the construction of aether (ether) theories in the nineteenth century.

The first key step in the unification of the theories of electricity and magnetism was a discovery made by Hans Christian Oersted whilst lecturing to students in 1820. He had placed a compass needle, pointing north-south, under a wire which ran in the same direction. Then he turned the current on, and to his astonishment the compass needle swung round through a quarter circle and pointed east-west. Oersted, who was influenced by romantic ideas of the unity of nature, responded with rhapsodies about 'the conflict of electricities', but he ought to have been deeply disturbed. The behaviour of his apparatus appeared to defy a basic principle of intelligibility—namely, the principle of symmetry. . . .

. . . The principle is that a symmetrical arrangement of causes cannot produce an asymmetrical effect, so we can ask, 'What must the intrinsic symmetry of a magnet be in order for its orientation at right-angles to the current to be the most symmetrical arrangement possible?' And the answer is, 'It must possess the intrinsic symmetry of a rotating cylinder' (see Figure 10-1).

If we follow this lead, it becomes natural to see a magnet as containing circulating electric currents, and this conveniently explains why you can never have an isolated magnetic pole:

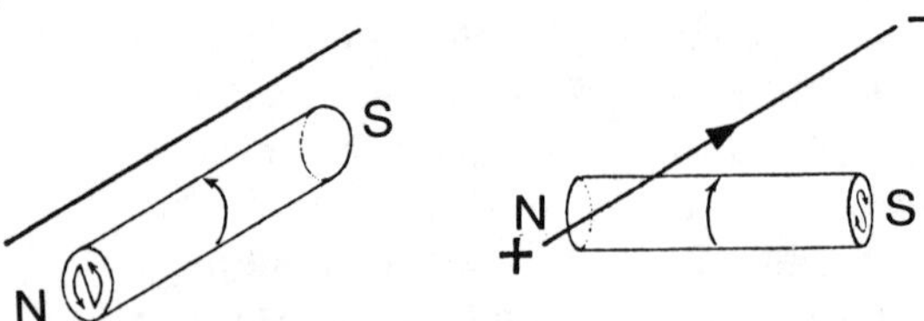

Figure 10-1 *Oersted's experiment reinterpreted. If the magnetic needle has the intrinsic symmetry of a rotating cylinder then, when the current flows in the wire, the second arrangement is more symmetrical than the first.*

a 'north pole' is a current circulating anticlockwise; a 'south pole' is a current circulating clockwise, and which you 'see' depends on where you stand. This suggestion means that 'magnetic poles' can disappear from the vocabulary of physics. It would in many ways be theoretically tidier to abolish magnetic fields too, and replace references to magnetism by descriptions of the interactions of electric currents. . . . The connection between electric and magnetic fields led to relativity theory.

At first, despite the impressive series of fundamental discoveries he made, [Michael] Faraday's ideas on fields of force were not taken seriously by theoreticians because of his informal, non-mathematical presentation. However, . . . James Clerk Maxwell . . . from 1855 onwards codified and subtly changed the message. Like many other British physicists of the period, Maxwell believed that space was not empty but filled with an all-pervasive aether, which he took to be governed by the laws of Newtonian mechanics. Maxwell's strategy was to cast Faraday's ideas into mathematical form by devising a mechanical analogue for them, thus showing that the theory of the electromagnetic field was consistent with Newtonian mechanics, and hence making it probable that the 'aether' had a mechanical structure obeying Newtonian principles.

. . . According to the wave theory, light is a 'transverse' vibration in an 'aether' which possesses the characteristics of an elastic solid. Given the high velocity of light this aether had to be conceived as combining such apparently incompatible properties as extremely high rigidity and negligible density. And, since there are no light-like 'compression' waves, analogous to sound waves, the aether also had to be absolutely incompressible, whilst offering no perceptible resistance to bodies moving through it. Maxwell's initiative was to try to link the idea of this 'luminiferous' aether with Faraday's notion of an electromagnetic field. He imagined that each magnetic line of force was enveloped by a sleeve-like magnetic vortex, and so that the vortices could rotate in the same sense around neighboring lines of force he separated them by layers of elastic electrical particles (see Figure 10-2). On this theory if the electrical particles are put in motion they cause the vortices adjacent to them to be put into rotation and thus generate magnetic field lines. Since the aetherial electrical particles were supposed to be elastic they could be seen as distorted by pressure or tension, which would bring about magnetic effects identical to those of an electric current. This structure allowed for a wave motion to be propagated through the aether, and Maxwell was able to calculate the velocity of these waves, which turned out to be the same as that of light.

By filling all of space with a medium which could be the basis for the transmission of causal influences, the aether theory re-established the primacy of 'action by contact', thus satisfying one of the requirements of a mechanistic world picture. The theory had other attractions too. There were some who were delighted by the idea that science had revealed the existence of an invisible reality, and were quick to draw parallels with religion. In addition, many of these late nineteenth-century scientists took an interest in spiritualism, and speculated that the 'aether' might provide not only the connection between 'mind' and 'body', but between this and the 'unseen' world. They also conceived of themselves as 'gentlemen', for whom science was an ethical pursuit, aimed at 'understanding' rather than material gain, and they took as much interest in the problems of integrating their physics into a world picture which supported 'spiritual values' as they took in the technical problems of physics itself.

As far as purely physical theory is concerned, the idea of an aether led to three main lines of development. First of all there were those who attempted to make the aether mechanistically intelligible by populating physical space with bizarre . . . constructions of cogs and flywheels, jointed rods and gyroscopes. Secondly, there were those who . . . tried to develop a completely 'electro-magnetic' picture of the world, in which mechanics itself would be 'explained'. To similar effect others tried to treat 'matter' as vortices in the aether, and 'inertia' as the interaction of these vortices with the field. A third line was suggested by Heinrich Hertz, the discoverer of radio waves, who argued that Maxwell's theory was simply Maxwell's equations, which did not need to be encumbered with such postulates as the 'aether'.

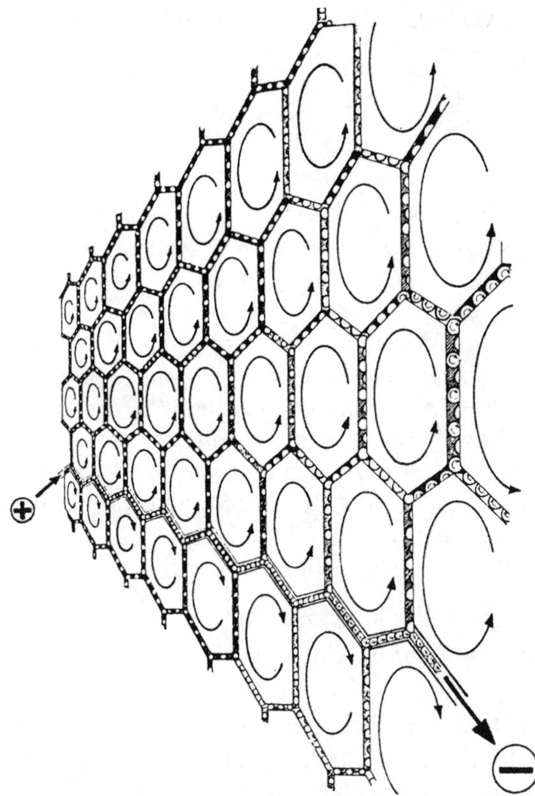

Figure 10-2 *Maxwell's mechanical aether model: a honeycomb of sleeves of magnetic vortices separated by elastic electrical particles.*

Whichever way the Faraday-Maxwell theory was taken, however, it had one subtly devastating consequence. . . . In Newtonian mechanics all states of uniform straightline motion are mechanically equivalent to 'rest in Absolute Space'. In the Faraday-Maxwell theory, however, you only get magnetic fields appearing when electrical charges *actually move*—relative to the aether frame, as Maxwell would have it, but in any case relative to a privileged frame of reference which has no parallel in mechanics. An almost invisible, but fatal, crack had appeared in the classical framework.

Pierre Duhem

Mechanical Models versus Mathematical Reasoning

Pierre Duhem, an early twentieth-century French physical chemist, historian, and philosopher of science, wrote extensively on both medieval science and on nineteenth-century science. One of his major concerns for the modern period was with national differences in scientific styles. In the following selection he focuses on the difference between British preferences for concrete mechanical models (he said that reading Maxwell's theories was like "walking into a factory") and continental preferences for abstract mathematical theories. It is important to note, however, that even continental mathematical theories depended on common assumptions about rigid bodies, the independence of space and time, and the distinction between matter and energy which had been formulated in connection with classical mechanics.

English Physics and the Mechanical Model

In the treatises on physics published in England, there is always one element which greatly astonishes the French student; that element which nearly invariably accompanies the exposition of a theory, is the model. Nothing helps us better understand how very different from ours is the manner in which the English mind proceeds in the construction of science than this use of the model.

Two electrically charged bodies are before us, and the problem is to give a theory of their mutual attractions or repulsions. The French or German physicist . . . will by an act of thought postulate in the space outside these bodies that abstraction called a material point and, associated with it, that other abstraction called an electric charge. He then tries to calculate a third abstraction: the force to which the material point is subjected. He gives formulas which permit one to determine the magnitude and direction of this force for each possible position of this material point. From these formulas he deduces a series of consequences: he shows clearly that in each point of space the force is directed along the tangent to a certain line called the line of force, and that all the lines of force cross normally (at right angles) certain surfaces—viz., equipotential surfaces—whose equations he gives. In particular, he shows that they are normal to the surfaces of the two charged conductors which are included among the number of equipotential surfaces. He calculates the force to which each element of these surfaces is subjected. Finally, he integrates all these elementary forces according to the rules of statics; he then knows the laws of the mutual actions of the two charged bodies.

This whole theory of electrostatics constitutes a group of abstract ideas and general propositions, formulated in the clear and precise language of geometry and algebra, and connected with one another by the rules of strict logic. This whole fully satisfies the reason of a French physicist and his taste for clarity, simplicity, and order.

The same does not hold for an Englishman. These abstract notions of material points, force, line of force, and equipotential surface do not satisfy his need to imagine concrete, material, visible, and tangible things. "So long as we cling to this mode of representation," says an English physicist, "we cannot form a mental representation of the phenomena which are really happening." It is to satisfy this need that he goes and creates a model.

The French or German physicist conceives, in the space separating two conductors, abstract lines of force having no thickness or real existence; the English physicist materializes these lines and thickens them to the dimensions of a tube which he will fill with vulcanized rubber. In place of a family of lines of ideal forces, conceivable only by reason, he will have a bundle of elastic strings, visible and tangible, firmly glued at both ends to the surfaces of the two conductors, and, when stretched, trying both to contract and to expand. When the two conductors approach each other, he sees the elastic strings drawing closer together; then he sees each of them bunch up and grow large. Such is the famous model of electrostatic action imagined by Faraday and admired as a work of genius by Maxwell and the whole English school.

The employment of similar mechanical models, recalling by certain more or less rough analogies the particular features of the theory being expounded, is a regular feature of the English treatises on physics. Here is a book (O. Lodge) intended to expound the modern theories of electricity and to expound a new theory. In it there are nothing but strings which move around pulleys, which roll around drums, which go through pearl beads, which carry weights; and tubes which pump water while others swell and contract; toothed wheels which are geared to one another and engage hooks. We thought we were entering the tranquil and neatly ordered abode of reason, but we find ourselves in a factory.

The use of such mechanical models, very far from facilitating the understanding of a theory by a French reader, requires him, in many cases, to make a serious effort to grasp the operation of what is often a very complicated apparatus, as described to him by the English author. Quite an effort is required in order to recognize the analogies between the properties of this apparatus and the propositions of the theory that is being illustrated. This effort is often much greater than the one the Frenchman needs to make in order to understand in its purity the abstract theory which it is claimed the model embodies.

The Englishman, on the other hand, finds the use of the model so necessary to the study of physics that to his mind the sight of the model ends up by being confounded with the very understanding of the theory. It is curious to see this confusion formally accepted and proclaimed by one who is the highest expression of the English scientific genius, one who, famous for a long time under the name of William Thomson, has been raised to the peerage with the title of Lord Kelvin. In his *Lectures on Molecular Dynamics*, he says:

"My object is to show how to make a mechanical model which shall fulfill the conditions required in the physical phenomena that we are considering, whatever they may be. At the time when we are considering the phenomenon of elasticity in solids, I want to show a model of that. At another time, when we have vibrations of light to consider, I want to show a model of the action exhibited in that phenomenon. We want to understand the whole about it; we only understand a part. It seems to me that the test of 'Do we or do we not understand a particular subject in physics?' is 'Can we make a mechanical model of it?' I have an immense admiration for Maxwell's mechanical model of electromagnetic induction. He makes a model that does all the wonderful things that electricity does in inducing currents, etc., and there can be no doubt that a mechanical model of that kind is immensely instructive and is a step towards a definite theory of electromagnetism." . . .

The English physicist does not, therefore, ask any metaphysics to furnish the elements with which he can design his mechanisms. He does not aim to know what the irreducible properties of the ultimate elements of matter are. W. Thomson, for example, never asks himself such philosophical questions as: Is matter continuous or is it formed of individual elements? Is the volume of one of the ultimate elements of matter variable or invariable? What is the nature of an atom's actions: are they efficacious at a distance or only by contact? These questions do not even enter his mind, or else, when they are presented to him, he pushes them away as otiose and injurious to scientific progress. For instance, he says: "The idea of an atom has been so constantly associated with incredible assumptions of infinite strength, absolute rigidity, mystical actions at a distance, and indivisibility, that chemists and many other reasonable naturalists of modern times, losing all patience with it, have dismissed it to the realms of metaphysics, and made it smaller than 'anything we can conceive.' But if atoms are inconceivably small, why are not all chemical actions infinitely swift? Chemistry is powerless to deal with this question, and many others of paramount importance, if barred by the hardness of its fundamental assumptions, from contemplating the atom as a real portion of matter occupying a finite space, and forming a not immeasurably small constituent of any palpable body."

The bodies with which the English physicist constructs his models are not abstract conceptions elaborated by metaphysics. They are concrete bodies, similar to those surrounding us; namely, bodies that are solid or liquid, rigid or flexible, flowing or viscous; and with solidity, fluidity, rigidity, flexibility, and viscosity it is not necessary to understand abstract properties defined in terms of a certain cosmology. These properties are nowhere defined, but imagined by means of observable examples: rigidity calls up the image of a block of steel; flexibility, that of a silk thread; viscosity, that of glycerine. In order to express in a more tangible manner the concrete character of the bodies with which he builds his mechanisms, W. Thomson is not afraid to designate them by the most everyday names: he calls them bell-cranks, cords, jellies. He could not indicate more clearly that what he is concerned with are not combinations intended to be conceived by reason, but mechanical contrivances intended to be seen by the imagination.

Neither could he warn us more clearly that the models he proposes should not be taken as *explanations* of natural laws; anyone who should attribute such a meaning to them would be exposed to strange surprises.

Bertrand Russell

The ABC of Relativity

Bertrand Russell was both one of the most brilliant British analytic philosophers of this century and one of the most effective popularizers of philosophical ideas. In 1925 he wrote The ABC of Relativity *to explain to a lay audience some of the fundamental implications of relativity theory. In this selection he addresses the inseparability of space and time which seems so at odds with our common sense.*

Until the advent of the special theory of relativity, no one had thought that there could be any ambiguity in the statement that two events in different places happened at the same time. It might be admitted that, if the places were very far apart, there might be difficulty in finding out for certain whether the events were simultaneous, but every one thought the meaning of the question perfectly definite. It turned out, however, that this was a mistake. Two events in distant places may appear simultaneous to one observer who has taken all due precautions to insure accuracy (and, in particular, has allowed for the velocity of light), while another equally careful observer may judge that the first event preceded the second, and still another may judge that the second preceded the first. This would happen if the three observers were all moving rapidly relatively to each other. It would not be the case that one of them would be right and the other two wrong: they would all be equally right. The time-order of events is in part dependent upon the observer; it is not always and altogether an intrinsic relation between the events themselves. Einstein has shown, not only that this view accounts for the phenomena, but also that it is the one which ought to have resulted from careful reasoning based upon the old data. In actual fact, however, no one noticed the logical basis of the theory of relativity until the odd results of experiment had given a jog to people's reasoning powers. . . .

This question of time in different places is perhaps, for the imagination, the most difficult aspect of the theory of relativity. . . . It seems obvious that we can speak of the positions of the planets at a given instant. The Newtonian theory enables us to calculate the distance between the earth and (say) Jupiter at a given time by the Greenwich clocks; this enables us to know how long light takes at that time to travel from Jupiter to the earth—say half an hour; this enables us to infer that half an hour ago Jupiter was where we see it now. All this seems obvious. But in fact it only works in practice because the relative velocities of the planets are very small compared with the velocity of light. When we judge that an event on the earth and an event on Jupiter have happened at the same time—for example, that Jupiter eclipsed one of his moons when the Greenwich clocks showed twelve midnight—a person moving relatively to the earth would judge differently, assuming that both he and we had made the proper allowance for the velocity of light. And naturally the disagreement about simultaneity involves a disagreement about periods of time. If we judged that two events on Jupiter were separated by twenty-four hours, another person might judge that they were separated by a longer time, if he were moving rapidly relatively to Jupiter and the earth.

The universal cosmic time which used to be taken for granted is thus no longer admissible. For each body, there is a definite time-order for the events in its neighbourhood; this may be called the 'proper' time for that body. Our own experience is governed by the proper time for our own body. As we all remain very nearly stationary on the earth, the proper times of different human beings agree, and can be lumped together as terrestrial time. But this is only the time appropriate to *large* bodies on the earth. For beta particles (electrons) in laboratories, quite different times would be wanted; it is because we insist upon using our own time that these particles seem to increase in mass with rapid motion. From their own point of view, their mass remains constant, and it is we who suddenly grow thin or corpulent. The history of a physicist as observed by a beta particle would resemble Gulliver's travels.

The question now arises: what really is measured by a clock? When we speak of a clock in the theory of relativity, we do not mean only clocks made by human hands: we mean anything which goes through some regular periodic performance. The earth is a clock, because it rotates once in every twenty-three hours and fifty-six minutes. An atom is a clock, because it emits light-waves of very definite frequencies; these are visible as bright lines in the spectrum of the atom. The world is full of periodic occurrences, and fundamental mechanisms, such as atoms, show an extraordinary similarity in different parts of the universe. Any one of these periodic occurrences may be used for measuring time; the only advantage of humanly manufactured clocks is that they are specially easy to observe. However, some of the others are more accurate. Nowadays the short radio waves emitted under certain conditions by caesium atoms and ammonia molecules are being used to establish standards of time measurement more uniform than those based on the earth's rotation. But the question remains: If cosmic time is abandoned, what is really measured by a clock in the wide sense that we have just given to the term?

Each clock gives a correct measure of its own 'proper' time, which, as we shall see presently, is an important physical quantity. But it does not give an accurate measure of any physical quantity connected with events on bodies that are moving rapidly in relation to it. It gives one datum towards the discovery of a physical quantity connected with such events, but another datum is required, and this has to be derived from measurement of distances in space. Distances in space, like periods of time, are in general not objective physical facts, but partly dependent upon the observer. How this comes about must now be explained.

First of all, we have to think of the distance between two events, not between two bodies. This follows at once from what we have found as regards time. If two bodies are moving relatively to each other—and this is really always the case—the distance between them will be continually changing, so that we can only speak of the distance between them at a given time. If you are in a train travelling towards Edinburgh, we can speak of your distance from Edinburgh at a given time. But, as we said, different observers will judge differently as to what is the 'same' time for an event in the train and an event in Edinburgh. This makes the measurement of distances relative, in just the same way as the measurement of times has been found to be relative. We commonly think that there are two separate kinds of interval between two events, an interval in space and an interval in time: between your departure from London and your arrival in Edinburgh, there are four hundred miles and ten hours. We have already seen that another observer will judge the time differently; it is even more obvious that he will judge the distance differently. An observer on the sun will think the motion of the train quite trivial, and will judge that you have travelled the distance travelled by the earth in its orbit and its diurnal rotation. On the other hand, a flea in the railway carriage will judge that you have not moved at all in space, but have afforded him a period of pleasure which he will measure by his 'proper' time, not by Greenwich Observatory. It cannot be said that you or the sun-dweller or the flea are mistaken: each is equally justified and is only wrong if he ascribes an objective validity to his subjective measures. The distance in space between two events is, therefore, not in itself a physical fact. But, as we shall see, there is a physical fact which can be inferred from the distance in time together with the distance in space. This is what is called the 'interval' in space-time.

Taking any two events in the universe, there are two different possibilities as to the relation between them. It may be physically possible for a body to travel so as to be present at both events or it may not. This depends upon the fact that no body can travel as fast as light. Suppose, for example, that it were possible to send out a flash of light from the earth and have it reflected back from the moon. (This is an experiment which has actually been performed, not with light, but with radar waves, which travel at the same speed.) The time between the sending of the flash and the return of the reflection would be about two and a half seconds. No body could travel so fast as to be present on the earth during any part of those two and a half seconds and also present on the moon at the moment of the arrival of the flash, because in order to do so the body would have to travel faster than light. But theoretically a body could be present on the earth at any time before or after those two and a half seconds and also present on the moon at the time when the flash arrived. When it is physically impossible for a body to travel so as to be present at both events, we shall say that the interval* between the two events is 'space-like'; when it is physically possible for a body to be present at both events, we shall say that the interval between the two events is 'time- like.' When the interval is 'space-like,' it is possible for a body to move in such a way that an observer on the body will judge the two events to be simultaneous. In that case, the 'interval' between the two events is what such an observer will judge to be the distance in space between them. When the interval is 'time-like,' a body can be present at both events; in that case, the 'interval' between the two events is what an observer on the body will judge to be the time between them, that is to say, it is his 'proper' time between the two events. There is a limiting case between the two, when the two events are parts of one light-flash—or, as we might say, when the one event is the seeing of the other. In that case, the interval between the two events is zero.

There are thus three cases. (1) It may be possible for a ray of light to be present at both events; this happens whenever one of them is the seeing of the other. In this case the interval between the two events is zero. (2) It may happen that no body can travel from one event to the other, because in order to do so it would have to travel faster than light. In that case, it is always physically possible for a body to travel in such a way that an observer on the body would judge the two events to be simultaneous. The interval is what he would judge to be the distance in space between the two events. Such

* I shall define 'interval' in a moment.

an interval is called 'space-like.' (3) It may be physically possible for a body to travel so as to be present at both events; in that case, the interval between them is what an observer on such a body will judge to be the time between them. Such an interval is called 'time-like.'

The interval between two events is a physical fact about them, not dependent upon the particular circumstances of the observer.

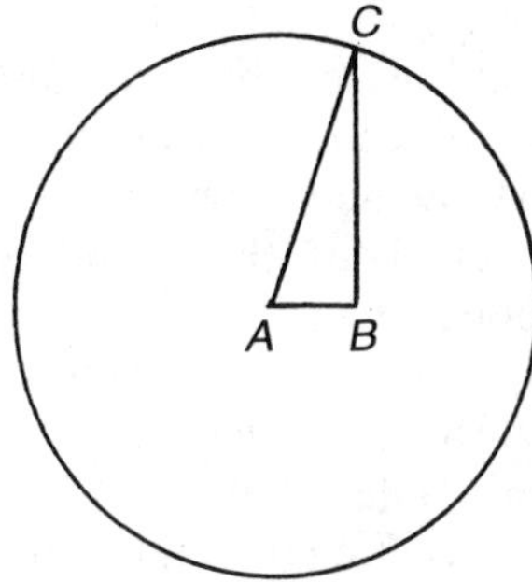

There are two forms of the theory of relativity, the special and the general. The former is in general only approximate, but becomes very nearly exact at great distances from gravitating matter. Whenever gravitation may be neglected, the special theory can be applied, and then the interval between two events can be calculated when we know the distance in space and the distance in time between them, estimated by any observer. If the distance in space is greater than the distance that light would have travelled in the time, the separation is space-like. Then the following construction gives the interval between the two events: Draw a line AB as long as the distance that light would travel in the time; round A describes a circle whose radius is the distance in space between the two events; through B draw BC perpendicular to AB, meeting the circle in C. Then BC is the length of the interval between the two events.

When the distance is time-like, use the same figure, but let AC be now the distance that light would travel in the time, while AB is the distance in space between the two events. The interval between them is now the time that light would take to travel the distance BC.

Although AB and AC are different for different observers, BC is the same length for all observers, subject to corrections made by the general theory. It represents the one interval in 'space-time' which replaces the two intervals in space and time of the older physics.

Hugh L. Rodgers

Charles A. Beard, the "New Physics," and Historical Relativity

Relativity theory formed an intellectual prop for arguments widely separated from the domain of physics. It was used by art critics, politicians, and, as Hugh Rodgers explains, by social scientists in general and historians in particular to justify abandoning all attempts to establish an "objective" understanding of human actions. Rodgers focuses on Charles Beard, one of the most influential American historians of the early twentieth century.

Writing in the middle of the 1930s, Tobias Dantzig summarized the profound transformation wrought by Einstein in basic scientific assumptions by noting that modern science differed from its classical predecessor in recognizing "the anthropomorphic origin and nature of human knowledge." It was "inherently impossible" to establish "whether any universe of human discourse possessed objective reality" at all. One would never be able to assert that he had "exhaustively determined all (his) limitations and discounted all (his) foibles." The new science of Einstein was skeptical of objectivity and frankly admitted the subjective nature of observations and thus of knowledge itself. . . .

The academic disciplines were affected by the new ideas. Some writers spoke of the relativity of human nature. Political scientists found metaphysical implications in the quantum theory and the doctrine of relativity. The immediate successor of Charles Beard as president of the American Political Science Association, William Bennett Munro of Harvard, entitled his presidential address in 1927, "Physics and Politics: An Old Analogy Revisited." Declaring outmoded the physics of Bagehot's day, Munro announced that the revolution produced by relativity in the physical world "must inevitably carry its echoes into other fields of human knowledge." He proceeded to apply the new physics to political science with rather startling results.

Historians were also affected by the new concepts although it required the better part of the decade . . . to adapt relativity to historiography. . . .

The papers read at the annual meeting of the American Historical Association in 1927 reflected an active interest in problems of method, relation, outlook, and philosophy of history. In his presidential address that year Henry Osborn Taylor called the concept of relativity in physics, together with its dethronement of solid matter, a representation of the modern temper. Such concepts might not satisfy the spirit of another age, he stated, making the doctrine of relativity itself relative to its historical setting.

Taylor went on to remark that written history was also a product of the "intellectual conditions (which are the actualities) of the time of its composition." But Taylor did not stop after pointing out the relativity of conditions. He approached a subjectivist position by defining knowledge as experience; each man had a different experience, therefore, different knowledge. This was the disturbing "human equation" which made the study of the facts dependent upon the observer's ability and bias.

Carl Becker magnified the note of subjectivism sounded by Taylor when he informed the American Historical Association in 1931 that every man was his own historian. Becker's position was pushed to its logical conclusion the next year by Edward Maslin Hulme, who told the Pacific Coast Branch of the American Historical Association that detachment and objectivity in the writing of history was a myth. Not only time and place, but the personal characteristics of the historian colored his work. Noting that even physicists now took the personal equation into account, Hulme concluded: "Historic truth is relative and subjective, not absolute and objective." The cause of "scientific" history seemed doomed.

It remained for the economic historian Charles Woolsey Cole to demonstrate a direct connection between the new physics and historiography. He did so in an essay published in 1933 entitled "The Relativity of History." As in the past when Newton and Darwin had provided useful patterns of thought, Cole suggested the time had come once more to draw new points of view from the "austere domain of physics." Cole found in the new physics three principles: relativity, indeterminism, and discontinuity. Upon elaborating these principles, Cole showed that Einstein had made it clear the physical facts of time, distance, motion, and measurement had no meaning. "They are not absolute. They take on significance only when related to some frame of reference." The same thing was true of historical facts. They are made meaningful "when by order, selection, and interpretation they are related to a frame of reference." Facts, in other words, had the meaning which historians bestowed on them.

Historians thus moved from absolute to relative truth and from pretensions of detachment to admissions of subjectivity. The debate about the nature of historical knowledge was the context in which Charles A. Beard's doctrine of historical relativity was formed. Naturally, ascertaining just when and under what circumstances an individual develops a particular line of thought is difficult at best. Important clues to the progress of Beard's thought, however, are provided in his book reviews and shorter essays, especially if they are approached in chronological order.

No overt hint of Beard's later suspicion of objectivity is apparent in his works prior to the First World War. Beard's associate, James Harvey Robinson, emphasized in 1904 that historical matter could be truly and impartially stated. Robinson and Beard hoped to make the morning paper more understandable with their *Development of Modern Europe* which they pointedly subtitled, *An Introduction to the Study of Current History*. But granting that this aim was held, the authors asserted there would be no "distortion of the facts in order to bring them into relation to any particular conception of the present." The new history, pragmatic and utilitarian, did not aim at the rejection of objectivity.

Beard held that the economic interpretation of history was a most excellent aid in reaching the goal of objectivity. In his study of the Constitution, Beard criticized the scientific historians for being overly concerned with mere classifying and arranging of facts. He proposed to explain their "proximate or remote causes and relations." Beard accepted the causal nature of history while objecting to the narrowness of the Rankean type of history writing. Beard was even confident that the historian might discover a law which would "reduce history to order" just as natural scientists had found such laws governing the physical world.

Beard was an evolutionist in 1920, although not necessarily an optimistic one. He wrote of a fast changing world in which the confusing elements of history constantly evolved "together in terrible fascination." The years from 1922 to 1925 found Beard becoming skeptical about the possibility of objectivity when writing about those elements. Instead of brashly talking about explaining proximate and remote causes, Beard now found that selection of material was a most "delicate" matter. "Each of us," he admitted, "must perforce see what is behind his own eyes." . . .

In July 1926, Beard had occasion to make his first direct reference in print to the new physics when he advised historians that research into the origins of World War I would be a complicated job. One should not dash in lightly and start at once to pontificate, just as he should not rush into "the matter of 'relativity' or the 'quantum theory' " with a deficient knowledge of mathematics. For, as Beard confessed later that year to the American Political Science Association, the historian can never present all the facts. He must select a few from the number that had survived the destructive work of time and circumstance. "And any selection," Beard warned, "is an interpretation."

By the end of 1926 Beard had moved some distance from his pre-1920 position. The historian, he now held, never used all the facts. Those he selected were picked according to the historian's system of values. The philosophy of values in part depended upon the historian's position in time and place. Impartiality in the writing of history was a pretension. Beard's attack on objective history became sharper. It was the individual historian, Beard insisted, who selected, interpreted, and made the facts come alive "for those for whom he is writing in the age for which he is writing." The dictum of writing history as it actually happened he blasted as a "clever saying of Ranke's," labeling it the "fruit of vain imagination." He reiterated that it was the historian of a particular time and place who assigned value to the facts.

The years of the mid-1920s during which the gradual shift in Beard's thinking occurred coincided with the publication of several works of theoretical science and numerous popularizations of them. A new edition of Sir James Hopwood Jeans' *The Mathematical Theory of Electricity and Magnetism* (1925), the appearance of Sir Arthur Stanley Eddington's *The Internal Constitution of the Stars* (1926), and two editions of Ernest William Hobson's *The Domain of Natural Science* (1923 and 1926) are examples of these publications. Beard was familiar with these works. He considered the Jeans and Eddington volumes as belonging to a group of books which had "had the most influence on thought and action during the last half century." Beard cited the Hobson book several times in his own study of the nature of the social sciences—a work marking a critical milestone in Beard's development as theorist.

The Jeans and Eddington books are specialized scientific treatises and are not directed to a popular audience. Both deal with Einstein's work and the relativity and quantum theories. Hobson attempted in his volume to explain developments in natural science and theoretical mathematics to the interested layman. Natural science, in Hobson's view, was not concerned with the nature of reality or with schemes of causation and determination. Science, like other branches of knowledge, was subjective. Its data were not absolute fact but consisted of conceptions of phenomena perceived by the senses. Hobson pointed out that whereas in the Newtonian system the scientist described the phenomena of motion by using an absolute frame of reference, in the Einstein system all frames of reference were "on a parity," thus making for complete relativity. Hobson cautioned against extending Einstein's theory beyond the scope of physics.

A number of these topical scientific themes were touched upon by Bertrand Russell in his essay contributed to *Whither Mankind*, a volume Charles Beard edited in 1928. Russell, after summarizing Einstein's work, remarked that the "old glad certainty" of science was gone. Natural law did not consist of immutable principles but of human conventions and statistical averages. Physicists had demonstrated the absence of causation in the natural world.

Beard continued to take note of developments in natural science, although he complained in 1931 that physicists resembled Supreme Court Justices in the diversity of their opinions. Beard, by that time deeply concerned about the depression crisis, was searching for some illumination "on the strange map of life." He told an assembly of history teachers that year that historians and social scientists were being outstripped in this search by the natural scientists. Jeans had ranged over complex areas of time and space to conclude that the universe consisted, not of familiar Newtonian matter, but of pure thought. Whitehead, Beard revealed, had collapsed before God as the "ultimate limitation," while Eddington had admitted the absence of causality in the material world. The three philosopher-scientists agreed there was a basic reality of some kind underlying the appearance of life, but that it was unknowable for the most part.

In the bewildering struggle of the Titans of science what could a mere historian do? Beard queried. "He knows nothing of relativity, electrons, and symbolic logic." But he went on to declare that the only discipline able to shed light on mankind's path was history. History revealed to Beard "contingency, the possibility of choices ever present in the stream of life." Beard stressed this point again in 1933 when he emphasized that the events of history, "the unfolding of ideas and interests in time-motion," could never be brought into a deterministic pattern. It was therefore possible for man to find a way out of his crises.

Beard categorically proclaimed the end of determinism in the social sciences. The bondage of social scientists was the result of transferring to their field assumptions "once deemed applicable to the world of physics." Social processes could not be analyzed in the manner of engineering problems. For too long, Beard remonstrated early in 1933, social thinkers had been dominated by the idea that once all the facts had been assembled, conclusions would inevitably flow from them. Boldly Beard challenged, "*All* the facts? Obviously, an impossibility. The obtrusive facts? Obtruded on whose vision? The important facts? Important to whom and to what ends?"

The limitations of the observer implied in these last statements came to receive more emphasis in Beard's thought. Early in 1933, Beard decried "the coming crisis in empirical method." He found that even in chemistry and astrophysics where the observer might be supposed to stand outside the discrete things observed, the scientist brought to his study a set of ideas and emotions which colored his perception of phenomena. Much less possible was objectivity in the social sciences where the "thinker floats in the streams of facts, so-called, which he observes. His thoughts are parts of the thing thought about." Such statements indicate Beard's deepening concern about the subjective nature of knowledge.

The position which historical thinking and social science theory had reached by 1933 makes Beard's famous address to the American Historical Association that year seem anticlimactic. Nearly everything Beard said then had already been stated. This is true of his attack on scientific method and Rankean objectivity and of his use of the well-known phrase, "frame of reference." But the address did serve as a fitting denouement to the long discussion about philosophy and method in historiography.

Beard called upon his listeners to throw off the shackles of analogies drawn from Newtonian physics and Darwinian biology. At one point he attempted, like Taylor six years earlier, to confound the doctrine of relativity itself.f. But Beard went on to formulate some of the classic statements of that doctrine. The scientific method was extremely limited and could never produce a science of history. The historian was not a detached, objective observer. In selecting and arranging his facts, the historian was guided "inexorably by the frame of reference" in his mind. This frame of reference included "things deemed necessary, things deemed possible, and things deemed desirable." Within this frame three conceptions of history were possible: history as meaningless chaos, a cyclic interpretation, or, the view that history moves in some direction. There was no test of validity, no inherent logic of the data, to guide the historian in making his choice. The selection made constituted the historian's "act of faith" and conditioned the kind of history he would write.

Beard's remarks threw into clear relief the gulf which separated historians. . . . The prospect frightened some. The objectivists refused the new teaching and resolved to go down fighting with Rankean colors firmly nailed to the masthead. Beard clarified his position in the resulting conflict. He shrank from pushing relativism to its ultimate conclusion. He denied that history was chaos and asserted that there are not "as many schemes of reference as there are historians." Yet Beard would not retreat from his view that the idea of absolute truth in history was an illusion. Above all, Beard rejoiced that the pretension to objectivity had been "wrecked beyond repair."

Beard was encouraged to reject the theory of mechanistic causation in history because he knew the physicists were no longer sure about causation. Contemporary physics, Beard found, was "chary about the use of terms cause and effect." It merely described events in the order of their occurrence. What could not be done in the physical could not be done in the social sciences. The data of the social sciences, Beard insisted, could not be forced into a "closed circle of deterministic sequences."

The new science also aided Beard in his search for a solution to the impasse in empirical method. Much of the speculation in higher physics, Beard discovered, was a result of "subjective feeling rather than the outcome of objective observation." Modern scientists had unhesitatingly dismissed the old concept of complete objectivity. They no longer believed " 'that the eye of any beholder is disinterested.' " Such admissions of subjectivism by natural scientists reinforced Beard's skepticism about objectivity in historiography and encouraged him in his bold view of written history as the historian's subjective act of faith in the future.

There can be little doubt that Beard was cognizant of trends in contemporary physics and of their implications. All the ingredients of Beard's relativism—the attack upon objectivity, the denial of determinist schemes and thereby of a science of history, the frequent use of phrases like "frame of reference"—all point to the impact of the new physics. The very terminology of Beard's statements fits easily into the discourse of modern physics and its popularizers. To suggest that Beard drew upon contemporary physics does not, of course, exclude other and perhaps ultimately more important influences on his thought. The sources of Charles A. Beard's inspiration and even phraseology, however, were varied, and in any consideration of his relativism the contribution of Einstein's physics must be taken into account.

Werner Heisenberg

Philosophical Problems of Atomic Physics

Werner Heisenberg, born in 1901, received the Nobel Prize for his work in quantum mechanics and was the formulator of the so-called "uncertainty" or "indeterminacy" principle. He has written one of the key explorations of the broad implications of modern physics, Physics and Philosophy: The Revolution in Modern Science. *In the following lecture, delivered in 1959, Heisenberg discusses how quantum theory forces the abandonment of "materialist" philosophies and returns to ways of thinking that have strong links with those of both Plato and Aristotle.*

Modern physics, and in particular the quantum theory—the most important of Max Planck's discoveries—have posed a series of questions of a very general nature concerning not only specific problems of physics but also the methods used by the exact sciences and the very nature of matter itself. The questions have forced physicists to reconsider philosophical problems which had apparently already been solved once and for all by the strict teachings of classical physics.

There are two cycles of problems in particular which have been brought back to the attention of the scientists by Planck's quantum theory. The first of them is concerned with the nature of matter itself, or, to be more exact, is the old question posed by the Greek philosophers as to how the abundance of material phenomena could be reduced to simple principles and thus made more easily comprehensible. The second of these cycles is concerned with the question . . . as to how far it is possible to objectivize natural science experience, or any kind of material experience, i.e. to draw conclusions as to the nature of an objective process which takes place quite independently of the observer. . . . The question concerning the objective background of the phenomena has been brought up again by the quantum theory in a very surprising manner, this time from a different angle. For this reason we can therefore tackle the problem, from the point of view of modern natural science, from a different angle too.

First of all, let us look at those problems which arise in natural philosophy when one begins to search for a homogeneous and consistent set of principles by which one can judge material phenomena. The natural philosophers of ancient Greece, in the course of their reflections on the common basis of all visible phenomena, had already stumbled over the question of what constituted the smallest particle of matter. At the end of this glorious epoch in the history of the human intellect, two different schools of thought confronted one another as a result of all this reflection. These two schools have exercised the strongest possible influence on the later development of philosophical thought and have come to be known under the headings of materialism and idealism.

The atomic theory, which was initiated by Leucippos and Democritus, envisaged the smallest particle of matter as existence in its true sense. The smallest particles were considered to be indivisible and unchangeable, they were the eternal and the true basis of all matter and were thus called atoms and therefore did not require—indeed were quite incapable of—any further explanation. The only other properties they were considered to possess were geometrical ones. In the opinion of the philosophers they had a certain form, were separated one from another by empty space and brought about the abundance of phenomena by positioning themselves or moving in different ways in space. They had, however, neither colour, nor taste nor smell, nor did they have temperature or any other physical property known at the time. The properties of the things we can perceive were indirectly the result of the movements or position of the atoms. Just as comedy and tragedy are written with the same letters, so, according to the teachings of Democritus, could very different phenomena in the world be caused by the same atoms. The atoms were therefore the intrinsic, objectively real core of matter and thus of all material phenomena. They were, as already stated above, existence in its true sense, whereas the multitudude of different phenomena was merely the indirect result of the behaviour of the atoms. This school of thought is therefore termed materialism.

Plato, on the other hand, considered the smallest particles of matter to be nothing more than geometrical forms. Plato identified the smallest particles of matter with the regular bodies in geometry. Like Empedocles, he believed in the existence of four elements, namely earth, air, fire and water, and envisaged the smallest particle of the earth element as a cube, and the smallest particle of the element water as an icosahedron, whilst the smallest particle of the element fire was considered to be a tetrahedron and the smallest particle of the element air an octohedron. The form was considered to be characteristic for the properties of the element. However, in contrast to Democritus, Plato did not hold these particles to be indestructible and unchangeable. On the contrary, he believed that they could be reduced to triangles and built up again from triangles. For this reason he does not refer to them as atoms. The triangles themselves are no longer matter, since they are not three-dimensional, they do not occupy space. At the lower end of the chain of material structures as seen by Plato we therefore no longer find something material, but simply a mathematical form, or if you like, a purely intellectual entity. The most basic concept comprehensible to man was, in Plato's opinion, mathematical symmetry, form, an idea; and this school of thought is therefore termed idealism.

Strangely enough, this old question of materialism or idealism has been brought up again in a very specific form by modern atomic physics and by the quantum theory of Max Planck in particular. Until the time of the discovery of Planck's action quantum, the exact natural sciences of our time, i.e., physics and chemistry, were materialistically orientated. During the 19th century the chemical atom and its components, which we now call elementary particles, were considered to be the real basis of all matter, and existence in its true sense. The existence of atoms did not seem to require any further explanation or even to be capable of the same.

However, Planck discovered a certain trait of discontinuity in radiation phenomena, which seemed to be related to a surprising extent to the existence of the atom but which could not be explained by this.

This trait of discontinuity, which was discovered through Planck's quantum of action, seemed to indicate that this very characteristic, as well as the existence of the atom, could be common effects of a fundamental natural law, of a mathematical structure within Nature itself, the formulation of which could at the same time open the way for the discovery of those principles governing the structure of matter for which the Greeks had sought so long. Perhaps the existence of the atom was not the final link in the chain after all—perhaps this existence could be traced back, similar to the manner in which Plato had sought to explain the basis of all matter, to the effect of natural laws which could be mathematically formulated, i.e. to the effect of mathematical symmetry.

Planck's law of radiation differed in a very characteristic way from natural laws formulated in earlier times. When earlier natural laws, e.g. Newton's law of mechanics, contained so-called constants, then these constants referred to the properties of things, e.g. their mass, or the strength of a force acting between two bodies or some such similar property. Planck's quantum of action, however, which appears as the characteristic constant in his law of radiation, does not describe the properties of things, but a property of Nature itself. It sets a standard in Nature and at the same time shows that in those regions of Nature where Planck's quantum of action can be considered as a negligible quantity, as in all phenomena of daily life, natural phenomena take a different course to that taken in cases where they are of atomic magnitude, i.e. of the order of Planck's quantum of action. Whereas earlier laws of physics, e.g. Newton's laws of mechanics, apply in exactly the same manner to all orders of magnitude—the rotation of the moon round the earth is supposed to take place according to the same laws as the falling of an apple from a tree or the deflection of an a-particle that flies past the nucleus of an atom—Planck's law of radiation shows for the first time that there are scales and standards in Nature; that phenomena are not simply alike in all orders of magnitude.

Only a few years after Planck's discovery of the quantum of action, a second constant of scale came to be understood in its full importance. Einstein's special theory of relativity made it clear to physicists that the velocity of light did not refer to a property of a special substance known as ether, as had earlier been assumed from a study of electrodynamics, in the belief that this substance was responsible for the propagation of the light, but that we were confronted in this case with a property of time and space, i.e. with a general property of Nature itself, which has nothing to do with specific objects or things in Nature at all. The velocity of light can therefore also be considered to be one of Nature's constants of scale. Our common concepts of time and space can only be used for the description of phenomena in which the velocities occuring are relatively small in comparison with the velocity of light. The well-known paradoxes of the theory of relativity, on the contrary, are due to the fact that phenomena in which the velocities occurring are of more or less the same order as the velocity of light cannot be correctly interpreted with our usual concepts of time and space. I should merely like to quote the well-known paradox of the clock, i.e. the fact that time seems to pass more slowly for the observer who is moving fast than it does for the observer who is standing still. After the mathematical structure of the special theory of relativity had been elucidated, the implications of this mathematical formulation for the interpretation of the phenomena were soon analysed with such thoroughness that the characteristics of Nature associated with the new constant of scale, i.e. the velocity of light, could be completely understood. Admittedly the considerable discussion still going on about the theory of relativity goes to show that our ingrained conceptions still place a number of difficulties in the way of this comprehension, but all doubts could be quickly dispelled.

It was, however, much more difficult to comprehend the physical implications of Planck's quantum of action. In one of Einstein's works, dating from the year 1918, it was already shown that the laws of the quantum theory were probably of a statistical nature. The first attempt to state precisely the statistical nature of the quantum theoretical laws was made in 1924 by Bohr, Kramers and Slater. The connection between electromagnetic fields, which had been regarded in classical physics since the time of Maxwell as responsible for the propagation of the phenomena of light, and the discontinuous absorption and emission of light by the atom postulated by Planck were interpreted in the following manner: The electromagnetic field of waves, which is so patently responsible for the phenomena of interference and diffraction, only determines the degree of probability with which an atom emits or absorbs light energy in the form of quanta at the point in space in question. The electromagnetic field was thus no longer considered directly as a field of forces which acts on the electrical charges of the atom and thus causes movement; this action was considered to take place in a more indirect manner, whereby the field only determines the degree of probability with which emission or absorption takes place. This interpretation later on turned out to be not quite correct. The true connections were still more complicated and were correctly formulated by Bohr at a later date. However, the work of Bohr, Kramers and Slater did contain the decisive concept that natural laws do not determine the occurrence of an event but only the probability of its occurrence and that this probability must be connected to a field that conforms to a wave equation that can be formulated mathematically.

Thus a decisive step away from classical physics was taken and scientists fell back on a formulation which had already played an important role in the philosophy of Aristotle. The probability waves, as interpreted in the work of Bohr, Kramers and Slater, can be taken as a quantitative version of the concept of 'possibility', or (in the later Latin form) of 'potentia' in the philosophy of Aristotle. The idea that the phenomenon does not take place as a result of some inescapable coercion but that the possibility or the tendency for some phenomenon to occur itself possesses a degree of reality—a certain intermediate kind of reality, which is somewhere between the solid reality of matter and the mental reality of an idea or concept—plays a decisive role in the philosophy of Aristotle. It acquires a new form in the quantum theory inasmuch as this very concept of possibility is quantitatively formulated as probability and is subjected to mathematically calculable natural laws. Natural laws formulated in the language of mathematics no longer determine the phenomenon itself, but the possibility or probability that some phenomenon will take place.

This way of introducing the probability concept corresponded very accurately at first to the situation found during experimentation in the course of the study of atomic phenomena. If the physicist determines the strength of radioactive radiation by counting the number of times the radiation sets off the counting tube during the course of a certain interval of time, then he automatically assumes in doing so that the intensity of the radioactive radiation will regulate the probability that the counting tube will register. The exact intervals between the pulses do not really interest the physicist at all—they are, as he would say, "statistically distributed". The only factor of interest to the scientist is the average frequency of the pulses. It has already been established by numerous tests that this statistical interpretation reproduces the experimental situation very precisely. In cases where quantum mechanics permits quantitative statements to be made, as concerning the wavelengths of spectral lines or the binding energies of molecules, they have also been confirmed by the experiments. In short, there was no reason to doubt the validity of this theory. However, the question as to how the statistical interpretation would agree with the wealth of experience embodied in classical physics was rather more difficult. All experiments are based on the fact that there is a definite connection between observation and the physical phenomenon. If, for example, we measure a spectral line of a certain frequency with a diffraction grating, we automatically assume that the atoms of the radiating substance must have emitted light of just this frequency. Or again, if a photographic plate is darkened, then we automatically assume that rays or particles of matter have fallen on it at this point. Physics therefore makes use of the causal determination of phenomena in order to gather experimental experience and thus becomes apparently contrasted to the experimental situation in the atomic field and to the quantum theory where such a causal determination of the phenomenon does frequently not exist.

The inner contradiction which appears to occur here is removed in modern physics by establishing that the phenomena are only determined in so far as they can be described with the concepts of classical physics. The use of these concepts is, on the other hand, limited by the so-called indeterminacy relations. These contain quantitative information concerning the limits of the use of classical concepts. The physicist therefore knows which phenomena he can consider as determined and which he must consider as undetermined, and he can thus use a method which is free from all contradictions in observing them and interpreting their physical behaviour. Admittedly the question arises as to why it should be necessary here to keep to the concepts of classical physics, instead of converting the entire description of the physical behaviour to a new system of concepts based on the quantum theory.

First of all one must stress that the concepts of classical physics play a similar part in the interpretation of the quantum theory to that played by the a priori forms of perception used by Kant in his philosophy. Just as Kant explained the concepts of space and time or causality as a priori, since they formed the preconditions for all forms of experience and could not therefore be considered as the result of experience, so are the concepts of classical physics an a priori basis for all experience obtained about atomic phenomena, as we can only carry out experiments in the atomic field with the aid of these classical concepts.

Admittedly, Kant's "a priori" loses a certain claim to be absolute, a claim which it certainly made in his philosophy, as a result of such an interpretation. Whereas Kant was still able to assume that our a priori forms of perception, i.e. space and time, would form the basis of physics for all time, we now know that this is by no means the case. For example, the complete independence of space and time which we take for granted in our perception does not in fact exist in Nature if we observe very accurately. Our forms of perception, although they are a priori, do not agree with the results to be obtained only with the aid of the most sensitive technical equipment about processes or phenomena which take place at a velocity close to that of light. Our statements about space and time must therefore vary, according to whether we mean the a priori forms of perception inherent in our human nature or the pattern of order which exists in Nature, quite independent of all human observation, and in which all the objective phenomena of the world appear to be spread out. In a similar manner, classical physics forms the a priori basis for atomic physics and quantum theory, but it does not apply everywhere with equal validity, i.e. there are many different types of phenomena which cannot be described in detail using the concepts of classical physics.

In these fields of atomic physics, of course, much of the old conventional physics becomes lost. Not only the applicability of the concepts and laws of such physics, but also the entire concept of reality on which the exact natural sciences right down to our modern atomic physics were based. The term concept of reality is here meant to denote the idea that there are objective phenomena which take place in time and space in a certain manner, irrespective of whether they are observed or not. In atomic physics observations can no longer be objectivized in this simple manner, i.e. they can no longer be traced back to an objective and describable course of events in time and space. Here we find a consequence of the fact that natural science is not concerned with Nature itself, but with Nature as man describes and understands it. This does not mean that an element of subjectivity is introduced into natural science—no one claims that the processes and phenomena that take place in the world are dependent on our observation—but attention is brought to the fact that natural science stands between man and Nature and that we cannot dispense with the aid of perceptual concepts or other concepts inherent in the nature of man. This characteristic of quantum theory makes it difficult to accept the system laid down in the philosophy of materialism and to consider the smallest particles of matter, the elementary particles, as existence in the true sense. For these elementary particles are, if quantum theory is correct, no longer real in the same sense as are things of our everyday life, such as the trees and stones. They rather appear to be abstractions taken from the observation material, which is real in the true sense. . . .

. . . We cannot therefore escape the conclusion that our earlier concept of reality is no longer valid in the field of atomic physics and that we will become entangled in complicated abstractions if we regard the atom as existence in the true sense. The concept of existence in the true sense has really been discredited by modern physics and the fundamental teaching of materialistic philosophy must be modified at this point.

John Lukacs

Heisenberg's Recognitions: The End of the Scientific World View

John Lukacs is an historian of twentieth-century European culture who brings a strongly religious perspective to his discussion of the intellectual implications of quantum mechanics. Like many theologians, he views the indeterminacy principle as a warrant for a renewed emphasis on free will, and like many humanistically inclined students of history, he welcomes the new challenge to traditional notions of objectivity.

Let me . . . insist that what follows is not the breathless attempt of an enthusiastic historian to hitch his wagon to Heisenberg's star, or to jump on Heisenberg's bandwagon. . . . Rather, the contrary: my wagon is self-propelled, and a Heisenberg bandwagon does not exist (at least in the United States, among one hundred people who know the name of Einstein, not more than one may know of Heisenberg). It is the philosophical, rather than the experimental, part of Heisenberg's physics that I am qualified to discuss; my principal interest . . . springs from the condition that among the physicists of this century who have made excursions into philosophy I have found Heisenberg's philosophical exposition especially clear, meaningful and relevant . . .; and I have drawn upon some of his writings . . . because I want to present some of his courageous epistemological recognitions in a form which every English-speaking historian may read and understand easily. I have arranged these matters in order to sum them up in the form of ten propositions, the phrasing, the selection, and the organization of which is entirely my own: it is but their illustrations which come from the sphere of physics, described . . . by Heisenberg. . . . They are illustrations in the literal sense: they are intended to illustrate, to illuminate new recognitions, certain truths, in the assertion of which this writer, as indeed any historian in the twentieth century, is no longer alone.

First: there is no scientific certitude. Atomic physics found that the behavior of particles is considerably unpredictable: but, what is more important, this uncertainty is not "the outcome of defects in precision or measurement but a principle that could be demonstrated by experiment." Physicists have now found that while they can reasonably predict the average reactions of great numbers of electrons in an experiment, they cannot predict what a single electron will do, and not even when it will do it. The implications of this are, of course, the limitations of measurement; of accuracy; of scientific predictability—all fundamental shortcomings of "classical," or Newtonian, physics—they suggest the collapse of absolute determinism even in the world of matter.

Second: the illusory nature of the ideal of objectivity. In quantum mechanics the very act of observing alters the nature of the object, "especially when its quantum numbers are small." Quantum physics, Heisenberg says, "do not allow a completely objective description of nature." "As it really happened" (or "as it is really happening") is, therefore, an incomplete statement in the world of matter, too. . . . "In our century," Heisenberg wrote . . .,"it has become clear that the desired objective reality of the elementary particle is too crude an over-simplification of what really happens. . . ." "We can no longer speak of the behaviour of the particle independently of the process of observation. As a final consequence, the natural laws formulated mathematically in quantum theory no longer deal with the elementary particles themselves but with our knowledge of them." . . .

In biology, too, "it may be important for a complete understanding that the questions are asked by the species man which itself belongs to the genus of living organisms, in other words, that we already know what life is even before we have defined it scientifically.'' The recognition of personal participation is inescapable.

Third: the illusory nature of definitions. It seems that the minds of most physicists during the present interregnum still clung to the old, ''logical'' order of things: they were always giving names to newly discovered atomic particles, to such elements of the atomic kernel that did not ''fit.'' Yet the introduction of the name ''wavicle'' does preciously little to solve the problem of whether light consists of waves or of particles; and it may be that the continuing nominalistic habit of proposing new terms (sometimes rather silly-sounding ones, such as ''neutrino'') suggests that illusion of the modern mind which tends to substitute vocabulary for thought, tending to believe that once we name or define something we've ''got it.'' Sometimes things may get darker through definitions, Dr. Johnson said: and Heisenberg seems to confirm the limited value of definitions even in the world of matter:

> Any concepts or words which have been formed in the past through the interplay between the world and ourselves are not really sharply defined with respect to their meaning; that is to say, we do not know exactly how far they will help us in finding our way in the world. We often know that they can be applied to a wide range of inner or outer experience but we practically never know precisely the limits of their applicability. This is true even of the simplest and most general concepts like ''existence'' and ''space and time''. . . . The words ''position'' and ''velocity'' of an electron, for instance, seemed perfectly well defined as to both their meaning and their possible connections, and in fact they were clearly defined concepts within the mathematical framework of Newtonian mechanics. But actually they were not well defined, as is seen from the relations of uncertainty. One may say that regarding their position in Newtonian mechanics they were well defined, but in their relation to nature they were not.

Fourth: the illusory nature of the absolute truthfulness of mathematics. The absoluteness of mathematical ''truth'' was disproven by Gödel's famous theorem in 1931, but even before that, in the 1920's, physicists were beginning to ask themselves this uneasy question; as Heisenberg put it:

> Is it true that only such experimental situations can arise in nature as can be expressed in the mathematical formalism? The assumption that this was actually true led to limitations in the use of those concepts that had been the basis of physics since Newton. One could speak of the position and of the velocity of an electron as in Newtonian mechanics and one could observe and measure these quantities. But one could not fix both these quantities simultaneously with an arbitrarily high accuracy. . . . One had learned that the old concepts fit nature only inaccurately.

Mathematical truth is neither complete nor infinite (the velocity of light added to the velocity of light may amount to the velocity of light; on the other end of the physical scale there can be no action smaller than the quantum of action; and under certain physical conditions two by two do not always amount to four). Quantum theory found, too, that certain mathematical statements depend on the time element: Heisenberg realized that p times q is not always the equivalent of q times p in physics (when, for example, p means momentum and q position). What this suggests is that certain basic mathematical operations are not independent of human concepts of time and perhaps not even of purpose. That certain quantities do not always obey arithmetical rules was suggested already in the 1830's by the Irish mathematical genius Hamilton; and the Englishman Dirac, still to some extent influenced by nominalism, tried in the 1920's to solve this problem by asserting the necessity to deal with a set of so-called ''Q numbers'' which do not always respond to the rules of multiplication. But perhaps the ''problem'' may be stated more simply: the order in which certain mathematical (and physical) operations are performed affects their results.

Fifth: the illusory nature of ''factual'' truth. Change is an essential component of all nature: this ancient principle reappears within quantum physics. We have seen that the physicist must reconcile himself to the condition that he cannot exactly determine both the position and the speed of the atomic particle. He must reconcile himself, too, to the consequent condition that in the static, or factual, sense a basic unit of matter does not exist. It is not measurable; it is not even ascertainable; it is, in a way, a less substantial concept than such "idealistic" concepts as "beauty" or "mind." We can never expect to see a static atom or electron, since they do not exist as "immutable facts"; at best, we may see the trace of their motions. Einstein's relativity theory stated that matter is transmutable, and that it is affected by time; but the full implications of this condition were not immediately recognized, since they mean, among other things, that the earlier watertight distinctions between "organic" and "inorganic" substances no longer hold. "A sharp distinction between animate and inanimate matter," writes Heisenberg, "cannot be made." "There is only one kind of matter, but it can exist in different discrete stationary conditions." Heisenberg doubts "whether physics and chemistry will, together with the concept of evolution, some day offer a complete description of the living organism."

Sixth: the breakdown of the mechanical concept of causality. We have seen how, for the historian, *causa* must be more than the *causa efficiens* [Aristotle's efficient cause; see Chapter 1], and that the necessarily narrow logic of mechanical causality led to deterministic systems that have harmed our understanding of history, since in reality, through life and in history this kind of causation almost always "leaks." But now not even in physics is this kind of causation universally applicable: it is inadequate, and moreover, "fundamentally and intrinsically undemonstrable." There is simply no satisfactory way of picturing the fundamental atomic processes of nature in categories of space and time and causality. The multiplicity and the complexity of causes reappears in the world of physical relationships, in the world of matter.

Seventh: the principal importance of potentialities and tendencies. Quantum physics brought the concept of potentiality back into physical science—a rediscovery, springing from new evidence, of some of the earliest Greek physical and philosophical theories. Heraclitus was the first to emphasize this in the reality of the world: his motto, "Everything Moves," "imperishable change that renovates the world"; he did not, in the Cartesian and Newtonian manner, distinguish between being and becoming; to him fire was *both* matter and force. Modern quantum theory comes close to this when it describes energy, according to Heisenberg, anything that moves: "it may be called the primary cause of all change, and energy can be transformed into matter or heat or light." To Aristotle, too, matter was not by itself a reality but a *potentia* [potentiality], which existed by means of form: through the processes of nature the Aristotelian "essence" passed from mere possibility through form into actuality. When we speak of the temperature of the atom, says Heisenberg, we can only mean an expectation, "an objective tendency or possibility, a *potentia* in the sense of Aristotelian philosophy." An accurate description of the elementary particle is impossible: "the only thing which can be written down as description is a probability function"; the particle "exists" only as a possibility, "a possibility for being or a tendency for being." But this probability is not merely the addition of the element of "chance," and it is something quite different from mathematical formulas of probabilities. . . .

We have already met Heisenberg's question: "What happens 'really' in an atomic event?" The mechanism of the results of the observation can always be stated in the terms of the Newtonian concepts: "but what one deduces from an observation is a probability function . . . [which] does not itself represent a course of events in the course of time. It represents a tendency for events and our knowledge of events."

Eighth: not the essence of "factors" but their relationship counts. Modern physics now admits, as we have seen, that important factors may not have clear definitions: but, on the other hand, these factors *may* be clearly defined, as Heisenberg puts it, "with regard to their connections." These relationships are of primary importance: just as no "fact" can stand alone, apart from its associations with other "facts" and other matters, modern physics now tends to divide its world not into "different groups of objects but into different groups of connections." In the concepts of modern mathematics, too, it is being increasingly recognized how the functions of dynamic connections may be more important than the static definitions of "factors." Euclid had said that a point is something which has no parts and which occupies no space. At the height of positivism, around 1890, it was generally believed that an even more perfect statement would consist in exact definitions of "parts" and of "space." But certain mathematicians have since learned that this tinkering with definitions tends to degenerate into the useless nominalism of semantics, and consequently they do not bother with definitions of "points" or "lines" or "connection"; their interest is directed, instead, to the axiom that two points can be always connected by a line, to the relationships of lines and points and connections.

Ninth: the principles of "classical" logic are no longer unconditional: new concepts of truths are recognized. "Men fail to imagine any relation between two opposing truths and so they assume that to state one is to deny the other," Pascal wrote. Three centuries later Heisenberg wrote about some of C. F. von Weizaecker's propositions:

> It is especially one fundamental principle of classical logic which seems to require a modification. In classical logic it is assumed that, if a statement has any meaning at all, either the statement or the negation of the statement must be correct. Of "here is a table" or "here is not a table" either the first or the second statement must be correct. . . . A third possibility does not exist. It may be that we do not know whether the statement or its negation is correct; but in "reality" one of the two is correct.
>
> In quantum theory this law . . . is to be modified . . . Weizaecker points out that one may distinguish various levels of language. . . . In order to cope with [certain quantum situations] Weizaecker introduced the concept "degree of truth". . . . [By this] the term "not decided" is by no means equivalent to the term "not known." . . . There is still complete equivalence between the two levels of language with respect to the correctness of a statement, but not with respect to the incorrectness. . . .

Knowledge means not certainty, and a half-truth is not 50 percent truth; everyday language cannot be eliminated from any meaningful human statement of truth, including propositions dealing with matter; after all is said, logic is human logic, our own creation.

Tenth: at the end of the Modern Age the Cartesian partition falls away. Descartes's framework, his partition of the world into objects and subjects, no longer holds:

> The mechanics of Newton [Heisenberg writes] and all the other parts of classical physics constructed after its model started out from the assumption that one can describe the world without speaking about God or ourselves. This possibility seemed almost a necessary condition for natural science in general.
>
> But at this point the situation changed to some extent through quantum theory . . . we cannot disregard the fact [I would say: the condition] that science is formed by men. Natural science does not simply describe and explain nature; it is a part of the interplay between nature and ourselves; it describes nature as exposed to our method of questioning. This was a possibility of which Descartes could not have thought [?] but it makes the sharp separation between the world and the I impossible.
>
> . . . The Cartesian partition . . . has penetrated deeply into the human mind during the three centuries following Descartes and it will take a long time for it to be replaced by a really different attitude toward the problem of reality.

We cannot avoid the condition of our participation. . . . The recognition of this marks the beginning of a revolution not only in physical and philosophical but also in biological (and, ultimately, medical) concepts, springing from the empirical realization that there is a closer connection between mind and matter than what we have been taught to believe. Still, because of our interregnum, decades and disasters may have to pass until this revolution will bring its widely recognizable results. Yet we may at least look back at what we have already begun to leave behind.

After three hundred years the principal tendency in our century is still to believe that life is a scientific proposition, and to demonstrate how all of our concepts are but the products of complex mechanical causes that may be ultimately determinable through scientific methods. Thus Science, in Heisenberg's words, produced "its own, inherently uncritical"—and, let me add, inherently unhistorical—philosophy. But now "the scientific method of analysing, [defining] and classifying has become conscious"—though, let me add, far from sufficiently conscious—"of its limitations, which rise out of the [condition] that by its intervention science alters and refashions the object of investigation. In other words, methods and object can no longer be separated. *The scientific world-view has ceased to be a scientific view in the true sense of the word.*"

These are Heisenberg's italics. They correspond with the arguments . . . in which I have tried to propose the historicity of reality as something which is prior to its mathematicability. They represent a reversal of thinking after three hundred years: but, in any event, such recognitions involve not merely philosophical problems or problems of human perception but the entirety of human involvement in nature, a condition from which we, carriers of life in its highest complexity, cannot separate ourselves. The condition of this participation is the recognition of our limitations which is . . . our gateway to knowledge. "There is no use in discussing," Heisenberg writes, "what could be done if we were other beings than what we are." We must even keep in mind that the introduction of the "Cartesian" instruments such as telescopes and microscopes, which were first developed in the seventeenth century, do not, in spite of their many practical applications, bring us *always and necessarily* closer to reality—since they are interpositions, *our* interpositions, between our senses and the "object." We may even ask ourselves whether *our* task is still to "see" more rather than to see better, since not only does our internal deepening of human understanding now lag behind our accumulation of external information, but too, this external information is becoming increasingly abstract and unreal. Hence the increasing breakdown of internal communications: for, in order to see better, we must understand our own limitations better and also trust ourselves better. At the very moment of history when enormous governments are getting ready to shoot selected men hermetically encased in plastic bubbles out of the earth onto the moon, the importance of certain aspects of the "expanding universe" has begun to decline, and not only for humanitarian reasons alone; we are, again, in the center of the universe—inescapably as well as hopefully so.

Our problems—all of our problems—concern primarily human nature. The human factor is the basic factor. These are humanistic platitudes. But they have now gained added meaning, through the unexpected support from physics. It is thus that the recognitions of the human condition of science, and of the historicity of science—let me repeat that Heisenberg's approach is also historical—may mark the way toward the next phase in the evolution of human consciousness, in the Western world at least.

CHAPTER 11

Progress and the Rationality of Science

Introduction by Frederick Gregory

Until relatively recently, virtually all investigators of nature have been united by a common conviction, perhaps an unconscious presupposition, that nature could eventually be understood by human beings. Because of this assumption scientists from different times never questioned their readiness to force nature into the molds of rational order. For example, the early Greek astronomers demonstrated to their satisfaction that the apparent chaos in the motions of the planets was an illusion which could be made to succumb to rational explanation. Centuries later Galileo concluded that God must have been a geometer when he created the cosmos, because nature lent itself so beautifully to mathematical explanation. And the nineteenth-century physicist John Tyndall exhibited the same bold confidence in reason's capacity to describe the physical world completely when he observed that every new discovery that fit into a scientific theory strengthened the theory and helped it approach certainty.

Closely associated with this belief was another, a belief in the progress of humankind. Since nature was susceptible of rational analysis, then good science was a matter of proper reasoning. Proper reasoning involved at least two components, which, if there were deficiencies in either, could frustrate genuine progress. First, the initial axioms or assumptions on which scientific reasoning was based could not be arbitrary. They had to be true, and their truth had to be guaranteed in some manner. Either they were self-evident, divinely revealed, clear and distinct, or they possessed some other unambiguous criterion of truth. Second, the deductions and inferences made from the initial assumptions had to be carried out according to accepted rules of logical manipulation.

In medieval times many of the assumptions that had been made about nature since the days of the Greeks acquired the endorsement of the Latin Church, and with that came an impressive guarantee of their truth and reliability. Some seemed as obvious as the first principles of Euclidean geometry; for example, everyone accepted that in nature the shortest distance between two points was a straight line, and all but a very few accepted the immobility of the earth. In the middle ages progress did not occur in natural science as much as it did in theology; that is, progress did not result from changing the fundamental assumptions on which the perception of nature was based as much as it did from refinements in theological reasoning.

With the Scientific Revolution of the seventeenth century the notion of progress in science, and a concomitant progress in Western civilization, came into its own. New assumptions, with new grounds for justifying their truth, and a new ideal model for reasoning replaced the older appeals to the authority of the Church and Aristotelian logic. Careful and unbiased observation began more and more to affect what was taken to be unalterably true in nature, and, because it was assumed that observed relationships could be expressed in quantitative terms, mathematical deduction became the model for proper rational analysis. The guarantee of truth now lay in the empirical confirmation of predicted results, not in the *a priori* dictates of revelation or tradition.

The possibility of progress in science became more enticing. Because it was carried out in the spirit of open inquiry, the combination of careful neutral observation and mathematical explanation seemed to offer the prospect of the final and correct explanation of natural phenomena. Although many conceded that the number of things to be explained would most likely always exceed human conquest, at least the route along which progress toward certainty could be realized was clear.

In the aftermath of the magnificent achievements of the Scientific Revolution, the theme of progress was heard ever more frequently. Not only did the eighteenth century produce writers like Condorcet (1743–1794), who saw the world moving from darkness toward an ever more enlightened state, but the post-Napleonic era gave rise in France to what became known as positive philosophy, a system of thought that purported to represent the culmination of the development of humankind through its several stages of progress. With German idealistic philosophy, the same period produced a veritable celebration of progress, in which improvement was seen as a constitutent feature of the universe itself.

In this context it was perhaps to be expected that several thinkers would begin to reflect philosophically about natural science. The English were especially prolific in this regard during the first half of the nineteenth century. Besides John Herschel (1792–1871), whose *A Preliminary Discourse on the Study of Natural Philosophy* is excerpted in this chapter, John Stuart Mill (1806–1873) and William Whewell (1794–1866) also wrote at length about the nature of scientific reasoning. Herschel and Mill in particular defended the implication contained in the thought of their countryman, Francis Bacon, who, two centuries earlier, had spelled out how the scientist should proceed as an unbiased observer of nature. Always they assumed that one could arrive at scientific laws by inducing them from observed facts. Their validity was assured if they could be empirically verified. At the bottom of this attitude toward nature, which remained extremely powerful throughout the nineteenth century and on into the twentieth century, lay the conviction that, in the final analysis, laws and even theories were obtained from experience. Laws were discovered, not invented, and the truth they expressed was totally independent of the scientist who discovered them. In 1867 a German scientist said it perhaps most succinctly: "The natural scientist does not give in to the belief that he has created the law. He feels in his innermost being that the facts impose it on him." [1]

Events in physics in the twentieth century undermined this confident and progressive route to scientific certainty as much as any other single cause. With the development of relativity and quantum theory, the older meaning of the rationality of science was thrown into question, and new practical considerations took its place. Nature's rationality was preserved at the level of mathematical description, but the price for this consistency in explanation was that scientists had to remain content with the mathematics; that is, as soon as they began to ask what the mathematical description described they were enmeshed in inconsistency and contradiction. The more that quantum theory was developed the less it seemed that it was obtained from the physical world of experience and the more it appeared to depend on the properties of complex mathematical systems. Physicists began to doubt that their task was solely that of the discoverer. They could no longer resist the notion that they were, at least in part, inventing the laws that governed scientific phenomena.

It was inevitable that the philosophical understanding of natural science would be affected by developments such as these. New interpretations began to appear just before and just after the revolutionary developments of the twentieth century. Some suggested that scientific theories were simply economies of thought, others that they represented conventions, still others that they were nothing more than the instruments scientists used to manipulate nature. In none of these philosophies of science did careful observation of nature serve as a sufficient guarantee of a theory's ultimate truth. Even philosophers who wished to preserve the image of science as an eminently rational activity surrendered the old claim that there was gradual progress toward truth.

Some contemporary philosophers have argued not only that the history of science contains episodes which reveal the direct involvement of irrational factors in the formation and selection of scientific theories, but that the choices between competing theories *must* be irrational. With this conclusion philosophy of science has reached a level of relativism that makes many uncomfortable, for it implies that there can be no progress in science like that which many have assumed differentiates science from other intellectual pursuits. But until philosophers agree that the methods of the scientist can guarantee that the results are progressing toward some final truth, or until there is a consensus about a new and different sense in which progress should be defined, a dogmatic rejection of all relativism in science will hardly suffice.

There is little doubt that when Galileo chose the Copernican world system over that of Ptolemy he thought he was choosing a system that was true over one that was false. The lessons of the history of science suggest, however, that things are more complicated than that. Decisions between competing scientific theories are most likely made on grounds that are more complex than the decision maker realizes. But if a scientific theory, or even the methodology of science itself, is not ultimately preferred because it represents or leads to completely objective truth, then the question immediately arises: What does determine these preferences? When, for example, Galileo embraced Copernicus over Ptolemy, or when Western societies prefer a "scientific" approach to nature over the role nature is assigned in Eastern contemplative perspectives, what factors actually govern these preferences? Whatever the complete answer to this question is, it must be acknowledged that the *values* of an individual, which obviously are affected by those of the society in which the individual lives, play a key role in determining how science is understood and the importance it is assigned. While it would be inappropriate to conclude that human beings are in no way dictated to by a nature that is beyond their reach, it would be equally foolish not to acknowledge that to a substantial degree humans help create the very reality they understand. These considerations do nothing to tarnish the image of natural science as the bold encounter of humanity with nature; on the contrary, they infuse into that image a new dimension of wonder and excitement.

[1]Jacob Moleschott, *Ursache und Wirkung in der Lehre vom Leben* (Giessen, 1867), 8.

Condorcet

Science and Progress

Before the French Revolution the mathematician Marie Jean Antoine Nicolas Caritat, Marquis de Condorcet, had been Secretary of the Academy of Sciences. During the Revolution, while hiding from the Jacobins, Condorcet hurriedly wrote down his famous sketch of history in ten epochs, the Outlines of an Historical View of the Progress of the Human Mind. *The conception of progress that Condorcet promulgates in the following selection, which is taken from the epoch concerned with humanity's future, illustrates why his sketch has been called "rationalism run riot, dominated by a simple-minded faith in science that confuses over and over again the improvement of techniques with advances in virtue and happiness."* [1]

The only foundation of faith in the natural sciences is the principle, that the general laws, known or unknown, which regulate the phenomena of the universe, are regular and constant; and why should this principle, applicable to the other operations of nature, be less true when applied to the development of the intellectual and moral faculties of man? In short, as opinions formed from experience, relative to the same class of objects, are the only rule by which men of soundest understanding are governed in their conduct, why should the philosopher be proscribed from supporting his conjectures upon a similar basis, provided he attribute to them no greater certainty than the number, the consistency, and the accuracy of actual observations shall authorise? . . .

The progress of the sciences secures the progress of the art of instruction, which again accelerates in its turn that of the sciences; and this reciprocal influence, the action of which is incessantly increased, must be ranked in the number of the most prolific and powerful causes of the improvement of the human race. At present, a young man, upon finishing his studies and quitting our schools, may know more of the principles of mathematics than Newton acquired by profound study, or discovered by the force of his genius, and may exercise the instrument of calculation with a readiness which at that period was unknown. The same observation, with certain restrictions, may be applied to all the sciences. . . .

. . . It is manifest that the improvement of the practice of medicine, become more efficacious in consequence of the progress of reason and the social order, must in the end put a period to transmissible or contagious disorders, as well to those general maladies resulting from climate, aliments, and the nature of certain occupations. Nor would it be difficult to prove that this hope might be extended to almost every other malady, of which it is probable we shall hereafter discover the most remote causes. Would it even be absurd to suppose this quality of melioration in the human species as susceptible of an indefinite advancement; to suppose that a period must one day arrive when death will be nothing more than the effect either of extraordinary accidents, or of the slow and gradual decay of the vital powers; and that the duration of the middle space, of the interval between the birth of man and this decay, will itself have no assignable limit? Certainly man will not become immortal; but may not the distance between the moment in which he draws his first breath, and the common term when, in the course of nature, without malady or accident, he finds it impossible any longer to exist, be necessarily protracted? We are now speaking of a progress that is capable of being represented with precision, by numerical quantities or by lines.

Lastly, may we not include in the same circle the intellectual and moral faculties? May not our parents, who transmit to us the advantages or defects of their conformation, and

[1]Peter Gay, *The Enlightenment: An Interpretation* (New York, 1969), II, 122.

from whom we receive our features and shape, as well as our propensities to certain physical affections, transmit to us also that part of organization upon which intellect, strength of understanding, energy of soul or moral sensibility depend? Is it not probable that education, by improving these qualities, will at the same time have an influence upon, will modify and improve this organization itself? Analogy, an investigation of the human faculties, and even some facts, appear to authorise these conjectures, and thereby to enlarge the boundary of our hopes.

. . . How admirably calculated is this view of the human race, emancipated from its chains, released alike from the dominion of chance, as well as from that of the enemies of its progress, and advancing with a firm and indeviate step in the paths of truth, to console the philosopher lamenting the errors, the flagrant acts of injustice, the crimes with which the earth is still polluted? It is the contemplation of this prospect that rewards him for all his efforts to assist the progress of reason and the establishment of liberty. He dares to regard these efforts as a part of the eternal chain of the destiny of mankind; and in this persuasion he finds the true delight of virtue, the pleasure of having performed a durable service, which no vicissitude will ever destroy in a fatal operation calculated to restore the reign of prejudice and slavery.

John Herschel

The Scientific Method

Through phrases such as "the actual structure or mechanism of the universe" and nature's "true causes," the nineteenth-century astronomer John Herschel reveals his adherence to the view that there is an objective truth toward which science strives. That observation and experience provide the route to this truth becomes quite clear from the central role assigned to inductive reasoning in Herschel's system. Notice that the inevitable inaccuracies within empirical observations do not speak against *the existence of objective scientific truth; rather, by averaging errors scientists actually come closer and closer to that truth.*

The immediate object we propose to ourselves in physical theories is the analysis of phenomena, and the knowledge of the hidden processes of nature in their production, so far as they can be traced by us. An important part of this knowledge consists in a discovery of the actual structure or mechanism of the universe and its parts, through which, and by which, those processes are executed; and of the agents which are concerned in their performance. Now, the mechanism of nature is for the most part either on too large or too small a scale to be immediately cognizable by our senses; and her agents in like manner elude direct observation, and become known to us only by their effects. It is in vain therefore that we desire to become witnesses to the processes carried on with such means, and to be admitted into the secret recesses and laboratories where they are effected. Microscopes have been constructed which magnify more than a thousand times in *linear* dimension, so that the smallest visible grain of sand may be enlarged to the appearance of one a thousand million times more bulky; yet the only impression we receive by viewing it through such a magnifier is, that it reminds us of some vast fragment of a rock, while the intimate structure on which depend its color, its hardness, and its chemical properties, remains still concealed: . . .

Now, nothing is more common in physics than to find two, or even many, theories maintained as to the origin of a natural phenomenon. . . . Are we to be deterred from framing hypotheses and constructing theories, because we meet with such dilemmas, and find ourselves frequently beyond our depth? Undoubtedly not. . . . Hypotheses, with respect to theories, are what presumed proximate causes are with respect to particular inductions: they afford us motives for searching into analogies; grounds of citation to bring before us all the cases which seem to bear upon them, for examination. A well imagined hypothesis, if it has been suggested by a fair inductive consideration of general laws, can hardly fail at least of enabling us to generalize a step farther, and group together several such laws under a more universal expression. But this is taking a very limited view of the value and importance of hypothesis: it may happen (and it has happened in the case of the undulatory doctrine of light) that such a weight of analogy and probability may become accumulated on the side of an hypothesis, that we are compelled to admit one of two things; either that it is an actual statement of what really passes in nature, or that the reality, whatever it be, must run so close a parallel with it, as to admit of some mode of expression common to both, at least in so far as the phenomena actually known are concerned. Now, this is a very great step, not only for its own sake, as leading us to a high point in philosophical speculation, but for its applications; because whatever conclusions we deduce from an hypothesis so supported must have at least a strong presumption in their favor: and we may be thus led to the trial of many curious experiments, and to the imagining of many useful and important contrivances, which we should never otherwise have thought of, and which, at all events, *if* verified in practice, are real additions to our stock of knowledge and to the arts of life.

In framing a theory which shall render a rational account of any natural phenomenon, we have *first* to consider the agents on which it depends, or the causes to which we regard it as ultimately referable. These agents are not to be arbitrarily assumed; they must be such as we have good inductive grounds to believe do exist in nature, and do perform a part in phenomena analogous to those we would render an account of; or such, whose presence in the actual case can be demonstrated by unequivocal signs. They must be *verae causae,* in short, which we can not only show to exist and to act, but the laws of whose action we can derive independently, by direct induction, from experiments purposely instituted; or at least make such suppositions respecting them as shall not be contrary to our experience, and which will remain to be verified by the coincidence of the conclusions we shall deduce from them, with facts. For example, in the theory of gravitation, we suppose an agent—*viz*. force, or mechanical power—to act on *any* material body which is placed in the presence of *any* other, and to urge the two mutually towards each other. This is a *vera causa;* for heavy bodies (that is, all bodies, but some more, some less) tend to, or endeavor to reach, the earth, and require the exertion of force to counteract this endeavor, or to keep them up. Now, that which opposes and neutralizes force *is* force. And again, a plumb-line, which when allowed to hang freely, always hangs perpendicularly, is found to hang observably aside from the perpendicular when in the neighborhood of a considerable mountain, thereby proving that a force is exerted upon it, which draws it towards the mountain. Moreover, since it is a fact that the moon does circulate about the earth, it must be drawn towards the earth by a force; for if there were no force acting upon it, it would go on in a straight line without turning aside to circulate in an orbit, and would, therefore, soon go away and be lost in space. This force, then, which we call the *force* of gravity, is a real cause.

We have next to consider the laws which regulate the action of these our primary agents; and these we can only arrive at in three ways: 1st, By inductive reasoning; that is, by examining all the cases in which we know them to be exercised, inferring, as well as circumstances will permit, its amount or intensity in each particular case, and then piecing together, as it were, these *disjecta membra* [scattered parts], generalizing from them, and so arriving at the laws desired; 2dly, By forming at once a bold hypothesis, particularizing the law, and trying the truth of it by following out its consequences, and comparing them with facts; or, 3dly, By a process partaking of both these, and combining the advantages of both without their defects, viz. by assuming indeed the laws we would discover, but so generally expressed, that they shall include an unlimited variety of particular laws;—following out the consequences of this assumption, by the application of such general principles as the case admits;—comparing them in succession with all the particular cases within our knowledge; and, lastly, *on this comparison,* so modifying and restricting the general enunciation of our laws as to *make the results agree*. . . .

In estimating, however, the value of a theory, we are not to look, *in the first instance,* to the question, whether it establishes satisfactorily, or not, a particular process or mechanism; for of this, after all, we can never obtain more than that indirect evidence which consists in its leading to the same results. What, in the actual state of science, is far more important for us to know, is whether our theory truly represent *all* the facts, and include *all* the laws, to which observation and induction lead. A theory which did this would, no doubt, go a great way to establish any hypothesis of mechanism or structure, which might form an essential part of it: but this is very far from being the case, except in a few limited instances; and, till it is so, to lay any great stress on hypotheses of the kind, except in as much as they serve to scaffold, for the erection of general laws, is to "quite mistake the scaffold for the pile." Regarded in this light, hypotheses have often an eminent use; and a facility in framing them, if attended with an equal facility in laying them aside when they have served their turn, is one of the most valuable qualities a philosopher can possess; while, on the other hand, a bigoted adherence to them, or indeed to peculiar views of any kind, in opposition to the tenor of facts as they arise, is the bane of all philosophy. . . .

When two theories run parallel to each other, and each explains a great many facts in common with the other, any experiment which affords a crucial instance to decide between them, or by which one or other must fall, is of great importance. In thus verifying theories, since they are grounded on general laws, we may appeal, not merely to particular cases, but to whole classes of facts; and we therefore have a great range among the individuals of these for the selection of some particular effect which ought to take place oppositely in the event of one of the two suppositions at issue being right and the other wrong. A curious example is given by M. Fresnel, as decisive, in his mind, of the question between the two great opinions on the nature of light, which, since the time of Newton and Huyghens, have divided philosophers.

When two very clean glasses are laid one on the other, if they be not perfectly flat, but one or both in an almost imperceptible degree convex or prominent, beautiful and vivid colors will be seen between them; and if these be viewed through a red glass, their appearance will be that of alternate dark and bright stripes. These stripes are formed *between* the two surfaces in apparent contact, as any one may satisfy himself by using, instead of a flat *plate* of glass for the upper one, a triangular-shaped piece, called a prism, like a three-cornered stick, and looking through the inclined side of it next to the eye by which arrangement the reflection of light from the upper surface is prevented from intermixing with that from the surfaces in contact. Now, the colored stripes thus produced are explicable on both theories, and are appealed to by both as strong confirmatory facts; but there is a difference in one circumstance according as one or the other theory is employed to explain them. In the case of the Huyghenian doctrine, the intervals between the bright stripes ought to appear *absolutely black;* in the other, *half bright,* when so viewed through a prism. This curious case of difference was tried as soon as the opposing consequences of the two theories were noted by M. Fresnel, and the result is stated by him to be decisive in favor of that theory which makes light to consist in the vibrations of an elastic medium.

Theories are best arrived at by the consideration of general laws; but most securely verified by comparing them with particular facts, because this serves as a verification of the whole train of induction, from the lowest term to the highest. But, then, the comparison must be made with facts purposely selected, so as to include every variety of case, not omitting extreme ones, and in sufficient number to afford every reasonable probability of detecting error. A single numerical coincidence in a final conclusion, however striking the coincidence or important the subject, is not sufficient. Newton's theory of sound, for example, leads to a numerical expression for the actual velocity of sound, differing but little from that afforded by the correct theory afterwards explained by La Grange, and (when certain considerations not contemplated by him are allowed for) agreeing with fact; yet this coincidence is no verification of Newton's view of the general subject of sound, which is defective in an essential point, as the great geometer last named has very satisfactorily shown. This example is sufficient to inspire caution in resting the verification of theories upon any thing but a very extensive comparison with a great mass of observed facts.

But, on the other hand, when a theory will bear the test of such extensive comparison, it matters little how it has been originally framed. However strange, and, at first sight, inadmissible, its postulates may appear, or however singular it may seem that such postulates should have been fixed upon,—if they only lead us, by legitimate reasonings, to conclusions in exact accordance with numerous observations purposely made, under such a variety of circumstances as fairly embrace the whole range of the phenomena which the theory is intended to account for,—we cannot refuse to admit them; or if we still hesitate to regard them as demonstrated truths, we cannot, at least, object to receive them as temporary substitutes for such truths, until the latter shall become known. If they suffice to explain all the phenomena known, it becomes highly improbable that they will not explain more; and if all their conclusions we have tried have proved correct, it is probable that others yet untried will be found so too; so that in rejecting them altogether, we should reject all the discoveries to which they may lead. . . .

The importance of obtaining exact physical data can scarcely be too much insisted on, for without them the most elaborate theories are little better than mere inapplicable forms of words. It would be of little consequence to be informed, abstractedly, that the sun and planets attract each other, with forces proportional to their masses, and inversely as the squares of their distances: but, as soon as we know the data of our system, as soon as we have an accurate statement (no matter how obtained) of the distances, masses, and actual motions of the several bodies which compose it, we need no more to enable us to predict all the movements of its several parts, and the changes that will happen in it for thousands of years to come; and even to extend our views backwards into time, and recover from the past, phenomena, which no observation has noted, and no history recorded, and which yet (it is possible) may have left indelible traces of their existence in their influence on the state of nature in our own globe, and those of the other planets.

But how, it may be asked, are we to ascertain *by* observation, data more precise than observation itself? How are we to conclude the value of that which we do not see, with greater certainty than that of quantities which we actually see and measure? It is the number of observations which may be brought to bear on the determination of data that enables us to do this. Whatever error we may commit in a single determination, it is highly improbable that we should always err the same way, so that, when we come to take an average of a great number of determinations (unless there be some constant cause which gives a bias one way or the other), we cannot fail, at length, to obtain a very near approximation to the truth, and even allowing a bias, to come much nearer to it than can fairly be expected from any single observation, liable to be influenced by the same bias.

This useful and valuable property of the average of a great many observations, that it brings us nearer to the truth than any single observation can be relied on as doing, renders it the most constant resource in all physical inquiries where accuracy is desired. And it is surprising what a rapid effect, in equalizing fluctuations and destroying deviations, a moderate multiplication of individual observations has. A better example can hardly be taken than the average height of the quicksilver in the common barometer, which measures the pressure of the air, and whose fluctuations are proverbial. Nevertheless, if we only observe it regularly every day, and, at the end of each month, take an average of the observed heights, we shall find the fluctuations surprisingly diminished in amount: and if we go on for a whole year, or for many years in succession, the annual averages will be found to agree with still greater exactness. This equalizing power of averages, by destroying all such fluctuations as are irregular or accidental, frequently enables us to obtain evidence of fluctuations really regular, periodic in their recurrence, and so much smaller in their amount than the accidental ones, that, but for this mode of proceeding, they never would have become apparent. Thus, if the height of the barometer be observed four times a day, constantly, for a few months, and the averages taken, it will be seen that a regular *daily* fluctuation, of very small amount, takes place, the quicksilver rising and falling twice in the four-and-twenty hours. . . .

In all cases where there is a direct and simple relation between the phenomenon observed and a single *datum* on which it depends, every single observation will give a value of this quantity, and the average of all (under certain restrictions) will be its exact value. We say, under certain restrictions; for, if the circumstances under which the observations are made be not alike, they may not all be equally favorable to exactness, and it would be doing injustice to those most advantageous, to class them with the rest. In such cases as these, as well as in cases where the *data* are numerous and complicated together so as not to admit of single, separate determination (a thing of continual occurrence), we have to enter into very nice, and often not a little intricate, considerations respecting the *probable* accuracy of our results, or the limits of error within which it is *probable* they lie. . . .

Now, this is a very similar case to that of an observer—an astronomer for example—who would determine the exact place of a heavenly body. He points to it his telescope, and obtains a series of results disagreeing among themselves, but yet all agreeing within certain limits, and only a comparatively small number of them deviating considerably from the mean of all; and from these he is called upon to say, definitively, what he shall consider to have been the most probable place of his star at the moment. Just so in the calculation of physical *data*; where no two results agree exactly, and where all come within limits, some wide, some close, what have we to guide us when we would make up our minds what to conclude respecting them? It is evident that any system of calculation that can be shown to lead of necessity to the most probable conclusion where certainty is not to be had must be valuable. However, as this doctrine is one of the most difficult and delicate among the applications of mathematics to natural philosophy, this slight mention of it must suffice at present.

Jacob Bronowski

Knowledge or Certainty

In this passage the mathematician Jacob Bronowski (1908–1974) investigates the claim that to achieve an exact knowledge of nature the investigator needs only to refine the detail of the observations. This was the assumption of nineteenth-century scientific materialists like Ludwig Büchner, who argued that the apparent mystery of human consciousness would be solved once scientists' ignorance of the "fineness" of matter was removed. Bronowski shows that developments in quantum physics in the twentieth century expose the impossibility of arriving at some final truth of nature through this approach.

One aim of the physical sciences has been to give an exact picture of the material world. One achievement of physics in the twentieth century has been to prove that that aim is unattainable.

Take a good, concrete object, the human face. I am listening to a blind woman as she runs her fingertips over the face of a man she senses for the first time, thinking aloud. 'I would say that he is elderly. I think, obviously, he is not English. He has a rounder face than most English people. And I should say he is probably Continental, if not Eastern-Continental. The lines in his face would be lines of possible agony. I thought at first they were scars. It is not a happy face' . . . .

We are aware that these pictures do not so much fix the face as explore it; that the artist is tracing the detail almost as if by touch; and that each line that is added strengthens the picture but never makes it final. We accept that as the method of the artist.

But what physics has now done is to show that that is the only method to knowledge. There is no absolute knowledge. And those who claim it, whether they are scientists or dogmatists, open the door to tragedy. All information is imperfect. We have to treat it with humility. That is the human condition; and that is what quantum physics says. I mean that literally. . . . The painter analyses the face, takes the features apart, separates the colours, enlarges the image. It is natural to ask, Should not the scientist use a microscope to isolate and analyse the finer features? Yes, he should. But we ought to understand that the microscope enlarges the image but cannot improve it: the sharpness of detail is fixed by the wavelength of the light. The fact is that at any wavelength we can intercept a ray only by objects about as large as a wavelength itself; a smaller object simply will not cast a shadow.

We are here face to face with the crucial paradox of knowledge. Year by year we devise more precise instruments with which to observe nature with more fineness. And when we look at the observations, we are discomfited to see that they are still fuzzy, and we feel that they are as uncertain as ever. We seem to be running after a goal which lurches away from us to infinity every time we come within sight of it.

The paradox of knowledge is not confined to the small, atomic scale; on the contrary, it is as cogent on the scale of man, and even of the stars. Let me put it in the context of an astronomical observatory. Karl Friedrich Gauss's observatory at Göttingen was built about 1807. Throughout his lifetime and ever since (the best part of two hundred years) astronomical instruments have been improved. We look at the position of a star as it was determined then and now, and it seems to us that we are closer and closer to finding it precisely. But when we actually compare our individual observations today, we are astonished and chagrined to find them as scattered within themselves as ever. We had hoped that the human errors would disappear, and that we would ourselves have God's view. But it turns out that the errors cannot be taken out of the observations. And that is true of stars, or atoms, or just looking at somebody's picture, or hearing the report of somebody's speech.

Gauss recognised this with that marvellous, boyish genius that he had right up to the age of nearly eighty at which he died. When he was only eighteen years old, when he came to Göttingen to enter the University in 1795, he had already solved the problem of the best estimate of a series of observations which have internal errors. He reasoned then as statistical reasoning still goes today.

When an observer looks at a star, he knows that there is a multitude of causes for error. So he takes several readings, and he hopes, naturally, that the best estimate of the star's position is the average—the centre of the scatter. So far, so obvious. But Gauss pushed on to ask what the *scatter* of the errors tells us. He devised the Gaussian curve in which the scatter is summarised by the deviation, or spread, of the curve. And from this came a far-reaching idea: the scatter marks an area of uncertainty. We are not sure that the true position is the centre. All we can say is that it lies *in the area of uncertainty,* and the area is calculable from the observed scatter of the individual observations.

In the years of the First World War, science was dominated at Göttingen as elsewhere by Relativity. But in 1921 there was appointed to the chair of physics Max Born, who began a series of seminars that brought everyone interested in atomic physics here. It is rather surprising to reflect that Max Born was almost forty when he was appointed. By and large, physicists have done their best work before they are thirty (mathematicians even earlier, biologists perhaps a little later). But Born had a remarkable personal, Socratic gift. He drew young men to him, he got the best out of them, and the ideas that he and they exchanged and challenged also produced his best work. Out of that wealth of names, whom am I to choose? Obviously Werner Heisenberg, who did his finest work here with Born. Then, when Erwin Schrödinger published a different form of basic atomic physics, here is where the arguments took place. And from all over the world people came to Göttingen to join in.

It is rather strange to talk in these terms about a subject which, after all, is done by midnight oil. Did physics in the 1920s really consist of argument, seminar, discussion, dispute? Yes, it did. Yes, it still does. The people who met here, the people who meet in laboratories still, only end their work with a mathematical formulation. They begin it by trying to solve conceptual riddles. The riddles of the sub-atomic particles—of the electrons and the rest—are mental riddles.

Think of the puzzles that the electron was setting just at that time. The quip among professors was (because of the way university time-tables are laid out) that on Mondays, Wednesdays, and Fridays the electron would behave like a particle; on Tuesdays, Thursdays, and Saturdays it would behave like a wave. How could you match those two aspects, brought from the large-scale world and pushed into a single entity, into this Lilliput, *Gulliver's Travels* world of the inside of the atom? That is what the speculation and argument was about. And that requires, not calculation, but insight, imagination—if you like, metaphysics. I remember a phrase that Max Born used when he came to England many years after, and that still stands in his autobiography. He said: 'I am now convinced that theoretical physics is actual philosophy'.

Max Born meant that the new ideas in physics amount to a different view of reality. The world is not a fixed, solid array of objects, out there, for it cannot be fully separated from our perception of it. It shifts under our gaze, it interacts with us, and the knowledge that it yields has to be interpreted by us. There is no way of exchanging information that does not demand an act of judgment. Is the electron a particle? It behaves like one in the Bohr atom. But de Broglie in 1924 made a beautiful wave model, in which the orbits are the places where an exact, whole number of waves closes round the nucleus. Max Born thought of a train of electrons as if each were riding on a crankshaft, so that collectively they constitute a series of Gaussian curves, a wave of probability. A new conception was being made, on the train to Berlin and the professorial walks in the woods of Göttingen: that whatever fundamental units the world is put together from, they are more delicate, more fugitive, more startling than we catch in the butterfly net of our senses.

All those woodland walks and conversations came to a brilliant climax in 1927. Early that year Werner Heisenberg gave a new characterisation of the electron. Yes, it is a particle, he said, but a particle which yields only limited information. That is, you can specify where it is at this instant, but then you cannot impose on it a specific speed and direction at the setting-off. Or conversely, if you insist that you are going to fire it at a certain speed in a certain direction, then you cannot specify exactly what its starting-point is—or, of course, its end-point.

That sounds like a very crude characterisation. It is not. Heisenberg gave it depth by making it precise. The information that the electron carries is limited in its totality. That is, for instance, its speed *and* its position fit *together* in such a way that they are confined by the tolerance of the quantum. This is the profound idea: one of the great scientific ideas, not only of the twentieth century, but in the history of science.

Heisenberg called this the Principle of Uncertainty. In one sense, it is a robust principle of the everyday. We know that we cannot ask the world to be exact. If an object (a familiar face, for example) had to be *exactly* the same before we recognised it, we would never recognise it from one day to the next. We recognise the object to be the same because it is much the same; it is never exactly like it was, it is tolerably like. In the act of recognition, a judgment is built in—an area of tolerance or uncertainty. So Heisenberg's principle says that no events, not even atomic events, can be described with certainty, that is, with zero tolerance. What makes the principle profound is that Heisenberg specifies the tolerance that can be reached. The measuring rod is Max Planck's quantum. In the world of the atom, the area of uncertainty is always mapped out by the quantum.

Yet the Principle of Uncertainty is a bad name. In science or outside it, we are not uncertain; our knowledge is merely confined within a certain tolerance. We should call it the Principle of Tolerance. And I propose that name in two senses. First, in the engineering sense. Science has progressed step by step, the most successful enterprise in the ascent of man, because it has understood that the exchange of information between man and nature, and man and man, can only take place with a certain tolerance. But second, I also use the word passionately about the real world. All knowledge, all information between human beings can only be exchanged within a play of tolerance. And that is true whether the exchange is in science, or in literature, or in religion, or in politics, or even in any form of thought that aspires to dogma. It is a major tragedy of my lifetime and yours that, here in Göttingen, scientists were refining to the most exquisite precision the Principle of Tolerance, and turning their backs on the fact that all around them tolerance was crashing to the ground beyond repair. . . .

When Hitler arrived in 1933, the tradition of scholarship in Germany was destroyed, almost overnight. Now the train to Berlin was a symbol of flight. Europe was no longer hospitable to the imagination—and not just the scientific imagination. A whole conception of culture was in retreat: the conception that human knowledge is personal and responsible, an unending adventure at the edge of uncertainty. . . .

Science is a very human form of knowledge. We are always at the brink of the known, we always feel forward for what is to be hoped. Every judgment in science stands on the edge of error, and is personal. Science is a tribute to what we can know although we are fallible.

Robert F. Baum

Popper, Kuhn, Lakatos: A Crisis of Modern Intellect

Robert Baum is a critic of the so-called irrationalist philosophers of science. In the following selection Baum sketches the explanations of the nature of scientific knowledge given by two contemporary thinkers who reject the classical view in which scientific truth progresses by accumulation toward final and complete comprehension. Baum is not satisfied with systems in which the "rationality of science" acquires different meanings at different times in history. Such systems produce, in his view, an intellectual anarchism that is incapable of a critical examination of social preferences.

A controversy agitating philosophers of science has made that sometimes narrow specialty a brilliant illumination of the crossroads to which modern intellect has come. The controversy has provided counter-culture anarchism on the one hand, and blind obedience to existing elites or experts on the other, with their most sophisticated rationales. It has also inspired what seems the most persuasive presentation possible of the rationalism that has characterized the West since Newton published his *Principia*. And in the end it seems likely to revive insights that were ancient before Copernicus was born.

The controversy began in 1962 with the publication of Thomas S. Kuhn's *The Structure of Scientific Revolutions*. That book made the autonomy or, as moderns typically construe that quality, the rationality of science a matter of debate. It raised the question, "Do scientific laws and theories chiefly articulate the peculiar interests and extra–scientific beliefs of scientists, or are they descriptions of reality based firmly on empirical fact?"

Non-Euclidean geometries, atomic theory, and Maxwell's field theory had disturbed the conventional modern view of science before 1900. Still, from Newton's time to the publication of Einstein's relativity paper of 1905 the question stated above would in most quarters have received a confident answer. That answer would have been: Unlike philosophy and theology, which deduce their conclusions from premises sanctioned only by authority or faith, science *induces* its laws and theories from empirical facts, and in consequence those laws and theories accurately describe reality. Einstein's paper, soon aided by the quanta hypotheses emanating from Max Planck's radiation studies, had on that long established confidence the effect of a large rock thrown into a quiet pool.

With Einstein's sharp modification or overthrow of what had seemed the most scrupulously induced and abundantly verified laws of all time, those of Newtonian physics, David Hume's simple but devastating criticism of laws induced from observed facts came once more into prominence. What reason, Hume had asked nearly two centuries earlier, had anyone to suppose that future observations would resemble past observations? Hence, what reason to place any trust in universal, predictive laws induced from observed facts?

Over generations, the accumulating success of Newtonian physics had seemed to Anglo-Saxon empiricists to make Hume's questions academic. Empiricists had completely failed to justify induction logically, but a triumphant science seemingly inducing its universal laws had appeared to guarantee induction's soundness. On the Continent, Kant had answered Hume with a far more knowledgeable conception of scientific method and laws than the empiricists' induction notion. But Kant's reply was so deeply implicated in Newtonian ideas of time and space, which Kant considered phenomenally final, that it fell with them. Within a decade, reflective people generally were wondering what scientific laws and theories really signified.

Concomitantly, an idea put forward somewhat earlier in France gained not a few adherents. According to the physicists Poincaré and Duhem, when science went beyond statements of empirical fact to construct laws and theories, these were simply "conventions," accepted for reasons in the main esthetic. Through many heads in the 1910's and 1920's there passed the unsettling thought that physical theory might really have no more claim to truth than metaphysics or theology, to which modern rationalists long had granted little claim at all.

No one of course denied modern science's contribution to such achievements as dynamos and X-rays, but just as Newtonian theory, now upset, had served to predict multitudes of planetary and other phenomena, so theories equally fallible might have served to produce dynamos and X-rays. In taking sun and star sights with their sextants, mariners were still assuming a geocentric universe and safely making port. In brief, it became apparent that a theory's truth, *i.e.*, its validity as a description of reality, could not be inferred from what was or seemed to be its practical success. Pretty quickly, reflective scientists and laymen retreated from the centuries old belief that scientific laws and theories constituted an accumulating store of finally proven, settled truth.

The embarrassment thus occasioned showed itself not only in the popularity of French conventionalism but also in the acceptance by a majority of physicists, with Einstein and Planck notable exceptions, of the view of science called instrumentalism. This view totally surrendered science's claim to truth. It declared scientific laws and theories mere instruments for technological manipulation of nature, tools that no more described nature than a pipewrench describes a pipe.

Instrumentalism thus exalted as the goal and finest product of science the technology and hardware that ordinarily interested men like Einstein and Planck only to the extent that technology and hardware might assist in the pursuit of truth. Had instrumentalism endured as the accepted view of science, it might well have diverted from science just the kind of men who usually have been its greatest practitioners. For through the arcane mazes of mathematics and laboratory apparatus such men have, since Copernicus' time, pursued truth in a spirit often comparable to that of saints at prayer. Happily for science, instrumentalism and conventionalism soon faced an attractive rival. This received its most notable expression in Karl (now Sir Karl) Popper's *Logik der Forschung* of 1934, translated in 1959 as *The Logic of Scientific Discovery.*

Popper wanted to assert the authority of empirical observation against conventionalism, which implied that scientific theories were accepted mainly by agreement. He also wanted to relieve science of the embarrassment inflicted by the induction notion, an embarrassment intensified by Logical Positivism, which considered induction the hallmark of science.

Stated as it was generally understood, Popper's argument ran:

Induction could not possibly be justified, because, as elementary logic tells us, the singular statements we base on observation can never justify universal statements. No number of true reports of white swans can ever justify "All swans are white." Hence science's universal theories were not inductions at all; they were bold hypotheses which, unlike metaphysical hypotheses, entailed empirically testable consequences. Tests could never finally prove theories true, since tests, too, were observations yielding only singular statements, but tests could often *falsify* theories, as one report of a black swan falsifies "All swans are white." Science therefore could and did progress toward truth by rejecting hypotheses that were falsified and by holding to those which the severest tests corroborated, meanwhile recognizing even these to be corrigible. Science did not accumulate a store of finally proven, settled truth, but by bold conjecture and self-criticism it could and did improve its theories' *approximation* to the truth.

Popper's verdict on induction (and Logical Positivism) could not rationally be challenged. At the same time, his conception of a science not absolutely true but both autonomous and rational, a science standing on logic and observation only, seemed satisfactorily to distinguish science from metaphysics and to justify belief in scientific progress. With Popper's *Logic,* modern rationalism had to some degree retreated—but to an apparently unassailable position. Falsificationism, as Popper's view came to be called, gave grounds for sober confidence in science as a purely logico–empirical and ever closer pursuit, though not final capture, of truth. Which brings us to 1962 and the heated controversy launched by Kuhn's *Structure of Scientific Revolutions*.

A professor of the history of science, Kuhn declared that falsificationism described, not the strategy actually employed by scientists, but one suggested by science textbooks. These, he wrote, pictured early science as a continuous effort toward modern science. Actually, he went on, the long history of mature sciences like astronomy or physics presented us with long periods of "normal science" wherein basic theories or, as he also called them, paradigms, like Ptolemy's geocentrism or Newton's theory of gravitation were contentedly accepted. While "protosciences" like psychology or sociology persisted in debating basic theory, agreement on basic theory typified mature sciences and enabled them to press forward into esoteric depths and details, aiming to bring additional areas of experience within the basic theories' compass. And here Kuhn advanced a major point:

In the resulting "normal" effort, which Kuhn called "puzzle-solving," scientists actually tested only their own ingenuity, not their basic theory. For examples, Kuhn cited Ptolemaic astronomers' ingenious eccentrics and epicycles, which for centuries had reconciled Ptolemaic geocentrism to observed planetary movements. Conversely, he pointed out that Newtonian astronomers had long recognized the perihelion of Mercury as an anomaly within Newtonian theory, but had taken it as a reflection on their puzzle-solving ingenuity rather than as a falsification of the theory. Instead of promptly rejecting basic theories that were contradicted by observation, as falsificationism envisioned, scientists normally displayed an ingenious loyalty to such theories.

Only, Kuhn continued, in periods of "extraordinary science" leading to scientific revolutions did the basic theories of mature sciences incur the criticism that Popper had taken as the distinctive mark of science generally. Yet even here, Kuhn declared, falsification did not occur. This because the new paradigms that then emerged differed so radically from the old ones as to be incommensurable with them.

Those astronomers, Kuhn wrote, who had called Copernicus insane for saying that the earth moved "were not either just wrong or quite wrong," because by "earth" they meant a stationary body. The earth *they* used in puzzle-solving *could* not move. In Kuhn's view, Copernicus' and other novel theories not only changed the meanings of words but in effect changed the world that presented itself for theoretical interpretation. Consequently, old theories were not in any real sense falsified; a historian could not find "a point at which resistance [to a new theory] becomes either illogical or unscientific." Instead, when scientists in a given science went over to a revolutionary basic theory they did so via a process not fully explicable in terms of observation and logic. The process was in part arbitrary and akin to religious conversion.

Kuhn's well documented book portrayed scientists as singleminded puzzle addicts, normally contented with whatever basic theory set up as solvable puzzles, and no more falsifying such theories than they proved them. Implicitly denying the autonomy of science, Kuhn announced that the explanation for scientists' choice of basic theories lay in the realm of sociology and individual psychology. Adding fuel to his fire, he suggested that the idea that successive theories carried science nearer to the truth might have to be given up.

Once more then, science, the beacon light of modern intellect, threatened to emerge as the stuff of dreams—or as a set of directions for puzzle-solving. Hydrogen bombs and voyages to the moon remained and could not be doubted. But, once more again, technological results cannot prove the truth of theories, and little thought of utilitarian ends had sustained the labor of the great architects of modern science. Rightly or wrongly, Copernicus, Kepler, Galileo, and their later peers had thought themselves in pursuit of a true description of reality. Most of them had plainly said so. Moreover, laymen had come to accord scientific theory much of the awed respect once given the pronouncements of prophets, oracles, and priests.

It should be evident, and Kuhn himself has come close to remarking it, that his argument projected into the field of science the historicism that after World War I began to relativize modern views of politics, art, and morals. Suspecting moral judgments like those too easily made by both sides in that war, historicist historians resolved to reject "all standards outside the object," *i.e.,* outside historical facts themselves, and to confine attention to facts alone. This positivistic rejection of philosophy and its standards inevitably made all men, ideas, and movements incomparable with each other, in Kuhn's phrase, incommensurable. Within historicism, not only moral but intellectual judgments became impossible. Everything was viewed on its own terms as no more than a product of its place and time. Kuhn viewed scientific theories in the same way and seemed to urge an equally unlimited relativism in science.

Kuhn's theses raised a host of issues in the philosophy of science and occasioned more than one formal symposium. Foremost among these was one held in London in 1965 and chaired by Popper. There, such luminaries in the field as Stephen Toulmin, John Watkins, Imre Lakatos, Paul Feyerabend, and others, as well as Popper and Kuhn themselves, presented papers.

Feyerabend, . . . whose command of much modern physics has not prevented him from embracing notions typical of the counterculture, explicitly welcomed the unlimited relativism which Kuhn had seemed to friend and foe alike to endorse. Mankind's progress, wrote

Feyerabend, easily assuming such progress's reality, results from allowing everyone to follow his own inclinations, and now in science as elsewhere that happy course is justified; nothing compels assent, hence everything deserves respect, even "the most outlandish products of the human brain." Formerly considered an austere and demanding mistress, science now stood revealed as "an attractive and yielding courtesan who tries to anticipate every wish of her lover."

In brief but adamant opposition to that, Popper's paper condemned Kuhnian relativism as part of a general relativism imperilling modern civilization. Lakatos observed that Kuhn, by spreading the impression that even in science no way existed to judge a theory except by the number and vocal energy of its supporters, had endorsed the credo of student revolutionaries, that truth lies in power. Popper and Lakatos clearly saw themselves as rational men confronted by a socially dangerous sophistry.

Now science's long immunity to historicist relativism has rested on science's presumed possession of a solid basis in empirical fact. However abstract an hypothesis might be, scientists have seemed able eventually to derive from it some definite predictions about the behavior of observable objects. And reports of that behavior, accepted as hard facts corroborating or contradicting the hypothesis, have been taken as a happily non-philosophical and therefore reliable standard for scientific judgments. In the modern view, philosophy itself, and politics and morality, could claim no such standard; empirical science could. . . .

Kuhn meanwhile has denied part of the relativism he has been charged with. He has written that in his view one scientific theory is not as good as another "for doing what scientists normally do," *i.e.,* for setting up puzzles and solving them. But in complete accord with our analysis above, he has added:

> Nevertheless, there is another step, or kind of step, which many philosophers of science wish to take and which I refuse. They wish, that is, to compare theories as representations of nature, as statements about "what is really out there." Granting that neither theory of a historical pair is [absolutely] true, they nevertheless seek a sense in which the later is a better approximation of the truth. I believe nothing of that sort can be found.

Also:

> The notion of a match between the ontology of a theory and its "real" counterpart in nature now seems to me illusive in principle.

Let us be clear about where such statements leave this undoubtedly learned educator and, presumably, his students. They leave them saying, not merely that Einstein's cosmology is no truer than Newton's, and Newton's no truer than Ptolemy's, but that Ptolemy's in turn is no truer than the theory that the earth is flat. With this, Kuhn might seem on the point of endorsing Feyerabend's anarchic notion that even "the most outlandish products of the human brain"—including not only flat earthism but, say, belief in witches and the Protocols of the Elders of Zion—deserve respect.

But Kuhn takes the somewhat different tack that has been steered by the historicism which, as earlier indicated, he projects into the field of science. Suspecting the normative judgments about historical events that philosophic or theological standards made possible, and taking historical events as the only realities, historicism ended by treating existing or "the current" events and trends themselves as standards. In politics, art, and morality historicists came to seek and defer to, not what they declared was good or true, but what was currently established. They ended by making what Burckhart once called the worst normative judgment of all, "the approval of the *fait accompli.*"

Kuhn appears to make just such a judgment when he urges that, however much existing science serves the merely puzzle-solving passion of scientists, however distant from real knowledge it persists in being, we not only should accept it as knowledge but should derive our very conception of reason from it. Instead of requiring science to measure up to reason, he would have us redefine reason by whatever a scientific elite, admittedly moved largely by "the value system, the ideology," current in science, currently accepts. He writes:

> To suppose, instead we possess criteria of rationality which are independent of our [the current?]

understanding of the scientific process is to open the door to cloud-cuckoo land.

That seems clearly to dismiss as "cloud-cuckoo" the land that most of us would call the land of reason. Is Kuhn not peremptorily dismissing the land wherein value systems and ideologies are critically examined, not accepted merely because they are current? And is not that land also the land wherein a tyrant is known to be a tyrant, even if currently successful, and wherein error is seen to be error, however firmly established? . . .

Taken as seriously as its roots in the traditional modern worldview deserve, Kuhn's sophistic seems sure increasingly to expose the inadequacy of modern rationalism and the unreality of the modern notion of a purely logico–empirical or autonomous, philosophy–free science. Intellectual (and social) anarchism on the one hand, and blind historicist subordination to existing authorities on the other, may well be that exposure's first consequences. But *their* consequences will no doubt, as in ages past, turn men once more to the reason that sets men apart from other animals.

Gary Zukav

Enlightenment and Modern Physics

More than one author has compared the approach characteristic of modern quantum physicists to that evident among the practitioners of Eastern religions. In both there is an acceptance of reality as something that can be encountered but which cannot be interpreted without contradiction. Gary Zukav underscores the impact that the values of a society exert on the manner in which nature is approached in the West as compared to the East. But he also argues that a comparison of the understanding of reality in the two groups reveals that there is more similarity than difference.

What does physics have in common with enlightenment? Physics and enlightenment apparently belong to two realms which are forever separate. One of them (physics) belongs to the external world of physical phenomena and the other of them (enlightenment) belongs to the internal world of perceptions. A closer examination, however, reveals that physics and enlightenment are not so incongruous as we might think. First, there is the fact that only through our perceptions can we observe physical phenomena. In addition to this obvious bridge, however, there are more intrinsic similarities.

Enlightenment entails casting off the bonds of concept ("veils of ignorance") in order to perceive directly the inexpressible nature of undifferentiated reality. "Undifferentiated reality" is the same reality that we are a part of now, and always have been a part of, and always will be a part of. The difference is that we do not look at it in the same way as an enlightened being. As everyone knows(?), words only *represent* (*re-present*) something else. They are not real things. They are only *symbols*. According to the philosophy of enlightenment, *everything* (every*thing*) is a symbol. The reality of symbols is an illusory reality. Nonetheless, it is the one in which we live.

Although undifferentiated reality is inexpressible, we can talk around it (using more symbols). The physical world, as it appears to the unenlightened, consists of many separate parts. These separate parts, however, are not really separate. According to mystics from around the world, each moment of enlightenment (grace/insight/samadhi/satori) reveals that everything—all the separate parts of the universe—are manifestations of the same whole. There is only *one* reality, and it is whole and unified. It is one. . . .

In short, both in the need to cast off ordinary thought processes (and ultimately to go "beyond thought" altogether), and in the perception of reality as one unity, the phenomenon of enlightenment and the science of physics have much in common. . . .

Classical science starts with the assumption of separate parts which together constitute physical reality. Since its inception, it has concerned itself with how these separate parts are related.

Newton's great work showed that the earth, the moon and the planets are governed by the same laws as falling apples. The French mathematician, Descartes, invented a way of drawing pictures of relationships between different measurements of time and distance. This process (analytic geometry) is a wonderful tool for organizing a wealth of scattered data into one meaningful pattern. Herein lies the strength of western science. It brings huge tracts of apparently unrelated experience into a rational framework of simple concepts like the laws of motion. The starting point of this process is a mental attitude which initially perceives the physical world as fragmented and different experiences as logically unrelated. Newtonian science is the effort to find the relationships between pre-existing "separate parts."

Quantum mechanics is based upon the opposite epistemological assumption. Thus, there are profound differences between Newtonian mechanics and quantum theory. . . .

David Bohm, Professor of Physics at Birkbeck College, University of London, proposes that quantum physics is, in fact, based upon a perception of a new order. According to Bohm, "We must turn physics around. Instead of starting with parts and showing how they work together (the Cartesian order) we start with the whole." . . .

Bohm's physics require, in his words, a new "instrument of thought." A new instrument of thought such as is needed to understand Bohm's physics, however, would radically alter the consciousness of the observer, reorientating it toward a perception of the "unbroken wholeness" of which everything is a form. . . .

The requirement for a new instrument of thought upon which to base Bohm's physics may not be as much of an obstacle as at first appears. There already exists an instrument of thought based upon an "unbroken wholeness." Furthermore, there exist a number of sophisticated psychologies, distilled from two thousand years of practice and introspection, whose sole purpose is to develop this thought instrument.

These psychologies are what we commonly call "Eastern religions." "Eastern religions" differ considerably among themselves. It would be a mistake to equate Hinduism, for example, with Buddhism, even though they are more like each other than either one of them is like a religion of the West. Nonetheless, all eastern religions (psychologies) are compatible in a very fundamental way with Bohm's physics and philosophy. All of them are based upon the experience of a pure, undifferentiated reality which is that-which-is.

While it would be naive to overstate the similarities between Bohm's physics and eastern philosophies, it would be foolish to ignore them. Consider, for example, the following sentences:

> The word "reality" is derived from the roots "thing" (*res*) and "think" (*revi*). "Reality" means "everything you can think about." This is not "that-which-is." No idea can capture "truth" in the sense of that-which-is.
>
> The ultimate perception does not originate in the brain or any material structure, although a material structure is necessary to manifest it. The subtle mechanism of knowing the truth does not originate in the brain.
>
> There is a similarity between thought and matter. All matter, including ourselves, is determined by "information." "Information" is what determines space and time.

Taken out of context, there is no absolute way of knowing whether these statements were made by Professor Bohm or a Tibetan Buddhist. In fact, these sentences were excerpted from different parts of two *physics* lectures that Professor Bohm gave at Berkeley in April, 1977. The first lecture was given on the campus to physics students. The second lecture was given in the Lawrence Berkeley Laboratory to a group of professional physicists. Two of these three statements were taken from the *second* lecture, the one given to the advanced physicists.

It is ironic that while Bohm's theories are received with some skepticism by most professional physicists, they would find an immediately sympathetic reception among the thousands of people in our culture who have turned their backs on science in their own quest for the ultimate nature of reality.

If Bohm's physics, or one similar to it, should become the main thrust of physics in the future, the dances of East and West could blend in exquisite harmony. Physics curricula of the twenty-first century could include classes in meditation.

The function of eastern religions (psychologies) is to allow the mind to escape the confines of the symbolic. According to this view, *everything* is a symbol, not only words and concepts, but also people and things. Beyond the confines of the symbolic lies that which is, pure awareness, the experience of the "suchness" of reality.

Nonetheless, every eastern religion resorts to the use of symbols to escape the realm of the symbolic. Some disciplines use symbols more than others, but all of them use symbols in one form or another. Therefore, the question arises, if pure awareness is considered distinct from the content of awareness, in what ways specifically does the content of awareness affect the realization of pure awareness? What types of content prompt the mind to leap forward. What enables it to activate the self-fulfilling capability to transcend itself.

It is very difficult to answer this question. Any answer is only a point of view. A point of view itself is limiting. To "understand" something is to give up some other way of conceiving it. This is another way of saying that the mind deals in forms of limitation. Nonetheless, there *is* a relationship between the content of awareness and the ability of the mind to transcend itself.

"Reality" is what we take to be true. What we take to be true is what we believe. What we believe is based upon our perceptions. What we perceive depends upon what we look for. What we look for depends upon what we think. What we think depends upon what we perceive. What we perceive determines what we believe. What we believe determines what we take to be true. What we take to be true is our reality.

The central focus of this process, initially at any rate, is "What we think." We at least can say that allegiance to a symbol of openness (Christ, Buddha, Krishna, "the infinite diversity of nature," etc.) seems to open the mind and that an open mind is often the first step in the process of enlightenment.

The psychological gestalt of physics has shifted radically in the last century to one of extreme openness. In the middle 1800s, Newtonian mechanics was at its zenith. There seemed to be no phenomenon which could not be explained in terms of mechanical models. All mechanical models were subject to long-established principles. The chairman of the physics department at Harvard discouraged graduate study because so few important matters remained unsolved.

In a speech to the Royal Institution in 1900, Lord Kelvin reflected that there were only two "clouds" on the horizon of physics, the problem of black-body radiation and the Michelson-Morley experiment. There was no doubt, said Kelvin, that they soon would be gone. He was wrong. Kelvin's two "clouds" signaled the end of the era that began with Galileo and Newton. The problem of black-body radiation led to Planck's discovery of the quantum of action. Within thirty years the entirety of Newtonian physics became a special limiting case of the newly developing quantum theory. The Michelson-Morley experiment foreshadowed Einstein's famous theories of relativity. By 1927, the foundations of the new physics, quantum mechanics and relativity, were in place.

In contrast to Kelvin's time, the allegiance of physicists today is to a symbol of extreme openness. Isidor Rabi, Nobel Prize winner and Chairman Emeritus of the Physics Department at Columbia University, wrote in 1975:

> I don't think that physics will ever have an end. I think that the novelty of nature is such that its variety will be infinite—not just in changing forms but in the profundity of insight and the newness of ideas . . .

Stapp wrote in 1971:

> . . . human inquiry can continue indefinitely to yield important new truths.

The "What we think" of physicists today is that the physics of nature, like human experience itself, is infinitely diverse. . . .

Buddhism is both a philosophy and a practice. Buddhist philosophy is rich and profound. Buddhist practice is called *Tantra*. *Tantra* is the Sanskrit word meaning "to weave." There is little that can be said about *Tantra*. It must be done.

Buddhist philosophy reached its ultimate development in the second century A.D. No one has been able to improve much on it since then. The distinction between Buddhist philosophy and *Tantra* is well defined. Buddhist philosophy can be intellectualized. *Tantra* cannot. Buddhist philosophy is a function of the rational mind. *Tantra* transcends rationality. The most profound thinkers of the Indian civilization discovered that words and concepts could take them only so far. Beyond that point came the actual doing of a practice, the experience of which was ineffable. This did not prevent them from progressively refining the practice into an extremely effective and sophisticated set of techniques, but it did prevent them from being able to describe the experiences which these techniques produce.

The practice of *Tantra* does not mean the end of rational thought. It means the integration of thought based on symbols into larger spectrums of awareness. (Enlightened people still remember their zip codes.)

The development of Buddhism in India shows that a profound and penetrating intellectual quest into the ultimate nature of reality can culminate in, or at least set the stage for, a quantum leap beyond rationality. In fact, on an individual level, this is one of the roads to enlightenment. Tibetan Buddhism calls it the Path without Form, or the Practice of Mind. The Path without Form is prescribed for people of intellectual temperament. The science of physics is following a similar path.

The development of physics in the twentieth century already has transformed the consciousness of those involved with it. The study of complementarity, the uncertainty principle, quantum field theory, and the Copenhagen Interpretation of Quantum Mechanics, produces insights into the nature of reality very similar to those produced by the study of eastern philosophy. The profound physicists of this century increasingly have become aware that they are confronting the ineffable.

Max Planck, the father of quantum mechanics, wrote:

> Science . . . means unresting endeavor and continually progressing development toward an aim which the poetic intuition may apprehend, but which the intellect can never fully grasp.

We are approaching the end of science. "The end of science" does not mean the end of the "unresting endeavor and continually progressing development" of more and more comprehensive and useful physical theories. (Enlightened physicists remember their zip codes, too.) The "end of science" means the coming of western civilization, in its own time and in its own way, into the higher dimensions of human experience.

CHAPTER 12

At the Brink of the Future

Introduction by John G. Burke

An astonishing number of scientifically based technologies have been developed in the twentieth century: synthetic fibers, frozen foods, plastics, television, antibiotic drugs, jet aircraft, atomic energy, and lasers, to name a few. The proliferation of new technologies has changed the way we live: our customs, our habits, our work, and our leisure. Two twentieth-century developments in particular, though—electronic computers and biotechnology—appear to many authorities to have the potential of producing radical change, even to the extent of transforming the human species. Pessimists predict that humans will become increasingly obsolescent in a computerized world, while optimists prophesy that the new race of humans fashioned by biotechnological advances will colonize the planets of our solar system and proceed on to inhabit the galaxy.

Peculiarly enough, biological technology began over three thousand years ago with attempts to bring about significant hereditary change in plants and animals by trial and error in order to enhance certain qualities thought to be desirable or to eliminate unwanted characteristics. Similarly, efforts to reduce by mechanical means the labor of computation or to develop methods by which industrial processes could be manipulated in response to coded information are centuries old. However, scientific knowledge began to be applied to both technologies only in the twentieth century.

Charles Babbage (1792–1871), an English mathematician, in the early nineteenth century envisaged a digital computer that would be mechanically operated; although some components were built, the projected machine was never completed. The first electronic computers using vacuum tubes were constructed in the 1940s in response to military demands for devices that would rapidly and accurately calculate the trajectories of long-range artillery shells under varying conditions—that is, exterior ballistics. The development of the transistor ushered in the second generation of computers, followed scarcely a decade later by the third generation, which employed integrated circuits. Guidance systems for intercontinental ballistic missiles and for space vehicles required continued miniaturization of computers and their components. Simultaneously, the need to monitor the precise geographical locations or the movements of the offensive and defensive weapons systems of both friends and potential foes prompted the design and construction of the fourth generation: very large-scale integrated computers. The fifth computer generation will, according to projections, incorporate completely new designs and be a thousand times more powerful than any now available.

The transfer of computer technology to the civilian sector has been astonishingly rapid; computers were deployed in manufacturing processes, finance, business, science, medicine, education, and the arts. One authority, Daniel Bell, views the computer as the central factor during the past three decades in changing the structure of work in our society. In earlier times the majority of employed people were engaged in producing goods; now they perform services, most notably the gathering, compilation, and distribution of information with the aid of computers.

The growing ubiquitous presence of computers has given rise to prolonged disputes among experts not only with respect to the impact on work, but also concerning the danger of the use of computers for the invasion of personal privacy and for the surveillance and control of the citizenry. The most heated argument, however, has to do with computer or artificial intelligence (AI), with whether or not computers now or in the future can match human intelligence. The controversy is far from trivial, because it involves the question of whether computers should be viewed as just another tool or technique to help decision makers or whether computers can or should be entrusted to make decisions without human intervention. These issues are discussed in the articles below by two computer scientists, Herbert A. Simon and Joseph Weizenbaum.

The contention that AI can be equated with human intelligence is based on the belief that the human brain (mind) is merely an information- or symbol-processing organ. René Descartes advanced the mechanistic view of man in the seventeenth century (see Chapter 6), although he made a careful distinction between brain and mind. However, more mundane mechanistic concepts have had much scientific support, notably by the eminent English biologist, Thomas Henry Huxley (1825–1895), in the nineteenth century and by the influential American psychologist, John B. Watson (1878–1958), in the early twentieth century. In opposition is the belief held currently by numerous scientists, including Weizenbaum, that the human brain is a unique, complex, and intricate organ, and that human intelligence incorporates attributes other than the ability to process information, including wisdom and intuition, which cannot be duplicated in a computer.

The controversy about intelligence and whether it consists of one or several inheritable traits may be settled sometime in the future by advances in biochemistry and molecular biology, aided to a large degree by super computers, which are necessary to identify genetic codes. Even well into the twentieth century, ideas about human heredity were primarily based on observation. For example, in the eighteenth century, a French scientist, Pierre de Maupertuis (1698–1759), found that polydactylism, the property of having more than the normal number of fingers or toes, appeared among a number of members of a family over several generations. Similarly, the famous Hapsburg protruding lower lip and jutting jaw were recognized features of many descendants of the royal family that ruled the Holy Roman Empire, Spain, and Austria for generations. And, about the turn of the twentieth century, Francis Galton (1822–1911) traced the lineage and accomplishments of numerous English families and concluded that intelligence was a hereditary characteristic.

Other studies incorporated measurements of brain sizes and weights and involved delineations of the facial features of criminals and the mentally defective. All fueled the eugenics movement, which came into prominence in the early twentieth century. The adherents in general had as their goal the improvement of the human species. However, there were quite different approaches to the solution. Some preached that miscegenation—intermarriage and procreation between races—should not only be discouraged but prohibited, because otherwise the human gene pool would be irreparably impaired. Others successfully advocated the prevention of the marriage of the mentally defective, with the result that laws were enacted to that end in many jurisdictions. Still others counseled that the progressive improvement of the human species and the prevention of its deterioration could be achieved only by the mating of superior men and women over many generations.

Hermann J. Muller (1890–1967), a biologist who was awarded the Nobel prize in 1946 for his demonstration (in 1926) that X-rays cause genetic mutations, was a foremost partisan of the latter approach. Muller assumed that great men and women—artists, scientists, writers, statesmen, or philosophers—were exceptional because of their genetic constitutions. Their outstanding talents, he believed, could be passed on to future generations by preserving the male sperm and artificially inseminating suitable females. The criteria to be used in selecting appropriate male donors, Muller thought, would be first, that they possessed a highly developed social feeling or fraternal love, and second, that they have superior intelligence, which he equated with analytical ability or reason. There were, he admitted, possibilities of error in the selection process. Genetic merit had to be distinguished from environmental good fortune, and possibly a generation had to pass before an individual could be judged as having been truly exceptional. Muller maintained, however, that any latent unfavorable traits that appeared in the offspring could be sifted out in later generations. And, he hoped that people who were not genetically well endowed would consider it their duty to refrain from having children.

Not all biologists, however, were as sanguine as Muller about the future uses of genetic knowledge. Excerpted below are two somewhat fanciful essays by the British physiologist, J. S. B. Haldane (1892–1964). Writing in the 1920s, he predicted *in vitro* fertilization; he also foresaw some of the moral and ethical dilemmas to which such developments could lead.

Meanwhile, the science of genetics was experiencing rapid development. Scientists determined that chromosomes, which are components of cell nuclei, were linked to heredity, and that the chromosomes contained DNA—deoxyribonucleic acid. Beginning in the 1950s genetics received tremendous impetus from the discovery of the structure of the DNA molecule and from the finding that each of the two strands composing it contain certain regions or sequences of nucleotides, which encode specific hereditary information. By 1968 the work of many investigators led to the breaking of the genetic code. In the early 1970s scientists took the initial step in modern biotechnology by producing recombinant DNA. In this technique strands of bacterial DNA are modified, yielding new molecular configurations and hence new functions, and then reinserted into bacterial cells.

Biotechnology is still in its infancy, and the initial results of this new applied science are just beginning to appear: the identification of chromosome or single-gene disorders, new pharmaceutical drugs, and new plant varieties incorporating such desired qualities as resistance to frost. Most experts at present think that the creation of new genetic strains of higher animals by biotechnological means is only a very remote possibility. However, there are both scientists and nonscientists who believe that humans should, as H. J. Muller preached, control their evolutionary future. And, the startling achievements of scientifically based technology in the twentieth century demonstrate that the prediction of future developments is a perilous undertaking. As an authority on science and a science fiction writer, Arthur C. Clarke suggests in the selection below that "the only way of discovering the limits of the possible is to venture a little way past them into the impossible." At the very least, the history of science should make us aware that science has a future as well as a past.

Arthur C. Clarke
The Hazards of Prophecy

Arthur C. Clarke (b. 1917) is best known for his award-winning science fiction and for 2001: A Space Odyssey, *but he has also chaired the British Interplanetary Society and has written books on space flight and lunar exploration. In the 1940s Clarke was the first to propose the idea of communications satellites. He was also one of the first to recognize the significance of computers. In* Profiles of the Future *(1963) he discusses previous attempts by scientists to foresee the future and analyzes where they went wrong.*

With few exceptions, scientists seem to make rather poor prophets; this is rather surprising, for imagination is one of the first requirements of a good scientist. Yet, time and again, distinguished astronomers and physicists have made utter fools of themselves by declaring publicly that such-and-such a project was impossible. . . .

. . . Apparently competent men . . . have been proved utterly wrong, sometimes while the ink was scarcely dry from their pens. On careful analysis, it appears that these debacles fall into two classes, which I will call "failures of nerve" and "failures of imagination."

The failure of nerve seems to be the more common; it occurs when *even given all the relevant facts* the would-be prophet cannot see that they point to an inescapable conclusion. Some of these failures are so ludicrous as to be almost unbelievable. . . .

The most famous, and perhaps the most instructive, failures of nerve have occurred in the fields of aero- and astronautics. At the beginning of the twentieth century, scientists were almost unanimous in declaring that heavier-than-air flight was impossible, and that anyone who attempted to build airplanes was a fool. The great American astronomer, Simon Newcomb, wrote a celebrated essay which concluded:

> The demonstration that no possible combination of known substances, known forms of machinery and known forms of force, can be united in a practical machine by which man shall fly long distances through the air, seems to the writer as complete as it is possible for the demonstration of any physical fact to be.

Oddly enough, Newcomb was sufficiently broad minded to admit that some wholly new discovery—he mentioned the neutralization of gravity—might make flight practical. One cannot, therefore, accuse him of lacking imagination; his error was in attempting to marshal the facts of aerodynamics when he did not understand that science. His failure of nerve lay in not realizing that the means of flight were already at hand.

For Newcomb's article received wide publicity at just about the time that the Wright brothers, not having a suitable antigravity device in their bicycle shop, were mounting a gasoline engine on wings. When news of their success reached the astronomer, he was only momentarily taken aback. Flying machines *might* be a marginal possibility, he conceded—but they were certainly of no practical importance, for it was quite out of the question that they could carry the extra weight of a passenger as well as that of a pilot. . . .

As far as the general public is concerned, the idea of space flight as a serious possibility first appeared on the horizon in the 1920's, largely as a result of newspaper reports of the work of the American Robert Goddard and the Rumanian Hermann Oberth. . . . For a sample of the kind of criticism the pioneers of astronautics had to face, I present this masterpiece from a paper published by one Professor A. W. Bickerton, in 1926. . . .

> This foolish idea of shooting at the moon is an example of the absurd length to which vicious specialisation will carry scientists working in

thought-tight compartments. Let us critically examine the proposal. For a projectile entirely to escape the gravitation of the earth, it needs a velocity of 7 miles a second. The thermal energy of a gramme at this speed is 15,180 calories. . . . The energy of our most violent explosive—nitroglycerine—is less than 1,500 calories per gramme. Consequently, even had the explosive nothing to carry, it has only one-tenth of the energy necessary to escape the earth. . . . Hence the proposition appears to be basically impossible. . . .

. . . When Lunik II lifted thirty-three years after Professor Bickerton said it was impossible, most of its several hundred tons of kerosene and liquid oxygen never got very far from Russia—but the half-ton payload reached the Mare Imbrium. . . .

The lesson to be learned from these examples is one that can never be repeated too often, and is one that is seldom understood by laymen—who have an almost superstitious awe of mathematics. But mathematics is only a tool, though an immensely powerful one. No equations, however impressive and complex, can arrive at the truth if the initial assumptions are incorrect. It is really quite amazing by what margins competent but conservative scientists and engineers can miss the mark, when they start with the preconceived idea that what they are investigating is impossible. When this happens, the most well-informed men become blinded by their prejudices and are unable to see what lies directly ahead of them. What is even more incredible, they refuse to learn from experience and will continue to make the same mistake over and over again. . . .

When the existence of the 200-mile-range V-2 was disclosed to an astonished world, there was considerable speculation about intercontinental missiles. This was firmly squashed by Dr. Vannevar Bush, the civilian general of the United States scientific war effort, in evidence before a Senate committee on December 3, 1945. Listen:

> There has been a great deal said about a 3,000 miles high-angle rocket. In my opinion such a thing is impossible for many years. The people who have been writing these things that annoy me, have been talking about a 3,000 mile high-angle rocket shot from one continent to another, carrying an atomic bomb and so directed as to be a precise weapon which would land exactly on a certain target, such as a city.
>
> I say, technically, I don't think anyone in the world knows how to do such a thing, and I feel confident that it will not be done for a very long period of time to come. . . . I think we can leave that out of our thinking. I wish the American public would leave that out of their thinking.

A few months earlier (in May 1945) Prime Minister Churchill's scientific advisor Lord Cherwell had expressed similar views in a House of Lords debate. This was only to be expected, for Cherwell was an extremely conservative and opinionated scientist who had advised the government that the V-2 itself was only a propaganda rumor. . . .

Of the many lessons to be drawn from this slice of recent history, the one that I wish to emphasize is this. Anything that is theoretically possible will be achieved in practice, no matter what the technical difficulties, if it is desired greatly enough. It is no argument against any project to say: "The idea's fantastic!" Most of the things that have happened in the last fifty years have been fantastic, and it is only by assuming that they will continue to be so that we have any hope of anticipating the future.

To do this—to avoid that failure of nerve for which history exacts so merciless a penalty—we must have the courage to follow all technical extrapolations to their logical conclusion. Yet even this is not enough. . . . To predict the future we need logic; but we also need faith and imagination which can sometimes defy logic itself.

The second kind of prophetic failure is less blameworthy, and more interesting. It arises when all the available facts are appreciated *and* marshaled correctly—but when the really vital facts are still undiscovered, and the possibility of their existence is not admitted.

A famous example of this is provided by the philosopher Auguste Comte, who in his *Cours de Philosophie Positive* (1835) attempted to define the limits within which scientific knowledge must lie. In his chapter on astronomy he wrote these words concerning the heavenly bodies:

> We see how we may determine their forms, their distances, their bulk, their motions, but we

> can never know anything of their chemical or mineralogical structure; and much less, that of organised beings living on their surface. . . . We must keep carefully apart the idea of the solar system and that of the universe, and be always assured that our only true interest is in the former. Within this boundary alone is astronomy the supreme and positive science that we have determined it to be . . . the stars serve us scientifically only as providing positions with which we may compare the interior movements of our system.

In other words, Comte decided that the stars could never be more than celestial reference points, of no intrinsic concern to the astronomer. Only in the case of the planets could we hope for any definite knowledge, and even that knowledge would be limited to geometry and dynamics. Comte would probably have decided that such a science as "astrophysics" was *a priori* impossible.

Yet within half a century of his death, almost the whole of astronomy *was* astrophysics, and very few professional astronomers had much interest in the planets. Comte's assertion had been utterly refuted by the invention of the spectroscope, which not only revealed the "chemical structure" of the heavenly bodies but has now told us far more about the distant stars than we know of our planetary neighbors.

Comte cannot be blamed for not imagining the spectroscope; *no one* could have imagined it, or the still more sophisticated instruments that have now joined it in the astronomer's armory. But he provides a warning that should always be borne in mind; even things that are undoubtedly impossible with existing or foreseeable techniques may prove to be easy as a result of new scientific breakthroughs. From their very nature, these breakthroughs can never be anticipated; but they have enabled us to bypass so many insuperable obstacles in the past that no picture of the future can hope to be valid if it ignores them.

Another celebrated failure of imagination was that persisted in by Lord Rutherford, who more than any other man laid bare the internal structure of the atom. Rutherford frequently made fun of those sensation mongers who predicted that we would one day be able to harness the energy locked up in matter. Yet only five years after his death in 1937, the first chain reaction was started in Chicago. What Rutherford, for all his wonderful insight, had failed to take into account was that a nuclear reaction might be discovered that would release more energy than that required to start it. . . .

The example of Lord Rutherford demonstrates that it is not the man who knows most about a subject, and is the acknowledged master of his field, who can give the most reliable pointers to its future. Too great a burden of knowledge can clog the wheels of imagination; . . .

. . . There is one form of mental exercise that can provide good basic training for would-be prophets: Anyone who wishes to cope with the future should travel back in imagination a single lifetime—say to 1900—and ask himself just how much of today's technology would be, not merely incredible, but *incomprehensible* to the keenest scientific brains of that time.

1900 is a good round date to choose because it was just about then that all hell started to break loose in science. As James B. Conant has put it:

> Somewhere about 1900 science took a *totally* unexpected turn. There had previously been several revolutionary theories and more than one epoch-making discovery in the history of science, but what occurred between 1900 and, say, 1930 was something different; it was a failure of a general prediction about what might be confidently expected from experimentation.

P. W. Bridgman has put it even more strongly:

> The physicist has passed through an intellectual crisis forced by the discovery of experimental facts of a sort which he had not previously envisaged, and which he would not even have thought possible.

The collapse of "classical" science actually began with Roentgen's discovery of X-rays in 1895; . . . X-rays—the very name reflects the bafflement of scientists and laymen alike—could travel through solid matter, like light through a sheet of glass. No one had ever imagined or predicted such a thing; that one would be able to peer into the interior of the human body—and thereby revolutionize medicine and surgery—was something that the most daring prophet had never suggested.

The discovery of X-rays was the first great breakthrough into the realms where no human mind had ever ventured before. Yet it gave scarcely a hint of still more astonishing developments to come—radioactivity, the internal structure of the atom, relativity, the quantum theory, the uncertainty principle. . . .

As a result of this, the inventions and technical devices of our modern world can be divided into two sharply defined classes. On the one hand there are those machines whose working would have been fully understood by any of the great thinkers of the past; on the other, there are those that would be utterly baffling to the finest minds of antiquity. And not merely of antiquity; there are devices now coming into use that might well have driven Edison or Marconi insane had they tried to fathom their operation.

Let me give some examples to emphasize this point. If you showed a modern diesel engine, an automobile, a steam turbine, or a helicopter to Benjamin Franklin, Galileo, Leonardo da Vinci, and Archimedes—a list spanning two thousand years of time—not one of them would have any difficulty in understanding how these machines worked. Leonardo, in fact, would recognize several from his notebooks. All four men would be astonished at the materials and the workmanship, which would have seemed magical in its precision, but once they had got over that surprise they would feel quite at home—as long as they did not delve too deeply into the auxiliary control and electrical systems.

But now suppose that they were confronted by a television set, an electronic computer, a nuclear reactor, a radar installation. Quite apart from the complexity of these devices, the individual elements of which they are composed would be incomprehensible to any man born before this century. Whatever his degree of education or intelligence, he would not possess the mental framework that could accommodate electron beams, transistors, atomic fission, wave guides and cathode-ray tubes.

The difficulty, let me repeat, is not one of complexity; some of the simplest modern devices would be the most difficult to explain. A particularly good example is given by the atomic bomb (at least, the early models). What could be simpler than banging two lumps of metal together? Yet how could one explain to Archimedes that the result could be more devastation than that produced by all the wars between the Trojans and the Greeks? . . .

The wholly unexpected discovery of uranium fission in 1939 made possible such absurdly simple (in principle, if not in practice) devices as the atomic bomb and the nuclear chain reactor. No scientist could ever have predicted them; if he had, all his colleagues would have laughed at him.

It is highly instructive, and stimulating to the imagination, to make a list of the inventions and discoveries that have been anticipated—and those that have not. Here is my attempt to do so.

All the items on the left have already been achieved or discovered, and all have an element of the unexpected or the downright astonishing about them. To the best of my knowledge, not one was foreseen very much in advance of the moment of revelation.

On the right, however, are concepts that have been around for hundreds or thousands of years. Some have been achieved; others will be achieved; others may be impossible. But which?

The Unexpected	*The Expected*
X-rays	automobiles
nuclear energy	flying machines
radio, TV	steam engines
electronics	submarines
photography	spaceships
sound recording	telephones
quantum mechanics	robots
relativity	death rays
transistors	transmutation
masers; lasers	artificial life
superconductors; superfluids	immortality
atomic clocks; Mössbauer effect	invisibility
determining composition of celestial bodies	levitation
dating the past (Carbon 14, etc.)	teleportation
detecting invisible planets	communication with dead
the ionosphere; van Allen Belts	observing the past, the future
	telepathy

The right-hand list is deliberately provocative; it includes sheer fantasy as well as serious scientific speculation. But the only way of discovering the limits of the possible is to venture a little way past them into the impossible. . . . As I glance down the left-hand column I am aware of a few items which, only ten years ago, I would have thought were impossible. . . .

Chart of the Future

The chart given [on the opposite page] is not, of course, to be taken too seriously, but it is both amusing and instructive to extrapolate the time scale of past scientific achievement into the future. If it does no more, the quick summary of what has happened in the *last* hundred-and-fifty years should convince anyone that no present-day imagination can hope to look beyond the years 2,100. I have not even tried to do so.

Herbert A. Simon

What Computers Mean for Humanity and Society

Herbert Simon, a professor of computer science and psychology, describes the present capabilities and future potentialities of the most complex device that has thus far been invented. Simon sees the computer as a benign device that will help us solve many social and economic problems, if appropriate legislative controls protect against its misuse. Far from the computer endangering us, Simon believes that it will eventually make a very important contribution to human progress by giving us knowledge of the human mind.

Energy and information are two basic currencies of organic and social systems. A new technology that alters the terms on which one or the other of these is available to a system can work on it the most profound changes. At the core of the Industrial Revolution, which began nearly three centuries ago, lay the substitution of mechanical energy for the energy of man and animal. It was this revolution that changed a rural subsistence society into an urban affluent one and touched off a chain of technological innovations that transformed not only production but also transportation, communication, warfare, the size of human populations, and the natural environment.

It is easy, by hindsight, to see how inexorably these changes followed one another, how "natural" a consequence, for example, suburbia was of cheap, privately owned transportation. It is a different question whether foresight could have predicted these chains of events or have aided in averting some of their more undesirable outcomes. The problem is not that prophets were lacking—they have been in good supply at almost all times and places. Quite the contrary, almost everything that has happened, and its opposite, has been prophesied. The problem has always been to pick and choose among the embarrassing riches of alternative projected futures; and in this, human societies have not demonstrated any large foresight. Most often we have been constrained to anticipate events just a few years before their occurrence, or even while they are happening, and to try to deal with them, as best we can, as they are engulfing us.

We are now in the early stages of a revolution in processing information that shows every sign of being as fundamental as the earlier energy revolution. Perhaps we should call it the Third Information Revolution. (The first produced written language, and the second, the printed book.) This third revolution, which began more than a century ago, includes the com-

THE PAST

Date	Transportation	Communication Information	Materials Manufacturing	Biology Chemistry	Physics
1800			Steam engines	Inorganic chemistry	Atomic theory
	Locomotive				
		Camera			
		Babbage calculator		Urea synthesized	
	Steamship	Telegraph	Machine tools		Spectroscope
1850					Conservation of energy
			Electricity	Organic chemistry	
		Telephone			
		Phonograph			Electromagnetism
		Office machines			Evolution
	Automobile		Diesel engine		
1900			Gasoline engine	Dyes	X-rays
	Airplane				Electron
		Vacuum tube	Mass production	Genetics	Radioactivity
				Vitamins	
			Nitrogen fixation	Plastics	
1910					
					Isotopes
		Radio			
					Quantum theory
				Chromosomes	
1920				Genes	
					Relativity
					Atomic structure
1930				Language of bees	
				Hormones	Indeterminacy
		TV			Wave mechanics
					Neutron
1940	Jet	Radar			
	Rocket				
	Helicopter				
		Tape recorders			
		Electronic computers	Magnesium from sea	Synthetics	Uranium fission
		Cybernetics	Atomic energy	Antibiotics	Accelerators
				Silicones	Radio astronomy
1950		Transistor	Automation		
	Satellite	Maser	Fusion bomb	Tranquilizers	I.G.Y.
	GEM	Laser			Parity overthrown

NOW

Date	Transportation	Communication Information	Materials Manufacturing	Biology Chemistry	Physics
1960	Spaceship	Communication Satellite		Protein structure	Nucleon structure

THE FUTURE

Date	Transportation	Communication Information	Materials Manufacturing	Biology Chemistry	Physics
	Space lab				
1970	Lunar landing	Translating machines			
	Nuclear rocket		Efficient electric storage	Cetacean languages	
1980	Planetary landings	Personal radio		Exobiology	Gravity waves
1990			Fusion power	Cyborgs	
		Artificial intelligence			
2000	Colonizing planets		"Wireless" energy	Time, perception enhancement	Sub-nuclear structure
		Global library	Sea mining		
2010	Earth probes	Telesensory devices			
			Weather control		
2020		Logical languages			Nuclear catalysts
	Interstellar probes	Robots		Control of heredity	
2030			Space mining		
		Contact with extra-terrestrials		Bioengineering	
2040				Intelligent animals	
			Transmutation		
2050	Gravity control			Suspended animation	
		Memory playback			
	"Space drive"		Planetary engineering		
2060		Mechanical educator			
					Space, time distortion
		Coding of artifacts		Artificial life	
2070					
			Climate control		
	Near-light speeds				
2080		Machine intelligence exceeds man's			
	Interstellar flight				
2090	Matter transmitter		Replicator		
	Meeting with extra-terrestrials	World brain		Immortality	
2100			Astronomical engineering		

puter but many other things as well. The technology of information comprises a vast range of processes for storing information, for copying it, for transmitting it from one place to another, for displaying it, and for transforming it.

Photography, the moving picture, and television gave us, in the course of a century, a whole new technology for storing and displaying pictorial information. Telegraphy, the telephone, the phonograph, and radio did the same for storing and transmitting auditory information. Among all of these techniques, however, the computer is unique in its capacity for manipulating and transforming information and hence in carrying out, automatically and without human intervention, functions that had previously been performable only by the human brain.

As with the energy revolution, the consequences of the information revolution spread out in many directions. First, there are the economic consequences that follow on any innovation that increases human productivity. . . . These are perhaps the easiest effects of technological change to predict. Second, there are consequences for the nature of work and of leisure—for the quality of life. Third, the computer may have special consequences for privacy and individual liberty. Fourth, there are consequences for man's view of himself, for his picture of the universe and of his place and goals in it. In each of these directions, the immediate consequences are, of course, the most readily perceived. . . . It is far more difficult to predict what indirect chains of effects these initial impacts will set off, for example, the chain that reaches from the steam engine through the internal-combustion engine to the automobile and the suburb.

Prediction is easier if we do not try to forecast in detail the time path of events and the exact dates on which particular developments are going to occur, but to focus, instead, upon the steady state toward which the system is tending. Of course, we are not so much interested in what is going to happen in some vague and indefinite future as we are in what the next generation or two holds for us. Hence, a generation is the time span with which I shall be concerned. . . .

Computer Capabilities

The computer is a device endowed with powers of utmost generality for processing symbols. It is remarkable not only for its capabilities but also for the simplicity of its underlying processes and organization. Of course, from a hardware standpoint it is not simple at all but is a highly sophisticated electronic machine. The simplicity appears at the level of the elementary information processes that the hardware enables it to perform, the organization for execution and control of those processes, and the programming languages in terms of which the control of its behavior is expressed. A computer can read symbols from an external source, output symbols to an external destination, store symbols in one or more memories, copy symbols, rearrange symbols and structures of symbols, and react to symbols conditionally—that is, follow one course of action or another, depending on what symbols it finds in its memory or in its input devices. . . .

There is great dispute among experts as to what the generality of the computer implies for its ability to behave intelligently. There is also dispute as to whether the computer, when it is behaving more or less intelligently, is using processes similar to those employed by an intelligent human being, or quite different processes. The views expressed here will reflect my own experience in research with computers and my interpretation of the scientific literature. First, no limits have been discovered to the potential scope of computer intelligence that are not also limits on human intelligence. Second, the elementary processes underlying human thinking are essentially the same as the computer's elementary information processes, although modern fast computers can execute these processes more rapidly than can the human brain. In the past, computer memories, even in large computers, have probably not been as capacious as human memory, but the scale of available computer memories is increasing rapidly, to the point where memory size may not be much longer an effective limit on the capacity of computers to match human performance. Any estimate of the potential of the computer in the near or distant future depends on one's agreement or disagreement with these assumptions. . . .

Of course humans, through the processes called learning, can improve their strategies by experience and instruction. By the same token, computers can be, and to some extent have been, provided with programs (strategies) for improving their own strategies. Since a computer's programs are stored in the same memories as data, it is entirely possible for programs to modify themselves—that is, to learn. . . .

How, then, have computers actually been used to date? At present, computers typically spend most of their time in two main kinds of tasks: carrying out large-scale engineering and scientific calculations and keeping the financial, production, and sales records of business firms and other organizations. . . . Now these tasks belong to the horseless-carriage phase of computer development. That is to say, they consist in doing things rapidly and automatically that were being done slowly and by hand (or by desk calculator) in the pre-computer era.

Such uses of computers do not represent new functions but only new ways of performing old functions. Of course, by greatly lowering the cost of performing them, they encourage us to undertake them on a larger scale than before. The increased analytic power provided by computers has probably encouraged engineers to design more complex structures (for example, some of the very tall new office buildings that have gone up in New York and Chicago) than they would have attempted if their analytic aids were less powerful. Moreover, by permitting more sophisticated analyses to be carried out in the design process, they have also brought about significant cost reductions in the designs themselves. In the same way, the mechanization of business record-keeping processes has facilitated the introduction of improved controls over inventories and cash flows, with resulting savings in costs. Thus, the computer not only reduces the costs of the information-processing operations that it automates but also contributes to the productivity of the activities themselves.

The remaining . . . computer uses are more sophisticated. Let us consider two different ways in which a computer can assist an engineer in designing electric motors. On the one hand, the engineer can design the motor using conventional procedures, then employ the computer to analyze the prospective operation of the design—the operating temperature, efficiency, and so on. On the other hand, the engineer can provide the computer with the specifications for the motor, leaving to the computer the task of synthesizing a suitable design. In the second, but not the first, case the computer, using various heuristic search procedures, actually discovers, decides upon, and evaluates a suitable design. In the same way, the role of the computer in managing inventories need not be limited to record-keeping. The computer program may itself determine (on the basis of usage) when items should be reordered and how large the orders should be. In these and many other situations, computers can provide not only the information on which decisions are made but can themselves make the decisions. Process-control computers, in automated or semiautomated manufacturing operations, play a similar role in decision-making. Their programs are decision strategies which, as the system's variables change from moment to moment, retain control over the ongoing process.

It is the capability of the computer for solving problems and making decisions that represents its real novelty and that poses the greatest difficulties in predicting its impact upon society. An enormous amount of research and developmental activity will have to be carried out before the full practical implications of this capability will be understood and available for use. In the single generation that modern computers have been in existence, enough basic research has been done to reveal some of the fundamental mechanisms. Although one can point to a number of applications of the computer as decision-maker that are already 20 or 25 years old, development and application on a substantial scale have barely begun.

Economic Effects of Computers

. . . Now the rate of technological change depends both upon the rate of discovery of new innovations and upon the availability of capital to turn them into bricks and steel (or wire and glass). In this process, computers compete with other forms of technology for the available capital. Hence, the process of computerization is simply a part, currently an important part, of the general process of technological change. [In the long run, computerization is, and has been, wholly compatible with full employment. (Summary of omitted material, by Herbert A. Simon.)]

In taking this very global and bird's-eye view of the economics of mechanization, we should not ignore the plight of the worker who is displaced by the computer. His plight is often genuine and serious, particularly if the economy as a whole is not operating near full employment, but even if it is. Society as a whole benefits from increased productivity, but often at the expense of imposing transient costs on a few people. But the sensible response to this problem is not to eschew the benefits of change; it is rather to take institutional steps to shift the burdens of the transition from the individual to society. Fortunately, our attitudes on these questions appear to be maturing somewhat, and our institutional practices improving, so that the widespread introduction of the computer into clerical operations over the past generation has not called forth any large-scale Ludditism. . . .

Control and Privacy

The potential of computers for increasing the control of organizations or society over their members and for invading the privacy of those members has caused considerable concern. The issues are important but are too complex to be discussed in detail here. I shall therefore restrict myself to a few comments which will serve rather to illustrate this complexity than to provide definitive answers.

A first observation is that our concern here is for competitive aspects of society, the power of one individual or group relative to others. Technologies tend to be double-edged in competitive situations, particularly when they are available to both competitors. For example, the computerization of credit information about individuals facilitates the assembly of such information from many sources, and its indefinite retention and accessibility. On the other hand, it also facilitates auditing such information to determine its sources and reliability. With appropriate legal rules of the game, an automated system can provide more reliable information than a more primitive one and can be surrounded by more effective safeguards against abuse. Some of us might prefer, for good reasons or bad, not to have our credit checked at all. But if credit checking is a function that must be performed, a strong case can be made for making it more responsible by automating it, with appropriate provision for auditing its operation.

Similarly, much has been said of the potential for embezzlement in computerized accounting systems, and cases have occurred. Embezzlement, however, was known before computers, and the computer gives auditors as well as embezzlers powerful new weapons. It is not at all clear which way the balance has been tilted.

The privacy issue has been raised most insistently with respect to the creation and maintenance of longitudinal data files that assemble information about persons from a multitude of sources. Files of this kind would be highly valuable for many kinds of economic and social research, but they are bought at too high a price if they endanger human freedom or seriously enhance the opportunities of blackmailers. While such dangers should not be ignored, it should be noted that the lack of comprehensive data files has never been the limiting barrier to the suppression of human freedom. The Watergate criminals made extensive, if unskillful, use of electronics, but no computer played a role in their conspiracy. The Nazis operated with horrifying effectiveness and thoroughness without the benefits of any kind of mechanized data processing.

Making the computer the villain in the invasion of privacy or encroachment on civil liberties simply diverts attention from the real dangers. Computer data banks can and must be given the highest degree of protection from abuse. But we must be careful, also, that we do not employ such crude methods of protection as to deprive our society of important data it needs to understand its own social processes and to analyze its problems.

Man's View of Man

Perhaps the most important question of all about the computer is what it has done and will do to man's view of himself and his place in the universe. The most heated attacks on the computer are not focused on its possible economic effects, its presumed destruction of job satisfactions, or its threats to privacy and liberty, but upon the claim that it causes people to be viewed, and to view themselves, as "machines."

To get at the real issues, we must first put aside one verbal confusion. All of us are familiar with a wide variety of machines, most of which predated the computer. Consequently, the word "machine" carries with it many connotations: of rigidity, of simplicity, of repetitive behavior, and so on. If we call anything a machine, we implicitly attribute these characteristics to it. Hence, if a computer is a machine, it must behave rigidly, simply, and repetitively. It follows that computers cannot be programmed to behave like human beings.

The fallacy in the argument, of course, lies in supposing that, because we have applied the term "machine" to computers, computers must behave like older forms of machines. But the central significance of the computer derives from the fact that it falsifies these earlier connotations. It can, in fact, be programmed to behave flexibly, in complex ways, and not repetitively at all. We must either get rid of the connotations of the term, or stop calling computers "machines."

There is a more fundamental question behind the verbal one. It is essentially the question that was raised by Darwinism, and by the Copernican revolution centuries earlier. The question is whether the dignity of man, his sense of worth and self-respect depends upon his being something special and unique in the universe. As I have said elsewhere:

> The definition of man's uniqueness has always formed the kernel of his cosmological and ethical systems. With Copernicus and Galileo, he ceased to be the species located at the center of the universe, attended by sun and stars. With Darwin, he ceased to be the species created and specially endowed by God with soul and reason. With Freud, he ceased to be the species whose behavior was—potentially—governable by rational mind. As we begin to produce mechanisms that think and learn, he has ceased to be the species uniquely capable of complex, intelligent manipulation of his environment.

What the computer and the progress in artificial intelligence challenge is an ethic that rests on man's apartness from the rest of nature. An alternative ethic, of course, views man as a part of nature, governed by natural law, subject to the forces of gravity and the demands of his body. The debate about artificial intelligence and the simulation of man's thinking is, in considerable part, a confrontation of these two views of man's place in the universe. It is a new chapter in the vitalism-mechanism controversy.

Issues that are logically distinct sometimes become stuck together with the glue of emotion. Several such issues arise here:

To what extent can human behavior be simulated by computer?

In what areas of work and life should the computer be programmed to augment or replace human activities?

How far should we proceed to explore the human mind by psychological research that makes use of computer simulation?

The first of these three issues will only be settled, over the years, by the success or failure of research efforts in artificial intelligence and computer simulation. Whatever our beliefs about the ultimate limits of simulation, it is clear that the current state of the art has nowhere approached those limits.

The second question will be settled anew each year by a host of individual and public decisions based on the changing computer technology, the changing economics of computer applications, and our attention to the social consequences of those applications.

The answer to the third question depends upon our attitudes. . . . One viewpoint is that knowledge can be dangerous—there are enough historical examples—and that the attempt to arrive at a full explanation of man's ability to think might be especially dangerous. A different point of view, closer to my own, is that knowledge is power to produce new outcomes, outcomes that were not previously attainable. To what extent these outcomes will be good or bad depends on the purposes they serve, and it is not easy, in advance, to predict the good and bad uses to which any particular technology will be put. Instead, we must look back over human history and try to assess whether, on balance, man's gradual emergence from a state of ignorance about the world and about himself has been something we should celebrate or regret. To believe that knowledge is to be preferred to ignorance is to believe that the human species is capable of progress and, on balance, has progressed over the centuries. Knowledge about the human mind can make an important contribution to that progress. It is a belief of this kind that persuades researchers in artificial intelligence that their endeavor is an important and exciting chapter in man's great intellectual adventure.

Joseph Weizenbaum

Artificial Intelligence

Joseph Weizenbaum, an eminent computer scientist, argues that the culture in which a human being lives contributes substantially to his or her intelligence. Therefore, the "intelligence" displayed by the most powerful and sophisticated computer cannot be equated with human intelligence.

Few "scientific" concepts have so thoroughly muddled the thinking of both scientists and the general public as that of the "intelligence quotient" or "I.Q." The idea that intelligence can be quantitatively measured along a simple linear scale has caused untold harm to our society in general, and to education in particular. It has spawned, for example, the huge educational-testing movement in the United States, which strongly influences the courses of the academic careers of millions of students and thus the degrees of certification they may attain. It virtually determines what "success" people may achieve in later life because, in the United States at least, opportunities to "succeed" are, by and large, open only to those who have the proper credentials, that is, university degrees, professional diplomas, and so on.

When modern educators argue that intelligence tests measure a subject's ability to do well in school, they mean little more than that these tests "predict" a subject's ability to pass academic-type tests. This latter ability leads, of course, to certification and then to "success." Consequently, any correlation between the results of such tests and people's "success," as that term is understood in the society at large, must necessarily be an artifact of the testing procedure. The test itself has become a criterion for that with which it is to be correlated! "Psychologists should be ashamed of themselves for promoting a view of general intelligence that has engendered such a testing program."

My concern here is that the mythology that surrounds I.Q. testing has led to the widely accepted and profoundly misleading conviction that intelligence is somehow a permanent, unalterable, and culturally independent attribute of individuals (somewhat like, say, the color of their eyes), and moreover that it may even be genetically transmittable from generation to generation.

The trouble with I.Q. testing is not that it is entirely spurious, but that it is incomplete. It measures certain intellectual abilities that large, politically dominant segments of western European societies have elevated to the very stuff of human worth and hence to the *sine qua non* of success. It is incomplete in two ways: first, in that it fails to take into account that human creativity depends not only on intellect but also crucially on an interplay between intellect and other modalities of thought, such as intuition and wisdom; second, in that it characterizes intelligence as a linearly measurable phenomenon that exists independent of any frame of reference.

Einstein taught us that the idea of motion is meaningless in and of itself, that we can sensibly speak only of an object's motion relative to some frame of reference, not of any *absolute* motion of an object. When, in speaking informally, we say that a train moved, we mean that it moved relative to some fixed point on the earth. We need not emphasize this in ordinary conversation, because the earth (or our body) is to us a kind of "default" frame of reference that is implicitly assumed and understood in most informal conversation. But a physicist speaking as a physicist cannot be so sloppy. His equations of motion must contain terms specifying the coordinate system with respect to which the motion they describe takes place.

So it is with intelligence too. Intelligence is a meaningless concept in and of itself. It requires a frame of reference, a specification of a domain of thought and action, in order to make it meaningful. The reason this necessity does not strike us when we speak of intelligence in ordinary conversation is that the required frame of reference—that is, our own cultural and social setting with its characteristic domains of thought and action—is so much with us that we implicitly assume it to be understood. But our culture and our social milieu are in fact neither universal nor absolute. It therefore behooves us, whenever we use the term "intelligence" as scientists or educators, to make explicit the domain of thought and action which renders the term intelligible.

Our own daily lives abundantly demonstrate that intelligence manifests itself only relative to specific social and cultural contexts. The most unschooled mother who cannot compose a single grammatically correct paragraph in her native language—as, indeed, many academics cannot do in theirs—constantly makes highly refined and intelligent judgments about her family. Eminent scholars confess that they don't have the kind of intelligence required to do high-school algebra. The acknowledged genius is sometimes stupid in managing his private life. Computers perform prodigious "intellectual feats," such as beating champion checker players at their own game and solving huge systems of equations, but cannot change a baby's diaper. How are these intelligences to be compared to one another? They cannot be compared.

Yet forms of the idea that intelligence is measurable along an absolute scale, hence that intelligences are comparable, have deeply penetrated current thought. This idea is responsible, at least in part, for many sterile debates about whether it is possible "in principle" to build computers more intelligent than man. Even as moderate and reasonable a psychologist as George A. Miller occasionally slips up, as when he says, "I am very optimistic about the eventual outcome of the work on machine solution of intellectual problems. Within our lifetime machines may surpass us in general intelligence."

The identification of intelligence with I.Q. has severely distorted the primarily mathematical question of what computers can and cannot do into the nonsensical question of "how much" intelligence one can, again "in principle," give to a computer. And, of course, the reckless anthropomorphization of the computer now so common, especially among the artificial intelligentsia, couples easily to such simpleminded views of intelligence. This joining of an illicit metaphor to an ill-thought-out idea then breeds, and is perceived to legitimate, such perverse propositions as that, for example, a computer can be programmed to become an effective psychotherapist. . . .

But, and this is the saving grace of which an insolent and arrogant scientism attempts to rob us, we come to know and understand not only by way of the mechanisms of the conscious. We are capable of listening with the third ear, of sensing living truth that is truth beyond any standards of provability. It is *that* kind of understanding, and the kind of intelligence that is derived from it, which I claim is beyond the abilities of computers to simulate.

We have the habit, and it is sometimes useful to us, of speaking of man, mind, intelligence, and other such universal concepts. But gradually, even slyly, our own minds become infected with what A. N. Whitehead called the fallacy of misplaced concreteness. We come to believe that these theoretical terms are ultimately interpretable as observations, that in the "visible future" we will have ingenious instruments capable of measuring the "objects" to which these terms refer. There is, however, no such thing as mind; there are only individual minds, each belonging, not to "man," but to individual human beings. I have argued that intelligence cannot be measured by ingeniously constructed meter sticks placed along a one-dimensional continuum. Intelligence can be usefully discussed only in terms of domains of thought and action. From this I derive the conclusion that it cannot be useful, to say the least, to base serious work on notions of "how much" intelligence may be given to a computer. Debates based on such ideas—e.g., "Will computers ever exceed man in intelligence?"—are doomed to sterility.

I have argued that the individual human being, like any other organism, is defined by the problems he confronts. The human is unique by virtue of the fact that he must necessarily confront problems that arise from his unique biological and emotional needs. The human individual is in a constant state of becoming. The maintenance of that state, of his humanity, indeed, of his survival, depends crucially on his seeing himself, and on his being seen by other human beings, as a human being. No other organism, and certainly no computer, can be made to confront genuine human problems in human terms. And, since the domain of human intelligence is, except for a small set of formal problems, determined by man's humanity, every other intelligence, however great, must necessarily be alien to the human domain.

I have argued that there is an aspect to the human mind, the unconscious, that cannot be explained by the information-processing primitives, the elementary information processes, which we associate with formal thinking, calculation, and systematic rationality. Yet we are constrained to use them for scientific explanation, description, and interpretation. It behooves us, therefore, to remain aware of the poverty of our explanations and of their strictly limited scope. It is wrong to assert that any scientific account of the "whole man" is possible. There are some things beyond the power of science to fully comprehend.

The concept of an intelligence alien to certain domains of thought and action is crucial for understanding what are perhaps the most important limits on artificial intelligence. But that concept applies to the way humans relate to one another as well as to machines and their relation to man. For human socialization, though it is grounded in the biological constitution common to all humans, is strongly determined by culture. And human cultures differ radically among themselves. Countless studies confirm what must be obvious to all but the most parochial observers of the human scene: "The influence of culture is universal in that in some respects a man learns to become like all men; and it is particular in that a man who is reared in one society learns to become in some respects like all men of his society and not like those of others." . . .

Every human intelligence is thus alien to a great many domains of thought and action. There are vast areas of authentically human concern in every culture in which no member of another culture can possibly make responsible decisions. It is not that the outsider is unable to decide at all—he can always flip coins, for example—it is rather that the *basis* on which he would have to decide must be inappropriate to the context in which the decision is to be made.

What could be more obvious than the fact that, whatever intelligence a computer can muster, however it may be acquired, it must always and necessarily be absolutely alien to any and all authentic human concerns? The very asking of the question, "What does a judge (or a psychiatrist) know that we cannot tell a computer?" is a monstrous obscenity. That it has to be put into print at all, even for the purpose of exposing its morbidity, is a sign of the madness of our times.

Computers can make judicial decisions, computers can make psychiatric judgments. They can flip coins in much more sophisticated ways than can the most patient human being. The point is that they *ought* not be given such tasks. They may even be able to arrive at "correct" decisions in some cases—but always and necessarily on bases no human being should be willing to accept.

There have been many debates on "Computers and Mind." What I conclude here is that the relevant issues are neither technological nor even mathematical; they are ethical. They cannot be settled by asking questions beginning with "can." The limits of the applicability of computers are ultimately statable only in terms of oughts. What emerges as the most elementary insight is that, since we do not now have any ways of making computers wise, we ought not now to give computers tasks that demand wisdom.

J. B. S. Haldane

Science and the Future

Son of a renowned British physiologist, J. B. S. Haldane (1892–1964) did pioneering work in mathematical population genetics during the 1920s that helped shape modern evolutionary theory. In a talk given at Cambridge University in 1923 and reprinted as Daedalus, or Science and the Future *(London, 1924), he first set forth the idea of "ectogenesis"—the* in vitro *fertilization and development of human ova (egg cells).*

Before we proceed to prophecy I should like to turn back to the past and examine very briefly the half dozen or so important biological inventions which have already been made. By a biological invention I mean the establishment of a new relationship between man and other animals or plants, or between different human beings, provided that such relationship is one which comes primarily under the domain of biology rather than physics, psychology or ethics. Of the biological inventions of the past . . . made before the dawn of history, I refer to the domestication of animals, the domestication of plants, the domestication of fungi for the production of alcohol. . . . And in our own day two more have been made, namely bactericide and the artificial control of conception.

The first point that we may notice about these inventions is that they have all had a profound emotional and ethical effect. . . .

The second point is perhaps harder to express. The chemical or physical inventor is always a Prometheus. There is no great invention, from fire to flying, which has not been hailed as an insult to some god. But if every physical and chemical invention is a blasphemy, every biological invention is a perversion. There is hardly one which, on first being brought to the notice of an observer from any nation which had not previously heard of their existence, would not appear to him as indecent and unnatural.

Consider so simple and time-honoured a process as the milking of a cow. The milk which should have been an intimate and almost sacramental bond between mother and child is elicited by the deft fingers of a milkmaid, and drunk, cooked, or even allowed to rot into cheese. We have only to imagine ourselves as drinking any of its other secretions, in order to realise the radical indecency of our relation to the cow.

No less disgusting a priori is the process of corruption which yields our wine and beer. But in actual fact the processes of milking and of the making and drinking beer appear to us profoundly natural; they have even tended to develop a ritual of their own whose infraction nowadays has a certain air of impropriety. There is something slightly disgusting in the idea of milking a cow electrically or drinking beer out of tea-cups. . . .

But to return, if I may use the expression, to the future, I am going to suggest a few obvious developments which seem probable in the present state of biological science, without assuming any great new generalizations of the type of Darwinism. I have the very best precedents for introducing a myth at this point, so perhaps I may be excused if I reproduce some extracts from an essay on the influence of biology on history during the 20th century which will (it is hoped) be read by a rather stupid undergraduate member of this university to his supervisor during his first term 150 years hence.

"As early as the first decade of the twentieth century we find a conscious attempt at the application of biology to politics in the so-called eugenic movement. A number of earnest persons, having discovered the existence of biology, attempted to apply it in its then very crude condition to the production of a race of supermen, and in certain countries managed to carry a good deal of legislation. They appear to have managed to prevent the transmission of a good deal of syphilis, insanity, and the like, and they certainly succeeded in producing the most violent opposition and hatred amongst the classes whom they somewhat gratuitously regarded as undesirable parents. . . . However, they undoubtedly prepared public opinion for what was to come, and so far served a useful purpose. Far more important was the progress in medicine which practically abolished infectious diseases in those countries which were prepared to tolerate the requisite amount of state interference in private life, and finally, after the league's ordinance of 1958, all over the world; . . .

But from a wider point of view the most important biological work in the first third of the century was in experimental zoology and botany. When we consider that in 1912 Morgan had located several Mendelian factors in the nucleus of Drosophila, and modified its sex-ratio, while Marmorek had taught a harmless bacillus to kill guinea-pigs, and finally in 1913 Brachet had grown rabbit embryos in serum for some days, it is remarkable how little the scientific workers of that time, and a fortiori the general public, seem to have foreseen the practical bearing of such results.

As a matter of fact it was not until 1940 that Selkovski invented the purple alga *Porphyrococcus fixator* which was to have so great an effect on the world's history. In the 50 years before this date the world's average wheat yield per nectar had been approximately doubled, partly by the application of various chemical manures, but most of all by the results of systematic crossing work with different races; there was however little prospect of further advance on any of these lines. *Porphyrococcus* is an enormously efficient nitrogen-fixer and will grow in almost any climate where there are water and traces of potash and phosphates in the soil, obtaining its nitrogen from the air. It has about the effect in four days that a crop of vetches would have had in a year. It could not, of course have been produced in the course of nature, as its immediate ancestors would only grow in artificial media and could not have survived outside a laboratory. Wherever nitrogen was the principal limiting factor to plant growth it doubled the yield of wheat, and quadrupled the value of grass land for grazing purposes. The enormous fall in food prices and the ruin of purely agricultural states was of course one of the chief causes of the disastrous events of 1943 and 1944. The food glut was also greatly accentuated when in 1942 the Q strain of *Porphyrococcus* escaped into the sea and multiplied with enormous rapidity. Indeed for two months the surface of the tropical Atlantic set to a jelly, with disastrous results to the weather of Europe. When certain of the plankton organisms developed ferments capable of digesting it the increase of the fish population of the seas was so great as to make fish the universal good that it is now, and to render even England self-supporting in respect of food. . . .

It was of course as the result of its invasion by *Porphyrococcus* that the sea assumed the intense purple colour which seems so natural to us, but which so distressed the more aesthetically minded of our great grand-parents who witnessed the change. It is certainly curious to us to read of the sea as having been green or blue. I need not detail the work of Ferguson and Rahmatullah who in 1957 produced the lichen which has bound the drifting sand of the world's deserts (for it was merely a continuation of that of Selkovski), nor yet the story of how the agricultural countries dealt with their unemployment by huge socialistic windpower schemes.

It was in 1951 that Dupont and Schwarz produced the first ectogenetic child. As early as 1901 Heape had transferred embryo rabbits from one female to another, in 1925 Haldane had grown embryonic rats in serum for ten days, but had failed to carry the process to its conclusion, and it was not till 1940 that Clark succeeded with the pig, using Kehlmann's solution as medium. Dupont and Schwarz obtained a fresh ovary from a woman who was the victim of an aeroplane accident, and kept it living in their medium for five years. They obtained several eggs from it and fertilized them successfully, but the problem of the nutrition and support of the embryo was more difficult, and was only solved in the fourth year. Now that the technique is fully developed, we can take an ovary from a woman, and keep it growing in a suitable fluid for as long as twenty years, producing a fresh ovum each month, of which 90 per cent can be fertilized, and the embryos grown successfully for nine months, and then brought out into the air. Schwarz never got such good results, but the news of his first success caused an unprecedented sensation throughout the entire world, for the birthrate was already less than the deathrate in most civilised countries. France was the first country to adopt ectogenesis officially, and by 1968 was producing 60,000 children annually by this method. In most countries the opposition was far stronger, . . .

As we know ectogenesis is now universal, and in this country less than 30 per cent of children are now born of woman. The effect on human psychology and social life of the separation of sexual love and reproduction which was begun in the 19th century and completed in the 20th is by no means wholly satisfactory. The old family life had certainly a good deal to commend it, and although nowadays we bring on lactation in women by injection of placentin as a routine, and thus conserve much of what was best in the former instinctive cycle, we must admit that in certain respects our great grandparents had the advantage of us. On the other hand it is generally admitted that the effects of selection have more than counterbalanced these evils. The small proportion of men and women who are selected as ancestors for the next generation are so undoubtedly superior to the average that the advance in each generation in any single respect, from the increased output of first-class music to the decreased convictions for theft, is very startling. Had it not been for ectogenesis there can be little doubt that civilization would have collapsed within a measurable time owing to the greater fertility of the less desirable members of the population in almost all countries.

It is perhaps fortunate that the process of becoming an ectogenetic mother of the next generation involves an operation which is somewhat unpleasant, though now no longer disfiguring or dangerous, and never physiologically injurious, and is therefore an honour but by no means a pleasure. Had this not been the case, it is perfectly possible that popular opposition would have proved too strong for the selectionist movement. As it was the opposition was very fierce, and characteristically enough this country only adopted its present rather stringent standard of selection a generation later than Germany, though it is now perhaps more advanced than any other country in this respect. The advantages of thorough-going selection, have, however, proved to be enormous. The question of the ideal sex ratio is still a matter of violent discussion, but the modern reaction towards equality is certainly strong."

Our essayist would then perhaps go on to discuss some far more radical advances made about 1990, but I have only quoted his account of the earlier applications of biology. The second appears to me to be neither impossible nor improbable, but it has those features which we saw above to be characteristic of biological inventions. If reproduction is once completely separated from sexual love mankind will be free in an altogether new sense. At present the national character is changing slowly according to quite unknown laws. The problem of politics is to find institutions suitable to it. In the future perhaps it may be possible by selective breeding to change character as quickly as institutions. I can foresee the election placards of 300 years hence, if such quaint political methods survive, which is perhaps improbable, "Vote for Smith and more musicians", "Vote for O'Leary and more girls", or perhaps finally "Vote for Macpherson and a prehensile tail for your great-grandchildren". We can already alter animal species to an enormous extent, and it seems only a question of time before we shall be able to apply the same principles to our own.

I suggest then that biology will probably be applied on lines roughly resembling the above. There are perhaps equally great possibilities in the way of the direct improvement of the individual, as we come to know more of the physiological obstacles to the development of different faculties. But at present we can only guess at the nature of these obstacles, and the line of attack suggested in the myth is the one which seems most obvious to a Darwinian. We already know however that many of our spiritual faculties can only be manifested if certain glands, notably the thyroid and sex-glands, are functioning properly, and that very minute changes in such glands affect the character greatly. As our knowledge of this subject increases we may be able, for example, to control our passions by some more direct method than fasting and flagellation, to stimulate our imagination by some reagent with less after-effects than alcohol, to deal with perverted instincts by physiology rather than prison. Conversely there will inevitably arise possibilities of new vices similar to but even more profound than those opened up by the pharmacological discoveries of the 19th century. . . .

I have tried to show why I believe that the biologist is the most romantic figure on earth at the present day. . . .

I do not say that biologists as a general rule try to imagine in any detail the future applications of their science. The central problems of life for them may be the relationship between the echinoderms and brachiopods, and the attempt to live on their salaries. They do not see themselves as sinister and revolutionary figures. They have no time to dream. But I suspect that more of them dream than would care to confess it.

I have given above a very small selection from my dreams. Perhaps they are bad dreams. It is of course almost hopeless to attempt any very exact prophecies as to how in detail scientific knowledge is going to revolutionize human life, but I believe that it will continue to do so, and even more profoundly than I have suggested. And though personally I am Victorian enough in my sympathies to hope that after all family life, for example, may be spared, I can only reiterate that not one of the practical advances which I have predicted is not already foreshadowed by recent scientific work. . . .

We must regard science from three points of view. First it is the free activity of man's divine faculties of reason and imagination. Secondly it is the answer of the few to the demands of the many for wealth, comfort and victory. . . . Finally it is man's gradual conquest, first of space and time, then of matter as such, then of his own body and those of other living beings, and finally the subjugation of the dark and evil elements in his own soul.

None of these conquests will ever be complete, but all, I believe will be progressive. The question of what he will do with these powers is essentially a question for religion and aesthetic. . . .

. . . Science is as yet in its infancy, and we can foretell little of the future save that the thing that has not been is the thing that shall be; that no beliefs, no values, no institutions are safe. So far from being an isolated phenomenon the late war is only an example of the disruptive results that we may constantly expect from the progress of science. The future will be no primrose path. It will have its own problems. Some will be the secular problems of the past, giant flowers of evil blossoming at last to their own destruction. Others will be wholly new. Whether in the end man will survive his accessions of power we cannot tell. But the problem is no new one. It is the old paradox of freedom re-enacted with mankind for actor and the earth for stage.

J. B. S. Haldane

The Last Judgment

In addition to being a renowned biologist, Haldane was also a popular writer who authored hundreds of provocative articles on science for various newspapers and magazines. In this influential essay he speculates about our ultimate destiny as a species and about a distant future when human survival requires tailoring the human organism to suit alien planetary settings.

In what follows I shall attempt to describe the most probable end of our planet as it might appear to spectators on another. I have been compelled to place the catastrophe within a period of the future accessible to my imagination. For I can imagine what the human race will be like in forty million years, since forty million years ago our ancestors were certainly mammals, and probably quite definitely recognizable as monkeys. But I cannot throw my imagination forward for ten times that period. Four hundred million years ago our ancestors were fish of a very primitive type. I cannot imagine a corresponding change in our descendants.

So I have suggested the only means which, so far as I can see, would be able to speed up the catastrophe. The account given here will be broadcast to infants on the planet Venus some forty million years hence. It has been rendered very freely into English, as many of the elementary ideas of our descendants will be beyond our grasp:

'It is now certain that human life on the Earth's surface is extinct, and quite probable that no living thing whatever remains there. The following is a brief record of the events which led up to the destruction of the ancient home of our species.

Eighteen hundred and seventy-four million years ago the Sun passed very close to the giant star 318.47.19543. The tidal wave raised by it in our sun broke into an incandescent spray. The drops of this spray formed the planets, of all of which the Earth rotated by far the most rapidly. . . . The liquid Earth spun round for a few years as a spheroid greatly expanded at the equator and flattened at the poles by its excessive rotation. Then the tidal waves raised in it by the Sun became larger and larger. Finally the crest of one of these waves flew off as the Moon. . . .

As the Moon raised large tides in the still liquid Earth the latter was slowed down by their braking action, for all the work of raising the tides is done at the expense of the Earth's rotation. But by acting as a brake on the Earth, the Moon was pushed forward along its course, as any brake is pushed by the wheel that it slows down. As it acquired more speed it rose gradually further and further away from the Earth, which has now a solid crust, and the month, like the day, became longer. When life began on the Earth the Moon was already distant, and during the sixteen hundred million years before man appeared it had only moved away to a moderate degree further. . . .

At this time the effect of tidal friction was to make each century, measured in days, just under a second shorter than the last. . . . As soon as the use of heat engines was discovered, man began to oxidize the fossil vegetables to be found under the Earth's surface. After a few centuries they gave out, and other sources of energy were employed. The power available from fresh water was small, from winds intermittent, and that from the Sun's heat only available with ease in the tropics. The tides were therefore employed, and gradually became the main source of energy. The invention of synthetic food led to a great increase in the world's population, and after the federation of the world it settled down at about twelve thousand million. As tide engines were developed, an ever-increasing use was made of their power; and before the human race had been in existence for a million years, the tide-power utilized aggregated a million million horsepower. The braking action of the tides was increased fiftyfold, and the day began to lengthen appreciably.

At its natural rate of slowing fifty thousand million years would have elapsed before the day became as long as the month, but it was characteristic of the dwellers on Earth that they never looked more than a million years ahead, and the amount of energy available was ridiculously squandered. By the year five million the human race had reached equilibrium; it was perfectly adjusted to its environment, the life of the individual was about three thousand years; and the individuals were 'happy', that is to say, they lived in accordance with instincts which were gratified. The tidal energy available was now fifty million million horsepower. Large parts of the planet were artificially heated. The continents were remodelled, but human effort was chiefly devoted to the development of personal relationships and to art and music, that is to say, the production of objects, sounds, and patterns of events gratifying to the individual.

Human evolution had ceased. Natural selection had been abolished, and the slow changes due to other causes were traced to their sources and prevented before very great effects had been produced. It is true that some organs found in primitive man, such as the teeth (hard, bone-like structures in the mouth), had disappeared. But largely on aesthetic grounds the human form was not allowed to vary greatly. The instinctive and traditional preferences of the individual, which were still allowed to influence mating, caused a certain standard body form to be preserved. The almost complete abolition of the pain sense which was carried out before the year five million was the most striking piece of artificial evolution accomplished. For us, who do not regard the individual as an end in itself, the value of this step is questionable.

Scientific discovery was largely a thing of the past, and men of a scientific bent devoted themselves to the more intricate problems of mathematics, organic chemistry, or the biology of animals and plants, with little or no regard for practical results. Science and art were blended in the practice of horticulture, and the effort expended on the evolution of beautiful flowers would have served to alter the human race profoundly. But evolution is a process more pleasant to direct than to undergo.

By the year eight million the length of the day had doubled, the Moon's distance had increased by 20 per cent, and the month was a third longer than it had been when first measured. It was realized that the Earth's rotation would now diminish rapidly, and a few men began to look ahead, and to suggest the colonization of other planets. The older expeditions had all been failures. . . .

. . . An expedition reached Mars successfully in the year 9,723,841, but reported that colonization was impracticable. The species dominant on that planet, which conduct its irrigation, are blind to those radiations which we perceive as light, and probably unaware of the existence of other planets; but they appear to possess senses unlike our own, and were able to annihilate this expedition and the only other which reached Mars successfully.

Half a million years later the first successful landing was effected on Venus, but its members ultimately perished owing to the unfavourable temperature conditions and the shortage of oxygen in its atmosphere. After this such expeditions became rarer.

In the year 17,846,151 the tide machines had done the first half of their destructive work. The day and the month were now of the same length. For millions of centuries the Moon had always turned the same face to the Earth, and now the Earth-dwellers could only see the Moon from one of their hemispheres. It hung permanently in the sky above the remains of the old continent of America. The day now lasted for forty-eight of the old days, so that there were only seven and a half days in the year. As the day lengthened the climate altered enormously. The long nights were intensely cold, and the cold was generally balanced by high temperatures during the day. . . .

. . .As the Earth's rotation slowed down, its equator contracted, causing earthquakes and mountain-building on a large scale. A good deal of land emerged from the oceans, especially the central Pacific. And with the lengthening of the nights snow began to be deposited on the uplands in fairly large amounts; near the poles the Sun occasionally failed to melt it during the day, and even where it was melted the subsoil was often permanently frozen. In spite of considerable efforts, ice-fields and giant glaciers had already appeared when the Moon ceased to rise and set. Above them permanent anticyclones once more produced storms in the temperate regions, and rainless deserts in the tropics.

The animals and plants only partially adapted themselves to the huge fluctuations of temperature. Practically all the undomesticated mammals, birds, and reptiles became extinct. Many of the smaller plants went through their whole life cycle in a day, surviving only as seeds during the night. But most of the trees became extinct except when kept warm artificially.

The human race somewhat diminished in numbers, but there was still an immense demand for power for heating and cooling purposes. The tides raised by the Sun, although they only occurred fifteen times per year, were used for these ends, and the day was thus still further lengthened.

The Moon now began once more to move relative to the Earth, but in the opposite direction, rising in the west and setting in the east. Very gradually at first, but then with ever-increasing speed, it began to approach the Earth again, and appear larger. By the year 25,000,000 it had returned to the distance at which it was when man had first evolved, and it was realized that its end, and possibly the Earth's, was only a few million years ahead. But the vast majority of mankind contemplated the death of their species with less aversion than their own, and no effective measures were taken to forestall the approaching doom. . . .

But if most men failed to look ahead, a minority felt otherwise, and expeditions to Venus became commoner. After 284 consecutive failures a landing was established, and before its members died they were able to furnish the first really precise reports as to conditions on that planet. Owing to the opaque character of our atmosphere, the light signals of the earlier expeditions had been difficult to pick up. Infrared radiation which can penetrate our clouds was now employed.

A few hundred thousand of the human race, from some of whom we are descended, determined that though men died, man should live for ever. It was only possible for humanity to establish itself on Venus if it were able to withstand the heat and want of oxygen there prevailing, and this could only be done by a deliberate evolution in that direction first accomplished on Earth. Enough was known of the causes responsible for evolution to render the experiment possible. The human material was selected in each generation. All who were not willing were able to resign from participation, and among those whose descendants were destined for the conquest of Venus a tradition and an inheritable psychological disposition grew up such as had not been known on Earth for twenty-five million years. The psychological types which had been common among the saints and soldiers of early history were revived. Confronted once more with an ideal as high as that of religion, but more rational, a task as concrete as and infinitely greater than that of the patriot, man became once more capable of self-transcendence. Those members of mankind who were once more evolving were not happy. They were out of harmony with their surroundings. Disease and crime reappeared among them. For disease is only a failure of bodily function to adjust itself to the environment, and crime a similar failure in behaviour. But disease and crime, as much as heroism and martyrdom, are part of the price which must be paid for evolution. The price is paid by the individual, and the gain is to the race. Among ourselves an individual may not consider his own interests a dozen times in his life. To our ancestors, fresh from the pursuit of individual happiness, the price must often have seemed too great, and in every generation many who have now left no descendants refused to pay it.

The modes of behaviour which our ancestors gradually overcame, and which only recur as the rarest aberrations among ourselves, included not only such self-regarding sentiments as pride and a personal preference concerning mating. They embraced emotions such as pity (an unpleasant feeling aroused by the suffering of other individuals). In a life completely dedicated to membership of a super-organism the one is as superfluous as the other, though altruism found its place in the emotional basis of the far looser type of society prevalent on Earth.

In the course of ten thousand years a race had been evolved capable of life at one-tenth of the oxygen pressure prevalent on Earth, and the body temperature had been raised by six degrees. The rise to a still higher temperature, correlated as it was with profound chemical and structural changes in the body, was a much slower process. Projectiles of a far larger size were dispatched to Venus. Of 1734, only 11 made satisfactory landings. The crews of the first two of these ultimately perished; those of the next eight were our ancestors. The organisms found on Venus were built of molecules which were mostly mirror images of those found in terrestrial bodies. Except as sources of fat they were therefore useless for food, and some of them were a serious menace. The third projectile to arrive included bacteria which had been synthesized on Earth to attack l-glucose and certain other components of the organisms on Venus. Ten thousand years of laboratory work had gone to their making. With their aid the previous life on that planet was destroyed, and it became available for the use of man and the sixty terrestrial species which he had brought with him.

The history of our planet need not be given here. After the immense efforts of the first colonizers, we have settled down as members of a super-organism with no limits to its possible progress. The evolution of the individual has been brought under complete social control, and besides enormously enhanced intellectual powers we possess two new senses. The one enables us to apprehend radiation of wavelengths between 100 and 1,200 metres, and thus places every individual at all moments of life, both asleep and awake, under the influence of the voice of the community. It is difficult to see how else we could have achieved as complete a solidarity as has been possible. We can never close our consciousness to those wavelengths on which we are told of our nature as components of a super-organism or deity, possibly the only one in space-time, and of its past, present, and future. It appears that on Earth the psychological equivalent of what is transmitted on these wavelengths included the higher forms of art, music, and literature, the individual moral consciousness, and, in the early days of mankind, religion and patriotism. The other wavelengths inform us of matters which are not the concern of all at all times, and we can shut them out if we so desire. Their function is not essentially different from that of instrumental radio-communication on Earth. The new magnetic sense is of less importance, but is of value in flying and otherwise in view of the very opaque character of our atmosphere. It would have been almost superfluous on Earth. We have also recovered the pain sense, which had become vestigial on Earth, but is of value for the survival of the individual under adverse circumstances, and hence to the race. So rapid was our evolution that the crew of the last projectile to reach Venus were incapable of fertile unions with our inhabitants, and they were therefore used for experimental purposes.

During the last few million years the Moon approached the Earth rather rapidly. . . . In the year 36,000,000 the Moon was at only a fifth of its distance from the Earth when history had begun. It appeared twenty-five times as large as the Sun, and raised the sea-level by some 200 metres about four times a year. The effects of the tidal strain raised in it by the Earth began to tell. Giant landslips were observed in the lunar mountains, and cracks occasionally opened in its surface. Earthquakes also became rather frequent on the Earth.

Finally the Moon began to disintegrate. It was so near to the Earth as to cover about a twentieth of the visible heavens when the first fragments of rock actually left its surface. The portion nearest to the Earth, already extensively cracked, began to fly away in the form of meteorites up to a kilometre in diameter, which revolved round the Earth in independent orbits. For about a thousand years this process continued gradually, and finally ceased to arouse interest on the Earth. The end came quite suddenly. It was watched from Venus, but the earlier stages were also signalled from the Earth. The depression in the Moon's surface facing the Earth suddenly opened and emitted a torrent of white-hot lava. As the Moon passed round the Earth it raised the temperature in the tropics to such an extent that rivers and lakes were dried up and vegetation destroyed.

The colour changes on Earth due to the

flowering of the plants which were grown on it for the pleasure of the human race, and which were quite visible from our planet, no longer occurred. Dense clouds were formed and gave some protection to the Earth. But above them the sea of flame on the Moon increased in magnitude, and erupted in immense filaments under the Earth's gravitation. Within three days the satellite had broken up into a ring of white-hot lava and dust. The last message received from the Earth stated that the entire human race had retired underground, except on the Antarctic continent, where however the ice-cap had already melted and the air temperature was 35°C. Within a day from the Moon's breakup the first large fragment of it had fallen on the Earth. The particles formed from it were continually jostling, and many more were subsequently driven down. Through the clouds of steam and volcanic smoke which shrouded the Earth our astronomers could see but little, but later on it became clear that its tropical regions had been buried many kilometres deep under lunar fragments, and the remainder, though some traces of the former continents remain, had been submerged in the boiling ocean. It is not considered possible that any vestige of human life remains, nor can our spectroscopes detect any absorption bands of chlorophyll which would indicate the survival of plants.

The majority of the lunar matter has formed a ring round the Earth, like those of Saturn, but far denser. It is not yet in equilibrium, and fragments will continue to fall on the Earth for about another thirty-five thousand years. At the end of that period the Earth, which now possesses a belt of enormous mountains in its tropical regions, separated from the poles by two rings of sea, will be ready for recolonization. Preparations are being made for this event. We have largely sorted out the useful elements in the outer five kilometres or so of our planet, and it is proposed, when the Earth is reoccupied, to erect artificial mountains on both planets which will extend above the Heaviside layer and enable continuous radio-communication instead of light signals to be used between the two.

The old human race successfully cultivated individual happiness and has been destroyed by fire from heaven. This is not a cause for great regret, since happiness does not summate. The happiness of ten million individuals is not a millionfold the happiness of ten. But the unanimous cooperation of ten million individuals is something beyond their individual behaviour. It is the life of a super-organism. If, as many of the Earth-dwellers hoped, the Moon had broken up quietly, their species might have lasted a thousand million years instead of thirty-nine million, but their achievement would have been no greater.

From the Earth it is proposed to colonize Jupiter. It is not certain that the attempt will succeed, for the surface temperature of that planet is 130°C, gravitation is three times as intense as that on Venus, and over twice that on Earth, while the atmosphere contains appreciable quantities of thoron, a radioactive gas. The intense gravitation would of course destroy bodies as large as our own, but life on Jupiter will be possible for organisms built on a much smaller scale. A dwarf form of the human race about a tenth of our height, and with short stumpy legs but very thick bones, is therefore being bred. Their internal organs will also be very solidly built. They are selected by spinning them round in centrifuges which supply an artificial gravitational field, and destroy the less suitable members of each generation. Adaptation to such intense cold as that on Jupiter is impracticable, but it is proposed to send projectiles of a kilometre in length, which will contain sufficient stores of energy to last their inhabitants for some centuries, during which they may be able to develop the sources available on that planet. It is hoped that as many as one in a thousand of these projectiles may arrive safely. If Jupiter is successfully occupied the outer planets will then be attempted.

About 250 million years hence our solar system will pass into a region of space in which stars are far denser than in our present neighbourhood. Although not more than one in ten thousand is likely to possess planets suitable for colonization, it is considered possible that we may pass near enough to one so equipped to allow an attempt at landing. If by that time the entire matter of the planets of our system is under conscious control, the attempt will stand some chance of success. Whereas the best time between the Earth and Venus was one-tenth of a terrestrial year, the time taken to reach another stellar system would be measured in hundreds or thousands of years, and only a very few projectiles per million would arrive safely. But in such a case waste of life is as inevitable as in the seeding of a plant or the discharge of spermatozoa or pollen. Moreover, it is possible that under the conditions of life in the outer planets the human brain may alter in such a way as to open up possibilities inconceivable to our own minds. Our galaxy has a probable life of at least eighty million million years. Before that time has elapsed it is our ideal that all the matter in it available for life should be within the power of the heirs of the species whose original home has just been destroyed. If that ideal is even approximately fulfilled, the end of the world which we have just witnessed was an episode of entirely negligible importance. And there are other galaxies.'

Epilogue

There are certain criteria which every attempt, however fantastic, to forecast the future should satisfy. In the first place, the future will not be as we should wish it. . . . Most of the great ideals of any given age are ignored by the men of later periods. They only interest posterity in so far as they have been embodied in art or literature. I have pictured a human race on the Earth absorbed in the pursuit of individual happiness; on Venus mere components of a monstrous ant-heap. My own ideal is naturally somewhere in between, and so is that of almost every other human being alive today. But I see no reason why my ideals should be realized. In the language of religion, God's ways are not our ways; in that of science, human ideals are the products of natural processes which do not conform to them.

Secondly, we must use a proper time-scale. The Earth has lasted between one and eight thousand million years. Recorded human history is a matter of about six thousand. This period bears the same ratio to the Earth's life as does a space of two or three days to the whole of human history. I have no doubt that in reality the future will be vastly more surprising than anything I can imagine. But when we once realize the periods of time which our thought can and should envisage we shall come to see that the use, however haltingly, of our imaginations upon the possibilities of the future is a valuable spiritual exercise. . . .

. . . Our private, national, and even international aims are restricted to a time measured in human life-spans.

> And yonder all before us lie
> Deserts of vast eternity. . . .

Man's little world will end. The human mind can already envisage that end. If humanity can enlarge the scope of its will as it has enlarged the reach of its intellect, it will escape that end. If not, the judgement will have gone out against it, and man and all his works will perish eternally.

Suggestions for Further Reading

Chapter 1: Greek Science and Society

Clagett, Marshall. *Greek Science in Antiquity*. New York: Abelard-Schumann, 1955.

Supplements Sambursky by following Greek scientific developments into the period of Roman domination in the Mediterranean basin.

Cohen, Morris R., and I. E. Drabkin, eds. *A Sourcebook in Greek Science*. Cambridge: Harvard University Press, 1966.

The most useful and extensive single-volume collection of primary materials related to Greek science.

Farrington, Benjamin. *Greek Science*. Baltimore: Penguin Books, 1953.

Although somewhat dated, remains the most outstanding general Marxist treatment of ancient science.

Lloyd, G. E. R. *Early Greek Science: Thales to Aristotle*. New York: W. W. Norton, 1970 and *Greek Science After Aristotle*. London: Chatto & Windus, 1973.

The most up-to-date survey of Greek science, in two volumes.

Neugelbauer, Otto. *The Exact Sciences in Antiquity*. 2d ed. Providence, R.I.: Brown University Press, 1957.

The standard treatment for Babylonian and Egyptian mathematics and astronomy.

Olson, Richard. *Science Deified and Science Defied*, Vol. 1. Berkeley: University of California Press, 1982.

Discusses the relationship of Ancient scientific developments to other aspects of ancient culture—especially to religion.

Ross, W. D. *Aristotle: A Complete Exposition of His Works and Thought*. 5th ed. Cleveland: World Publishing, 1953.

Wide coverage of Aristotelian science in a brief span.

Sambursky, S. *The Physical World of the Greeks*. New York: Macmillan, 1956.

Deals more extensively than Lloyd with late Atomist and Stoic traditions.

Solmsen, Frederick. *Aristotle's System of the Physical World*. Ithaca, New York: Cornell University Press, 1960.

Covers an important part of Aristotelian science in greater detail than Ross.

Van Der Waerden, B. L. *Science Awakening*, Vol. 1. New York: John Wiley & Sons, 1963 (which treats mathematics) and *Science Awakening*, Vol. 2. New York: Oxford University Press, 1974 (which treats astronomy).

Treats the same subjects as Neugelbauer, but at slightly greater length and with more concern for the cultural context.

Vlastos, Gregory. *Plato's Universe*. Seattle: University of Washington Press, 1975.

Treats Plato's science with great sympathy and grace.

Chapter 2: Medieval Contexts of Natural Knowledge

Crombie, A. C. *Medieval and Early Modern Science*, rev. ed. 2 vols. Garden City, N. Y.: Doubleday Anchor Books, 1959.

An older, comprehensive survey that offers a unified vision of medieval science.

Crombie, A. C. *The Science of Mechanics in the Middle Ages*. Madison: University of Wisconsin Press, 1959.

An accessible work that is strong on mechanics.

Grant, Edward. *Physical Science in the Middle Ages*. New York: John Wiley & Sons, 1971.

An excellent but relatively brief survey with a unified vision of medieval science.

Grant, Edward. *A Source Book in Medieval Science*. Cambridge: Harvard University Press, 1974.

A useful and representative collection of sources. Many English translations of medieval scientific treatises have also been published recently by the University of Wisconsin Press.

Leff, Gordon. *Paris and Oxford Universities in the Thirteenth and Fourteenth Centuries*. New York: John Wiley & Sons, 1968.

Stands out among studies of the context for medieval science because of its skillful combination of institutional, curricular, and doctrinal issues.

Lindberg, David C., ed. *Science in the Middle Ages*. Chicago: University of Chicago Press, 1978.

If you can have only one book on medieval science in your library, it should be this one. It contains essays on the cultural background to and institutional settings for medieval science as well as essays on the conceptual content of all important medieval sciences by the leading scholars working today. In addition, it offers a superb bibliography. Does not present a unified vision of medieval science.

Lindberg, David C., *Theories of Vision from Al-Kindi to Kepler*. Chicago: University of Chicago Press, 1976.

This study of optics is suitable for nonspecialists.

Nasr, Seyyed Hossein. *Science and Civilization in Islam*. Cambridge: Harvard University Press, 1968.

The most accessible general survey of Islamic science.

Pederson, Olaf, and Mogens Pihl. *Early Physics and Astronomy*. New York: American Elsevier, 1972.

Especially good on astronomy and mechanics.

Singer, Charles. *From Magic to Science*. New York: Dover Publications, 1958 and *A Short History of Anatomy and Physiology from the Greeks to Harvey*. New York: Dover, 1957.

No comprehensive surveys of medical and biological developments exist, but several of the sections in these books are very good.

Chapter 3: Scientific Imagination in the Renaissance

Introductory-level Surveys of Renaissance Science

Boas, Marie. *The Scientific Renaissance*. New York: Harper & Brothers, 1962.

Quite comprehensive.

Debus, Allen G. *Man and Nature in the Renaissance*. Cambridge: Cambridge University Press, 1978.

The most recent, and offers slightly more emphasis on magic and alchemy.

Wightman, W. P. D. *Science and the Renaissance*, 2 vols. Edinburgh and London, 1962.

The strongest on natural history and medicine, it includes an intensive inventory of primary sources.

Biographical Literature

Cardan, Jerome. *The Book of My Life*. New York: Dover, 1962.

An autobiography that offers marvelous insights into the variety of magical, mathematical, and medical influences on a single individual.

French, Peter J. *John Dee: The World of an Elizabethan Magus*. London: 1972.

Merejowski, Dimitri. *Romance of Leonardo Da Vinci*. New York: Modern Library, 1928.

An insightful fictional biography.

Montgomery, John Warwick. *The Cross and the Crucible: John Valentine Andrea (1586–1654)*. The Hague, Martinus Nijhoff, 1973.

Pagel, Walter. *Paracelsus: An Introduction to Philosophical Medicine in the Era of the Renaissance*. New York: 1958.

Rossi, Paolo. *Francis Bacon: From Magic to Science*. Chicago: University of Chicago Press, 1968.

Science and the Arts

Rossi, Paolo. *Philosophy, Technology, and the Arts in the Early Modern Era*. New York: Harper & Row, 1970.

Excellent.

The Mathematical Tradition of the Renaissance

Feingold, Mordechai. *The Mathematicians Apprenticeship: Science, Universities, and Society in England 1560–1640*. Cambridge: Cambridge University Press, 1984.

A work that supplements Taylor.

Rose, Paul L. *The Italian Renaissance of Mathematics*. Geneva: 1975.

Taylor, E. G. R. *The Mathematical Practitioners of Tudor and Stuart England: 1485–1714*. Cambridge: 1968.

Magic and Science

Shumaker, Wayne. *The Occult Sciences in the Renaissance*. Berkeley: University of California Press, 1972.

A good but not very sympathetic summary of astrological and magical doctrines in the Renaissance.

Westman, Robert S. *Hermeticism and the Scientific Revolution*. Los Angeles: Clark Library, 1977.

One of the most compelling arguments by a critic of Yates's work.

Yates, Frances. *Giordano Bruno and the Hermetic Tradition*. New York: Random House, 1964, and one should consult her *The Rosicrucian Enlightenment*. London: Routledge & Kegan Paul, 1972.

Chapter 4: Patronage and Printing

Dickens, A. G., ed. *The Courts of Europe. Politics, Patronage and Royalty, 1400–1800*. London: Thames & Hudson, 1977.

A series of commissioned articles on the question of courtly patronage. Accompanied by spectacular illustrations.

Eisenstein, Elizabeth L. *The Printing Press as an Agent of Change*. 2 vols. Cambridge: Cambridge University Press, 1979.

A detailed study of the impact of printing on early modern society. Includes a comprehensive bibliography of all aspects of the history of printing.

Elias, Norbert. *The Court Society*. Trans. Edmund Jephcott. New York: Pantheon, 1983.

A stimulating historical-sociological account of the culture of royal and aristocratic courts in the seventeenth and eighteenth centuries.

Evans, R. J. W. *Rudolf II and His World. A Study in Intellectual History, 1576–1612*. Oxford: Oxford University Press, 1973.

A valuable in-depth history of the court of one of the most complex of the Hapsburg monarchs.

Febvre, Lucien, and Henri-Jean Martin. *The Coming of the Book. The Impact of Printing, 1450–1800*. Trans. David Gerard. London: Verso Editions, 1984.

An original analysis of the importance of printing in the formation of early modern culture.

Lytle, Guy Fitch, and Stephen Orgel, eds. *Patronage in the Renaissance*. Folger Institute Essays. Princeton: Princeton University Press, 1981.

A superb collection of essays on various aspects of literary, artistic dramatic, political—but not scientific—patronage.

Martines, Lauro. *Power and Imagination. City-States in Renaissance Italy*. New York: Vintage, 1979.

An important political and social analysis of the emergence of Italian Renaissance Culture.

Chapter 5: The Reform of the Heavens

Kuhn, Thomas S. *The Copernican Revolution*. Cambridge: Harvard University Press, 1957.

Still the classic account of the Copernican Revolution, although many aspects of it are now disputed.

Lindberg, David C., and Ronald L. Numbers, eds. *God and Nature. Historical Essays on the Encounter between Christianity and Science*. Berkeley and Los Angeles: University of California Press, 1986.

An important collection of articles, containing several pertinent to this chapter, and in general disputing the notion of "warfare" between science and Christianity.

Redondi, Pietro. *Galileo Eretico*. Torino: Giulio Einaudi, 1983; French trans. *Galilee heretique*. Paris: Gallimard, 1985. Forthcoming in English translation, Princeton University Press.

A vividly written reinterpretation of the circumstances leading to the trial of Galileo which connects Galileo's troubles, not to the Copernican debates, but to Galileo's atomism and the Eucharist.

Thomas, Keith. *Religion and the Decline of Magic*. New York: Charles Scribner's Sons, 1971.

A seminal analysis of astrology, magic, and witchcraft in early modern England; a useful account to set alongside this chapter.

Wallace, William A. *Galileo and His Sources, The Heritage of the Collegio Romano in Galileo's Science*. Princeton: Princeton University Press, 1984.

A scholarly and provocative account of Galileo's science which links it directly to origins among the Roman Jesuits.

Westman, Robert S., ed. *The Copernican Achievement*. Berkeley and Los Angeles: University of California Press, 1975.

A collection of valuable scholarly papers devoted to different aspects of Copernicus's work and its reception.

Chapter 6: Mechanical and Mathematical Visions of the Universe

Cohen, I. Bernard. *Birth of a New Physics*. 2d ed. New York and London: W. W. Norton, 1985.

An excellent introduction to the revolution in astronomy and physics, starting from Aristotle and ending with Newton. This book features both up-to-date historical scholarship and an engaging style, stemming from its origin as an enrichment book for bright high-school students. Especially good on the role of mathematics and of scientific method.

Kline, Morris. *Mathematics in Western Culture*. Oxford and New York: Oxford University Press, 1964.

A very readable account of the history of mathematics and its involvement in all aspects of life, from art to physics. Chapters 11–18 cover the seventeenth century from "Projective geometry: a science born of art" through Descartes's philosophy, mathematics of change (the calculus), and the success of the quantitative approach to nature in the work of Newton.

Kuhn, T. S. *The Copernican Revolution: Planetary Astronomy in the Development of Western Thought*. Cambridge: Harvard University Press, 1957.

One of the great classic writings in the history of science. Especially recommended is the chapter on Copernicus himself, showing his use of past techniques to work out fully the implications of the "innovation" of setting the earth in motion. Rewarding of careful study.

Lindberg, David D., and Ronald L. Numbers, eds. *God and Nature: Historical Essays on the Encounter between Christianity and Science*. Berkeley and Los Angeles: University of California Press, 1986.

A collection of essays, including Robert S. Westman on Copernicus and the churches, William Shea on Galileo and the Church, Charles Webster on Puritanism and science, Richard Westfall on the rise of science and the decline of orthodox Christianity, and Margaret Jacob on Christianity and the Newtonian world view.

Thayer, H. S., ed. *Newton's Philosophy of Nature*. New York: Hafner, 1953.

Selections from Newton's own nontechnical writings, including an introduction. The materials on "The Method of Natural Philosophy" and "God and Natural Philosophy" are especially interesting.

Westfall, Richard S. *The Construction of Modern Science: Mechanisms and Mechanics*. New York: John Wiley & Sons, 1971.

A general history of seventeenth-century science, this book is especially valuable for its discussion of the way the "mechanical philosophy" was involved in the sciences. The chapters on "The Mechanical Philosophy" and "Mechanical Science" are the most relevant.

Willey, Basil. *The Seventeenth-Century Background: The Thought of the Age in Relation to Religion and Poetry*. New York: Doubleday, 1953.

Although originally published in 1935, this book still gives a nice overview of many leading philosophical ideas (Descartes, Hobbes, and Locke are subjects of good chapters); the discussion of poetry and of religion in relationship to the philosophy is thorough and interesting (Milton, Sir Thomas Browne, Joseph Glanville are some of the main authors discussed). Rewarding of careful study.

Chapter 7: Enlightenment and Industrialization

Science during the Enlightenment

Hankins, Thomas L. *Science and the Enlightenment*. Cambridge: Cambridge University Press, 1985.

Brief, comprehensive, and clearly written—a very good place for a beginner to start.

Rousseau, G. S., and Roy Porter, eds. *The Ferment of Knowledge: Studies in the Historiography of Eighteenth Century Science*. Cambridge: Cambridge University Press, 1980.

Contains a series of essays by many of the best of the current generation of scholars working in the field.

Wolf, Abraham. *A History of Science, Technology, and Philosophy in the Eighteenth Century*. 2 vols. New York: Macmillan, 1939.

Remains particularly useful on the history of scientific instruments and on technology.

Science and the Enlightenment as an Intellectual Movement

Cassirer, Ernest. *Philosophy of the Enlightenment*. Boston: Beacon Press, 1951.

Excellent, but emphasizes both the role of Newton and that of Kant more than most English or French scholars are inclined to.

Gay, Peter. *The Enlightenment, An Interpretation*. 2 vols. New York: Alfred Knopf, 1966, 1969.

More balanced than Cassirer, but is very difficult reading.

Smith, Preserved. *The Enlightenment: 1687–1776*. New York: Collier Books, 1962. A reissue of the superb 1934 original.

It expresses a kind of optimistic rationalist perspective that has generally been abandoned by scholars.

Biographies

Andrade, E. N. da C. *Sir Issac Newton, His Life and Works*. Garden City, N.Y.: Doubleday Anchor Books, 1954.

Clear, short, and very elementary.

Hankins, Thomas L. *Jean D'Alembert: Science and the Enlightenment*. Oxford: Oxford University Press, 1970.

McKie, Douglas. *Antoine Lavosier: Scientist, Economist, Social Reformer*. New York: Henry Schuman, 1952.

Schofield, Robert, ed. *A Scientific Autobiography of Joseph Priestly, 1733–1804: Selected Scientific Correspondence, with Commentary*. Cambridge: M.I.T. Press, 1966.

Thayer, H. S., ed. *Newton's Philosophy of Nature*. New York: Hafner, 1953.

A collection including several key selections from Newton's writing which are accessible to a lay audience.

Westfall, Richard A. *Never at Rest: A Biography of Isaac Newton*. Cambridge: Cambridge University Press, 1980.

Clear, long, and very comprehensive.

The Social Dimensions of Enlightenment Science

Gillispie, Charles Coulston. *Science and Polity in France at the End of the Old Regime*. Princeton: Princeton University Press, 1980.

McClellan, James E., III. *Science Reorganized: Scientific Societies in the Eighteenth Century*. New York: Columbia University Press, 1985.

Schofield, Robert. *The Lunar Society of Birmingham: A Social History of Provincial Science and Industry in Eighteenth Century England.* Oxford: Oxford University Press, 1963.

Science and Technology in the Eighteenth Century

Musson, A. E., ed. *Science, Technology, and Economic Growth in the Eighteenth Century*. London: Methuen, 1972.

Musson, A. E., and Eric Robinson. *Science and Technology in the Industrial Revolution*. Manchester: University of Manchester Press, 1969.

Chapter 8: Scientific Medicine and Social Statistics

Ackerknecht, Erwin. *Medicine at the Paris Hospital.* Baltimore: Johns Hopkins University Press, 1967.

Development of French medicine after the Revolution.

Faber, Knud. *Nosography.* Hoeber, 1923.

Unsurpassed on the history of clinical medicine in relation to science.

Ludmerer, Kenneth. *Learning to Heal.* New York: Basic Books, 1985.

Superb account of medical education.

Maulitz, Russell. *Morbid Appearances.* Cambridge: Cambridge University Press, 1987.

The rise of pathological anatomy.

Rosen, George. *History of Public Health.* MD Publications, 1958.

Still the best overview of public health and hygiene.

Rosenberg, Charles. *Cholera Years.* Chicago: University of Chicago, 1962.

Classic story of disease in the nineteenth century.

Stevens, Rosemary. *American Medicine and the Public Interest.* New Haven: Yale University Press, 1971.

Most useful work to date linking history with current circumstances.

Warner, John. *The Therapeutic Perspective.* Cambridge: Harvard University Press, 1986.

Excellent new study of role of therapeutics in nineteenth-century medical life.

Chapter 9: Darwinism as Science and Ideology

Allen, Garland. *Life Sciences in the Twentieth Century*. New York: Wiley History of Science Series, 1975. Cambridge University Press.

This well-written survey of the major trends in modern biology assumes that the reader has an elementary knowledge of the life sciences. It traces developments in biochemistry and molecular biology and describes the beginnings of the modern theory of heredity.

Bowler, Peter J. *Evolution: The History of an Idea*. Berkeley, Los Angeles, and London: University of California Press, 1984.

The author describes pre-Darwinian speculations about evolution, the origins of Darwin's thought, the immediate impact of Darwin's theory of evolution, and the twentieth-century debates which led to the modern synthesis in evolutionary theory. The book is a comprehensive survey incorporating recent scholarship.

Coleman, William. *Biology in the Nineteenth Century*. New York: Wiley History of Science Series, 1971. Cambridge University Press.

This book documents the major nineteenth-century theoretical advances in biology and describes the emergence of experimental biology at the end of the century. It is a highly readable exposition.

Darwin, Charles. *On the Origin of Species: A Facsimile of the First Edition*. Cambridge: Harvard University Press, 1964.

Darwin's major scientific work is eminently readable. His clear exposition and his patient presentation of the mass of evidence pointing to the evolution of species were undoubtedly responsible for the immediate impact of his theory on scientific and social thought.

Gould, Stephen J. *Ever Since Darwin*. New York: W. W. Norton, 1977.

The author describes the impact of Darwin's theory, particularly with reference to the sciences of anthropology and geology, and also traces its influence on social and political thought.

Greene, John C. *The Death of Adam*. Ames, Iowa: Iowa State University Press, 1959.

This book is a very clear and readable survey of the background of evolutionary ideas in science, philosophy, and religion in the century and a half before the publication of Darwin's theory.

Ruse, Michael. *The Darwinian Revolution*. Chicago: University of Chicago Press, 1979.

The author presents a comprehensive account of evolutionary thought in Great Britain from 1830 to 1875. It is a well-organized and very readable synthesis.

Chapter 10: Uncertainty in Physics and Society

Einstein, Albert, and Leopold Infled. *The Evolution of Physics: From Early Concepts to Relativity and Quanta*. New York: Simon and Schuster, 1938; 2d ed., 1960.

A very nice, readable popular introduction to physics, showing the development of some major ideas from Newton to the twentieth century. Sections 3 and 4, which deal, respectively, with relativity and quanta, are among the best popularizations for nonscientists.

Hoffman, Banesh. *The Strange Story of the Quantum: An Account for the General Reader of the Growth of the Ideas Underlying Our Present Atomic Knowledge*. 2d ed. New York: Dover, 1959.

A sometimes misleading but eminently readable history of quantum theory, which also nicely explains the ideas behind the theory for the nonscientist reader.

Huff, Darrell. *How to Lie with Statistics*. New York: Norton, 1954.

This beautifully written book is effortlessly easy to read. Its purpose is to teach readers how to avoid being misled by misuses of statistics. But it also gives the reader a nice feeling for statistical thinking in general. (Note: Because the book was published in 1954, adjust all dollar figures for inflation by multiplying them by seven.)

Kline, Morris. *Mathematics in Western Culture*. Oxford and New York: Oxford University Press, 1964.

The history of mathematics and its involvement in all aspects of life, from art to physics, are presented in this very readable account. Chapters 23–24 deal with the applications of statistics and probability to the sciences, while Chapter 26 deals with non-Euclidean geometry and 27 with relativity. Chapter 22 deals with the applications of statistical thinking to the study of human beings. The orientation of these chapters is historical and philosophical, not *technical, and they give a good view—for nonmathematicians!—of how and why mathematics is used in science.*

Kuhn, Thomas S. *The Structure of Scientific Revolutions*. 2d ed. Chicago: University of Chicago Press, 1970.

This book gives a general theory of scientific change. It is intended for readers with some background in the history of science and deals with a number of examples with which readers of the present volume are familiar, including the rise of relativity theory, the Copernican Revolution, and Lavoisier's revolution in chemistry. Through focusing on scientific communities, it treats science sociologically, and through focusing on the way individuals' perceptions change, it applies psychology to science. In addition, it raises philosophical questions about the nature of truth, explanation, and science.

Weaver, Warren. *Lady Luck: The Theory of Probability*. New York: Doubleday, 1963. Reprint. New York: Dover, 1982.

An excellent popular introduction to this topic. The most useful chapters for a philosophical understanding of probability theory and statistical thinking are both quite readable: Chapter 1, on logical versus probabilistic thinking, and Chapter 16, on the role of such thinking in modern science. Only elementary algebra is needed even for the more mathematical chapters, and Chapters 2 and 3, introducing the concept of probability, are highly recommended.

Williams, L. Pearce, ed. *Relativity Theory: Its Origins and Impact on Modern Thought*. New York: John Wiley & Sons, 1968. Reprint. New York: Robert Krieger Publishing, 1979.

A collection of readings, with a good introduction, on the origins of relativity theory (readings from Newton, Mach, Lorentz, Poincaré, Michelson and Morey, Einstein); some historical articles on relativity theory; and some examples of reactions to the theory, some by scientists, others by philosophers, artists, and journalists.

Chapter 11: Progress and the Rationality of Science

Barnes, Barry. *Scientific Knowledge and Sociological Theory*. Boston: Routledge & Kegan Paul, 1974.

One statement of the sociological analysis of science known as the "Edinburgh Program."

Blake, Ralph M., et al. *Theories of Scientific Method*. Seattle: University of Washington Press, 1966.

Thirteen studies of the methods used by scientists and commentators on science from the Renaissance through the nineteenth century.

Faust, David. *The Limits of Scientific Reasoning*. Minneapolis: University of Minnesota Press, 1984.

A careful analysis of the restricted capacity of human cognition from the viewpoint of an astute clinical practitioner.

Hanson, Norwood Russell. *Patterns of Discovery*. Cambridge: Cambridge University Press, 1958.

A classic analysis, through an examination of modern elementary particle theory, of how scientific theories are discovered.

Lakatos, Imre, and Alan Musgrave, eds. *Criticism and the Growth of Knowledge*. Cambridge: Cambridge University Press, 1970.

Individual analyses of Kuhn's Structure of Scientific Revolutions, *by Karl Popper and others of Kuhn's critics.*

Laudan, Larry. *Science and Values: The Aims of Science and their Role in Scientific Debate*. Berkeley and Los Angeles: University of California Press, 1984.

An argument that the adoption of one cognitive goal in science over another need not merely result from emotional preference, but can also follow from rational assessment.

Losee, John. *A Historical Introduction to the Philosophy of Science*. Oxford: Oxford University Press, 1972.

A philosophical examination of the historical development of thinking on the fundamental issues in philosophy of science from antiquity to the present.

Tweney, Ryan et al., eds. *On Scientific Thinking*. New York: Cambridge University Press, 1981.

A series of historical and contemporary readings selected to establish the idea that the "psychology of scientific thinking" has become a unique domain of scientific inquiry.

Chapter 12: At the Brink of the Future

Computers

Historical Perspective

Bernstein, Jeremy. *The Analytical Engine: Computers—Past, Present, and Future*. New York: William Morrow, 1981.

This nicely written book originated as a series of articles in the New Yorker. *It introduces nontechnical readers to how computers work, to the history of computers, and to some of the philosophical issues they raise.*

Westfall, Richard S. *The Construction of Modern Science: Mechanisms and Mechanics*. New York: John Wiley & Sons, 1971.

Chapters 2–5 give a good discussion of the seventeenth-century mechanical philosophy, with numerous examples showing its application to physics, chemistry, and biology in the work of such men as Kepler, Newton, Boyle, Descartes, Harvey, and Leeuwenhoek.

Artificial Intelligence

Grabiner, Judith V. "Computers and the Nature of Man: A Historian's Perspective on Controversies about Artificial Intelligence." *Bulletin of the American Mathematical Society* 15, (October 1986): 113–126.

This article reviews some recent work in AI and the claims made for it, seeing the basic issues as moral ones about the nature of human beings. It focuses on the debates about human nature rather than on technical developments.

McCorduck, Pamela. *Machines Who Think*. San Francisco: W. H. Freeman, 1979.

This is a lively popular history of the computer scientists who developed the field of Artificial Intelligence in the United States. Its provocative title indicates the point of view that it takes—great enthusiasm about the practice and promise of AI.

Computers and Society

Dertouzos, Michael L., and Joel Moses, eds. *The Computer Age: A Twenty-Year View*. Cambridge: M.I.T. Press, 1979.

Forester, Tom, ed. *The Information Technology Revolution*. Cambridge: M.I.T. Press, 1985.

These two anthologies of readings contain articles on all aspects of computers, technology, and society, from factory automation to computer crime, and from mathematical modeling to the politics of technical change. Very useful. Included is a very informative analysis by Daniel Bell, initiator of the term "information society," which describes the shift in the structure of work from manufacturing to service, from production to information, and documents the principal features of the new age.

Perrolle, Judith A. *Computers and Social Change: Information, Property, and Power*. Belmont, Calif.: Wadsworth Publishing, 1987.

The author, in addition to describing the ways in which industries that produce and distribute information have changed the economy, discusses some broader social, political, and ethical issues raised by the increasing use of computers.

Shore, John. *The Sachertorte Algorithm and Other Antidotes to Computer Anxiety*. New York: Viking, 1985.

A humane and literate introduction to how computers are programmed. This is not *a technical treatise on programming, but a repository of wisdom about how human beings interact with computers. Jargon like the term "algorithm" is illustrated and demystified (the title refers to the "program" his aunt gave him for making a cake).*

Weizenbaum, Joseph. *Computer Power and Human Reason*. San Francisco: W. H. Freeman, 1976.

This is the most thoughtful of the books on this subject. The introduction tells how Weizenbaum's concerns about the dehumanizing effects of computers arose through people's misinterpretation of his own AI work. Chapters 2–3 give a difficult and technical, but excellent, account of where the power of the computer comes from. The remainder of the book presents a reasoned discussion of what computers can and cannot do, and argues that since we cannot make computers wise, we should not give them tasks requiring wisdom. Highly recommended.

Biotechnology

Garber, Edward D., ed. *Genetic Perspectives in Biology and Medicine*. Chicago: University of Chicago Press, 1985.

This volume is a new edition of original articles by biologists and biochemists. It gives an excellent historical overview of twentieth-century developments, concentrating on advances in medical genetics.

Lappe, Marc. *Broken Code*. San Francisco: Sierra Book Club, 1984.

The author describes current trends in biotechnology, particularly research in recombinant DNA. He is interested primarily in problems that biotechnology poses for agriculture and the physical environment.

Zimmerman, Burke K. *Biofuture: Confronting the Genetic Era*. New York: Plenum Press, 1984.

This is a well-written popular account of the historical background of modern genetics, and it also describes the various directions in which biotechnology is advancing.

About the Contributing Editors

Senior Editor

John G. Burke Emeritus Professor of History and Emeritus Dean of the College of Letters and Sciences at the University of California, Los Angeles. A graduate of M.I.T., he received his Ph.D. from Stanford University. He is the author of *Origins of the Science of Crystals* (1966), editor of *The New Technology and Human Values* (1966, 1972), and his most recent book is *Cosmic Debris: Meteorites in History* (1986).

As Senior Editor for *Science and Culture in the Western Tradition,* Professor Burke coordinated the efforts of the contributing editors and made his own contributions, including material for Chapter 12 and the volume Introduction.

Mark B. Adams Associate Professor, Department of the History and Sociology of Science at the University of Pennsylvania. He received his A.B., A.M., and Ph.D. degrees from Harvard University. He has been Advisory Editor of *Isis* and *Mendel Newsletter* and has published widely on Darwinism, evolution, heredity, and eugenics, and on Soviet science. He has served as both Graduate and Undergraduate Departmental Chairperson.

Judith V. Grabiner Professor of Mathematics and the History of Science at Pitzer College in Claremont, California. She studied mathematics at the University of Chicago and the history of science at Harvard University, from which she received her Ph.D. As a National Science Foundation Faculty Development Fellow, she studied computer science at Indiana University. Her major research interests are in the history of mathematics, especially its significant role in modern science and philosophy, and computers and society. Her most important publications include *The Origins of Cauchy's Rigorous Calculus* (1981), many articles in historical and mathematical journals, and a paper, "The Centrality of Mathematics in Western Thought."

Frederick Gregory Associate Professor of the History of Science at the University of Florida. He studied the history of science at the University of Wisconsin and at Harvard University, where he received his Ph.D. He also holds a degree from Gordon-Conwell Theological Seminary, where his interest in the relation between science and religion was born. The focus of his research and publications has been the German scientific heritage of the eighteenth and nineteenth centuries. He has published on the history of mathematics, history of chemistry, history of medicine, and on the historical interaction of natural science with religious and philosophical thought. His *Scientific Materialism in Nineteenth Century Germany* (1977) is to be followed by a study of the German theological reaction to Darwin. Professor Gregory is also at work on a synthetic interpretation of the interaction between science and thought in the German Romantic period.

Russell C. Maulitz Lecturer in the History and Sociology of Science in Medicine at the School of Medicine, University of Pennsylvania. Dr. Maulitz is a practicing physician and a historian whose interests are focused on American and European scientific medicine in the nineteenth and twentieth centuries. He is author of *Morbid Appearances* (1987) and editor, with Diana Long, of *Grand Rounds* (1987).

Richard Olson Professor of History and Willard W. Kieth Fellow in the Humanities at Harvey Mudd College in Claremont, California. He studied physics and the history of science at Harvard University, from which he received his Ph.D. His special research interests include British natural philosophy in the eighteenth and nineteenth centuries and the relationships among religion, political ideology, and scientific activity in Western cultures from antiquity to the present. In addition to numerous articles in scholarly journals, his publications include *Scottish Philosophy and British Physics* (1975), an analysis of the Victorian scientific style, *Science Deified and Science Defied,* Volume I (1982), on the roles of science in Western culture from the Bronze Age to the seventeenth century, and an edited volume, *Science as Metaphor* (1971). He is currently working on the second volume of *Science Deified and Science Defied* to bring the story up to the early nineteenth century.

Robert S. Westman Professor of History at the University of California, Los Angeles. He studied the history of science at Imperial College and The Warburg Institute, University of London, and at the University of Michigan, where he earned his doctorate. His special research interests include the social and political relations among the disciplines in the sixteenth and seventeenth centuries, particularly astronomy, physics, theology, poetry, and art. He has also focused on science and the universities, scientific and magical knowledge, and aspects of the history of psychoanalysis. In addition to a number of significant articles in major scholarly journals, his most important publications include two edited volumes, *The Copernican Achievement* (1975) and, with David C. Lindberg, *Reappraisals of the Scientific Revolutions*. A major study, *The Copernicans: Universities, Courts and Interdisciplinary Conflict, 1540–1700,* is forthcoming.

Copyrights and Acknowledgments

CHAPTER 1

Page 2 Approximately forty lines abridged from *The Odyssey of Homer* trans. by Richard Lattimore. Copyright © 1965, 1967 by Richard Lattimore. Reprinted by permission of Harper & Row, Publishers, Inc.

6 From Henri Frankfort, et al., "Myth and Reality," *The Intellectual Adventure of Ancient Man*. University of Chicago Press, pp. 12–15. Copyright © 1946 by The University of Chicago. Reprinted by permission.

7 Reprinted from *The Republic of Plato,* translated by F. M. Cornford (1941) by permission of Oxford University Press.

10 Reprinted from "Metaphysics," translated by W. D. Ross from The Oxford Translation of Aristotle, W. D. Ross, ed., vol. 8, (1928) by permission of Oxford University Press.

12 Reprinted by permission of the publishers and The Loeb Classical Library from *Hippocrates,* Vol. II, W. H. S. Jones, translator, Harvard University Press, Copyright © 1923.

13 From B. L. Van Der Waerden, *Science Awakening,* trans. Arnold Dresden. Groningen, Holland: P. Noordhoff Ltd., 1954. Copyright © Martinus Nijhoff/ Dr. W. Junk Publishers, Dordrecht, The Netherlands. Reprinted by permission of Sijthoff Publishers and Martinus Nijhoff/Dr. W. Junk Publishers, The Netherlands.

15 From J. Ralph Lindgren, ed., *The Early Writings of Adam Smith*. New York: Augustus M. Kelley, Publisher, 1967, pp. 45–50 (edited). Reprinted with permission.

16 From Benjamin Farrington, *Greek Science: Its Meaning for Us,* Volume I. Published in one volume with Volume II (Pelican Books, 1953, rev. ed., 1961). Copyright © Benjamin Farrington 1944, 1949, 1961. London: Penguin Books Ltd, pp. 104–107. Reprinted with permission.

17 From *Magic, Reason and Experience: Studies in the Origin and Development of Greek Science,* by G. E. R. Lloyd. Copyright © 1979 Cambridge University Press. Reprinted with permission.

19 Richard Olson, *Science Deified and Science Defied*. Berkeley, Los Angeles, London: University of California Press, pp. 78–82. Copyright © 1982 by the Regents of the University of California. Reprinted with permission.

CHAPTER 2

27 From David C. Lindberg and Ronald L. Numbers, *God and Nature*. Berkeley, Los Angeles, London: University of California Press, pp. 29–41. Copyright © 1986 The Regents of the University of California. Reprinted with permission.

30 Brian Stock, "Science, Technology, and Economic Progress in the Early Middle Ages," in David C. Lindberg, ed., *Science in the Middle Ages,* 1978. University of Chicago Press, pp. 11–22 (edited). Copyright © 1978 by the University of Chicago Press. Reprinted by permission.

34 Lynn White, Jr., "Natural Science and Naturalistic Art in the Middle Ages," *American Historical Review* 52, (1947), pp. 46–51. Reprinted by permission of Lynn White, Jr.

37 From Joseph Ben-David, *The Scientist's Role in Society.* Englewood Cliffs, N.J.: Prentice-Hall, Inc., 1971, pp. 46–51. Copyright © Estate of Joseph Ben-David. Reprinted with permission from Estate of Joseph Ben-David.

41 From A. C. Crombie, *Augustine to Galileo,* Vol. I: *Science in the Middle Ages*. Cambridge: Harvard University Press, pp. 56–64. Copyright © 1952, 1959 by A. C. Crombie. Reprinted by permission of Harvard University Press and Heinemann Educational Books Ltd, London.

CHAPTER 3

49 Marshall Clagett, ed., *Critical Problems in the History of Science*. Madison, Wisconsin: University of Wisconsin Press, 1959, pp. 33–62. Reprinted with permission.

51 Frances A. Yates, "Cornelius Agrippa's Survey of Renaissance Magic," *Giordano Bruno and the Hermetic Tradition*. Chicago, London: University of Chicago Press, 1964, pp. 130–156 (edited); London: Routledge & Kegan Paul, Ltd. © 1964 by Frances A. Yates. Reprinted by permission of the University of Chicago Press and Routledge & Kegan Paul, Plc.

55 From *Man and Nature in the Renaissance* by A. G. Debus. Copyright © 1978 Cambridge University Press. Reprinted with permission.

57 From pp. 197–209 of Chapter VII, "The Uses of Mathematics," in *The Scientific Renaissance: 1450–1630* by Marie Boas. Copyright © 1962 by Marie Boas. Reprinted by permission of Harper & Row, Publishers, Inc. New York and William Collins Sons & Co., Ltd., London.

59 Francis Bacon, A Selection: *The New Atlantis,* pp. 446–457.

CHAPTER 4

67 Guy Fitch Lytle and Stephen Orgel, *Patronage in the Renaissance*. Copyright © 1981 by Princeton University Press. Excerpts, pp. 207–210, reprinted with permission of Princeton University Press.

68 Reprinted from *Painting and Experience in Fifteenth-Century Italy* by Michael Baxandall (1972), pp. 5–14 by permission of Oxford University Press.

72 Bruce T. Moran, "German Prince-Practitioners: Aspects in the Development of Courtly Science, Technology, and Procedures in the Renaissance, *Technology and Culture* 22, No. 2 (1981), pp. 253–274. Copyright © 1981 by The Society for the History of Technology. Reprinted by permission of the author and The University of Chicago Press.

252 Werner Heisenberg, ''Planck's Quantum Theory and the Philosophical Problems of Atomic Physics,'' *Science: Men, Methods, Goals* (edited). Copyright © 1986 W. A. Benjamin, Inc. Original text of lecture delivered in Geneva, reprinted from *Univeritas*, Vol. 3. No. 2, 1959.

257 Abridged and adapted from pp. 278–289 of John Lukacs, *Historical Consciousness: Or, The Remembered Past* by John Lukacs (New York: Harper & Row, Inc., 1968). Copyright © 1968 by John Lukacs. Reprinted by permission of Harper & Row, Publishers, Inc.

CHAPTER 11

266 Condorcet, ''Outlines of an Historical View of the Progress of the Human Mind.'' Philadelphia: Lang and Ustick, 1796.

267 Sir J. F. W. Herschel, *A Preliminary Discourse on the Study of Natural Philosophy*, London: Longran, Orme, Brown, Green, & Longmans. 1840.

271 Jacob Bronowski, *The Ascent of Man*. Boston: Little, Brown and Company, pp. 353–374 (edited). Copyright © 1973 by Jacob Bronowski. Reprinted with permission. Reproduced from *The Ascent of Man* by Jacob Bronowski with the permission of BBC Enterprises Ltd., London.

274 Robert F. Baum, ''Popper, Kuhn, Lakatos: A Crisis of Modern Intellect,'' *The Intercollegiate Review*, Spring 1974, pp. 99–110 (edited). Reprinted with permission.

279 Abridged from pp. 255–313 in *The Dancing Wu Li Masters* by Gary Zukav (Bantam edition). Copyright © 1979 by Gary Zukav. By permission of William Morrow & Company, Inc., New York.

CHAPTER 12

286 Excerpts from *Profiles of the Future: An Inquiry into the Limits of the Possible* by Arthur C. Clarke. Copyright © 1958 by Arthur C. Clarke. New York: Harper & Row, 1963. Reprinted by permission of the author and the author's agents, Scott Meredith Literary Agency, Inc., 845 Third Avenue, New York, NY 10022.

290 Excerpted from: Herbert A. Simon, ''What Computers Mean for Man and Society,'' *Science* 195 No. 4283, 1186–1191 (edited), 18 March 1977. Copyright © 1977 by AAAS. Published by The American Association for the Advancement of Science, Washington D.C. Reprinted with permission.

296 Joseph Weizenbaum, *Computer Power and Human Reason*. San Francisco: W. H. Freeman, 1976, pp. 196–201. © 1976 by W. H. Freeman and Company.

299 J. B. S. Haldane, *Daedalus or Science and the Future*. Copyright © 1924 by E. P. Dutton & Company. London, New York: E. P. Dutton & Company, 1924, pp. 42–49, 56–88. Reprinted with permission from Routledge & Kegan Paul Ltd.

303 J. B. S. Haldane, *The Last Judgment*. London: Chatto & Windus, 1927, pp. 49–66 (edited). Reprinted with permission from Harper & Row, New York.

176 Lemuel Shattuck, et al., *Report of the Sanitary Commission of Massachusetts* 1850. Cambridge: Harvard University Press, 1948, pp. 143–155.

179 Rudolf Virchow, *Collected Essays on Public Health and Epidemiology*, 1867, Vol. I. Nantucket, Mass.: Science History Publications, U.S.A. 1985, pp. 94–103 (edited). Reprinted with permission.

185 Max von Pettenkofer, "Cholera: How to Prevent and Resist It," trans. Thomas Whiteside Hime, 1883; rev., von Pettenkofer. London: Bailliere, Tindall & Cox, pp. 25–42.

189 Thomas McKeown, *The Role of Medicine: Dream, Mirage or Nemesis?*. Copyright © 1979 by Princeton University Press. Excerpts reprinted with permission of Princeton University Press.

192 Reprinted by permission of *Daedalus*, Journal of the American Academy of Arts and Sciences, Leighton E. Cluff, "America's Doctors, Medical Science, Medical Care," Vol. 115, No. 2, 1986, Cambridge, MA.

CHAPTER 9

209 From Peter J. Bowler, *Evolution: The History of an Idea*. © 1984 The Regents of the University of California. Berkeley: University of California Press, pp. 76–84 and 154–73 (edited). Reprinted with permission.

219 Charles Darwin, *On the Origin of Species: A Facsimile of the First Edition*. Cambridge: Harvard University Press, 1966, pp. 1–6, 126–130, 489–490 (edited).

223 Herbert Spencer, "Essays: Scientific, Political, and Speculative," Vol. 1. London: William and Norgate, 1868, pp. 1–3, 32–58, 384–397.

228 Trofim D. Lysenko, *The Situation in Biological Science*: Proceedings of the Lenin Academy of Agricultural Sciences of the U.S.S.R., Session July 31–August 7, 1948. Moscow: Foreign Languages Publishing House, 1949.

231 P. Kropotkin, *Mutual Aid: A Factor of Evolution* (1902). New York: McClure Phillips & Co., 1902.

CHAPTER 10

240 From Jonathan Powers, *Philosophy and the New Physics*. Copyright © 1982 Jonathan Powers. London: Methuen, 1982, pp. 51–58 (edited). Reprinted with permission.

242 Pierre Duhem, *The Aim and Structure of Physical Theory*, trans. Philip P. Wiener. Copyright 1954 Princeton University. © 1982 renewed by Princeton University Press. Excerpts, pp. 69–75, reprinted with permission of Princeton University Press.

244 Bertrand Russell, *The ABC of Relativity*, revised ed. © George Allen & Unwin Ltd, 1958, London. Copyright under the Berne Convention. New York: The New American Library of World Literature, Inc., pp. 33–41. Reprinted by permission of George Allen & Unwin Ltd, London.

247 Hugh L. Rodgers, "Charles A. Beard, the 'New Physics,' and Historical Relativity, " Allentown, PA: *The Historian* 30 (1968) pp. 545–560 (edited). Copyright © 1968 by Phi Alpha Theta. Reprinted by permission of the publisher.

127 From Étienne de Condillac, *La Logique,* trans. W. R. Albury. New York: Abaris Books, Inc., 1980, pp. 77–83. Reprinted by permission.

129 From René Descartes, *Treatise of Man*, trans. Thomas Steele Hall. Cambridge, Mass.: Harvard University Press. Copyright © 1972 by the President and Fellows of Harvard College. Reprinted by permission of Harvard University Press.

130 William Petty, "On the Value of People," Charles Henry Hull, ed., *The Economic Writings of Sir William Petty*, Vol. 1. Cambridge: Cambridge University Press, 1899, pp. 105–109. Printed with permission.

131 Thomas Hobbes, "Leviathan," from *The Great Books of the Western World*, Vol. 23, pp. 60–87. Reprinted by permission from Encyclopaedia Britannica, Inc.

134 From Sheldon S. Wolin, *Hobbes and the Epic Tradition of Political Theory*. University of California, Los Angeles: William Andrews Clark Memorial Library, 1970, pp. 38–42. Reprinted with permission.

136 Alexandre Koyre, "The Significance of the Newtonian Synthesis," from *Archives Internationales d'histoire des sciences* 3, 1950. Paris: Hermann, Maison d'editions, pp. 291–311.

CHAPTER 7

143 From pp. 62–79 from *Letters on England* by Voltaire, translated with an introduction by Leonard Tancock (Penguin Classics 1980). Copyright © 1980 Leonard Tancock. London: Penguin Books Ltd. Reproduced by permission of Penguin Books Ltd.

146 From James A. Secord, "Newton in the Nursery: Tom Telescope and the Philosophy of Tops and Balls, 1761–1838," *History of Science* 13 (1985), pp. 127–139. Bucks, England: Science History Publications. Reprinted with permission.

150 Article from *Impact of Science on Society*, Vol. VI:1. © Unesco 1955. Reproduced by permission of Unesco.

155 Charles C. Gillispie, "The Natural History of Trades," *Isis*, Vol. 48, 1957, pp. 121–134 (edited). Reprinted with permission of *Isis*, Philadelphia.

157 Neil McKendrick, "The Role of Science in the Industrial Revolution: A Study of Josiah Wedgwood as a Scientist and Industrial Chemist," Mikulas Teich and Robert Young, eds., *Changing Perspectives in the History of Science*. Copyright © Heinemann Educational Books 1973. Dordrecht, Holland, Boston: D. Reidel Publishing Company, 1973, pp. 279–318 (edited).

160 Neil McKendrick, "Josiah Wedgwood and Factory Discipline," *The Historical Journal*, Vol. 4 (1961). Copyright © 1961 Cambridge University Press. New York: Cambridge University Press, pp. 30–55 (edited). Reprinted with the permission of Cambridge University Press.

CHAPTER 8

170 Philippe Pinel, *The Clinical Training of Doctors: An Essay of 1793*, ed., trans. Dora B. Weiner. Baltimore: Johns Hopkins University Press, 1980, pp. 67–96 (edited). Reprinted with permission.

77 David Goodman, "Philip II's Patronage of Science and Engineering." *The British Journal for the History of Science* 16 (1983), pp. 49–56. Reprinted with permission.

80 Robert S. Westman, "The Astronomer's Role in the Sixteenth Century: A Preliminary Study." *The British Journal for the History of Science* 18 (1980). Oxford: Blackwell Scientific Publications, pp. 105–125.

85 From *The Printing Press as an Agent of Change,* Elizabeth L. Eisenstein. Copyright © 1979 Cambridge University Press. Reprinted with permission.

CHAPTER 5

93 "The Derivation and First Draft of Copernicus's Planetary Theory: A Translation of *The Commentariolus* with commentary," From Noel M. Swerdlow, "Symposium on Copernicus," *Proceedings of the American Philosophical Society,* Vol. 117, No. 6 (1973), p. 436. Reprinted with permission.

94 From Nicolaus Copernicus "De revolutionibus orbium coelestium." 1543. *Nicholas Copernicus Complete Works II: Nicholas Copernicus on the Revolutions,* Jerzy Dobrzcki, ed., Edward Rosen, trans. Warsaw: Polish Scientific Publishers, 1978. Baltimore, Maryland: The Johns Hopkins University Press. Reprinted by permission.

99 From Marie Boas and A. Rupert Hall, "Tycho Brahe's System of the World." *Occasional Notes of the Royal Astronomical Society,* III: 21 (1959). London: Royal Astronomical Society, pp. 257–263. Reprinted with permission from A. Rupert Hall and the Royal Astronomical Society.

102 From Owen Gingerich, "Ptolemy, Copernicus, Kepler." *The Great Ideas Today 1983*. Chicago: Encyclopaedia Britannica, Inc., 1983, pp. 170–179. Reprinted with permission.

107 Excerpts from *Discoveries and Opinions of Galileo* by Stillman Drake. Copyright © 1957 by Stillman Drake. Reprinted by permission of Doubleday & Company, Inc., New York.

111 From René Descartes, *Principles of Philosophy.* Trans., Valentine Rodger Miller and Reese P. Miller. Copyright © 1983/1984 by D. Reidel Publishing Co., Dordrecht, Holland. Reprinted with permission.

CHAPTER 6

120 Frederick Burwick, *Approaches to Organic Form.* Article "On the Nature of God's Existence, Wisdom and Power: The Interplay Between Organic and Mechanistic Imagery in Anglican Natural Theology, 1640–1740," by Richard Olson. Copyright © 1987 by D. Reidel Publishing Company, Dordrecht, Holland. Reprinted with permission.

123 Paolo Rossi, "Hermiticism, Rationality and the Scientific Revolution," in M. L. Righini Bonelli and William R. Shea, eds., *Reason, Experiment and Mysticism.* New York: Science History Publications, c/o Watson Publishing International, 1975, pp. 248–255 (edited). Reprinted with permission.

126 From pp. 1–11, *The Philosophical Works of Descartes,* Vol. I, Elizabeth S. Haldane and G. R. T. Ross, trans. Copyright © 1911, 1934 by the Cambridge University Press. New York: Cambridge University Press. Reprinted with the permission of Cambridge University Press.